Hydrologic Remote Sensing

Capacity Building for Sustainability and Resilience

Hydrologic Remote Sensing
Capacity Building for Sustainability and Resilience

Edited by
Yang Hong
Yu Zhang
Sadiq Ibrahim Khan

CRC Press
Taylor & Francis Group
Boca Raton London New York

CRC Press is an imprint of the
Taylor & Francis Group, an **informa** business

Cover figure credits:

(left insert) Courtesy of Sadiq Ibrahim Khan.; (right insert) From NASA's SMAP Mission website.

CRC Press
Taylor & Francis Group
6000 Broken Sound Parkway NW, Suite 300
Boca Raton, FL 33487-2742

© 2017 by Taylor & Francis Group, LLC
CRC Press is an imprint of Taylor & Francis Group, an Informa business

No claim to original U.S. Government works

Printed and bound in India by Replika Press Pvt. Ltd.

Printed on acid-free paper
Version Date: 20160411

International Standard Book Number-13: 978-1-4987-2666-5 (Hardback)

Library of Congress Cataloging-in-Publication Data
Names: Hong, Yang, 1973- editor. | Zhang, Yu, 1975- editor. | Khan, Sadiq Ibrahim, editor.
Title: Hydrologic remote sensing : capacity building for sustainability and resilience / editors, Yang Hong, Yu Zhang, and Sadiq Ibrahim Khan.
Description: Boca Raton : Taylor & Francis, CRC Press, 2016. | Includes bibliographical references and index.
Identifiers: LCCN 2016009009 | ISBN 9781498726665 (alk. paper)
Subjects: LCSH: Watershed management. | Hydrologic models. | Groundwater--Remote sensing. | Droughts--Remote sensing. | Floods--Remote sensing. | Sustainable development.
Classification: LCC TC413 .H935 2016 | DDC 363.34/630284--dc23
LC record available at https://lccn.loc.gov/2016009009

Visit the Taylor & Francis Web site at
http://www.taylorandfrancis.com

and the CRC Press Web site at
http://www.crcpress.com

Contents

SECTION I Remote Sensing Observations and Estimations

SECTION II Modeling, Data Assimilation, and Analysis

SECTION III Hydrologic Capacity Building for Improved Societal Resilience

Preface

Inadequate in situ observations have been historical barriers in the development of hydrologic science. Additionally, with the changing climate and ever accelerating human activities, it is anticipated that water-related disasters such as flood, landslide, and drought will be more severe and frequent, thus threatening more regions around the world. Flooding, which is considered as one of the most hazardous disasters in both rural and urban areas, accounts for around one-third of all global geophysical hazards and causes significant human suffering, loss of life, and property damage. The International Flood Network indicates that from 1995 to 2004, natural disasters caused 471,000 fatalities worldwide and economic losses totaling approximately $49 billion, out of which approximately 94,000 (20%) of the fatalities and $16 billion (33%) of the economic damages were attributed to floods alone. Comparing with fast-paced flood, drought is a slow-onset natural hazard that can nonetheless have wide and prolonged negative environmental and socioeconomic impacts.

As water security is one of the most critical international issues, the current challenges and opportunities in hydrologic science lie in (1) the capability of the hydrologic model in representing the physical hydrologic process, that is, the model structure; (2) model parameter optimization; and (3) accuracy, reliability, and adequacy of forcing data (e.g., in situ and remote sensing observation). Satellite remote sensing techniques bear the promising potential to overcome the limited spatial coverage of in situ observations, thus enabling better understanding of hydrologic cycles on regional and global scales. Compared to site-based measurements, remote sensing observations can better represent the spatial variability of the hydrologic components such as precipitation, soil moisture, evapotranspiration, and total water storage on a large scale (e.g., over the globe). Satellite remote sensing techniques are of particular importance for emerging regions that have inadequate or even no in situ observations. To date, remote sensing techniques have become a cost-effective approach to observe hydrologic variables on large scales and to further improve flood and drought prediction, especially for those emerging regions.

Motivated by innovative technologies such as satellite remote sensing and data assimilation, which can advance hydrologic science, this book particularly addresses the challenges and opportunities stated earlier and reviews multiple satellite remote sensing observations for water cycle in emerging regions and over the globe, the application of satellite remote sensing in hydrologic modeling and data assimilation, and the hydrologic capacity building from NASA and the HyDROS Group at the University of Oklahoma (http://hydro.ou.edu/) over the past decade. This book also intends to address the following questions:

1. What are the satellite missions for hydrologic remote sensing measurements?
2. How are the accuracy and uncertainty of the satellite remote sensing estimated in key hydrologic variables such as precipitation, soil moisture, and evapotranspiration?

3. How do satellite-driven hydrologic predictions perform in emerging regions?
4. Can remote sensing data complement or even replace in situ networks to force and calibrate hydrologic models, especially over vast and sparsely gauged basins or even ungauged basins?
5. Can spaceborne streamflow signals compensate for the uncertainty of spaceborne precipitation data to achieve hydrologic prediction skill comparable to results benchmarked with conventional observations?
6. What have we contributed to the hydrologic capacity building over the past decade in emerging regions and over the globe?

In this book, cutting-edge satellite remote sensing techniques and hydrologic, land surface, climate models, data assimilation methods, and capacity building over remote regions are introduced in three sections. Section I overviews satellite remote sensing observations of precipitation, soil moisture, evapotranspiration, reservoir, and total water storage. Section II reviews various applications of satellite remote sensing in hydrologic/land surface/climate modeling and ensemble square root filter (EnSRF) data assimilation in different regions. The first two sections collectively demonstrate the critical potential of satellite remote sensing, hydrologic modeling, and data assimilation techniques in improving societal resilience and environmental sustainability, as it is among the most cost-effective measures to reduce the devastation triggered by hydrologic hazards such as floods and landslides. Section III further introduces the development of hydrologic capacity building applications by collaborating with local stakeholders and decision makers, in order to enhance the preparedness and mitigation of water-related disasters.

In summary, this book presents a collection of recent innovative hydrology-relevant capacity building research conducted on emerging regions with inadequate observations, and hopefully provides a hydrologic research brochure for government officials, researchers, students, and practitioners in hydrology, remote sensing, meteorology, and climate science.

Yang Hong
National Weather Center

Yu Zhang
Risk Management Solutions Inc.

Sadiq Ibrahim Khan
NOAA National Water Center

Editors

Yang Hong is a professor of hydrometeorology and remote sensing in the School of Civil Engineering and adjunct professor in the School of Meteorology, University of Oklahoma. Currently, he also serves as honorary chair professor at Tsinghua University, Beijing, China. Previously, he was a research scientist at NASA's Goddard Space Flight Center and a postdoctoral researcher at University of California, Irvine.

Dr. Hong currently directs the HyDROS Lab (http://hydro.ou.edu) at the National Weather Center and also serves as the codirector of WaTER (Water Technology for Emerging Regions) Center. He is a faculty member of the Advanced Radar Research Center and is also an affiliated member of the Center for Analysis and Prediction of Storms at the University of Oklahoma. His areas of research span the wide range of hydrology–meteorology–climatology, with a particular interest in bridging the gap among water–weather–climate–human systems across scales in space and time. He has developed and taught subjects such as remote sensing retrieval and applications, advanced hydrologic modeling, climate change and natural hazards, engineering survey/measurement and statistics, land surface modeling, and data assimilation systems for hydrologic cycle and water systems under a changing climate.

Dr. Hong has served on several international and national committees, review panels, and editorial boards of several journals. He has served as chair of the AGU-Hydrology Section Technical Committee on Precipitation (2008–2012) and has been an editor of numerous journals. He is a two-time recipient of the NASA Group Achievement Award "For Significant Achievement in Systematically Promoting and Accelerating the Use of NASA Scientific Research Results for Societal Benefits" and "For Global Precipitation Measuring Mission Post-launch Performance" by the NASA Headquarter Administrator in 2008 and 2015, respectively. He also received the 2014 University Regents' Award for Superior Research "For Superior Accomplishments in Research and Creative Activity, and Professional and University Service."

He has extensively published in journals of remote sensing, hydrology, meteorology, and hazards and has disseminated several technologies to universities, governmental agencies, and private companies. Dr. Hong earned a PhD major in hydrology and water resources with a PhD minor in remote sensing and spatial analysis at the University of Arizona (2003), an MS (1999) in environmental sciences, and a BS (1996) in geosciences at Peking (Beijing) University, China.

Yu Zhang is currently a modeler at Risk Management Solutions, Inc, Newark, California. Prior to that, Dr. Zhang was a postdoctoral research associate in Civil and Environmental Engineering Department at Princeton University, New Jersey. Dr. Zhang earned her PhD in civil engineering from University of Oklahoma in December 2013 and her MS (2010) and BS (2008) degrees are in hydrology and water resources engineering, respectively, from Northwest A&F University, China. Dr. Zhang was a recipient of Best Student Paper Award at Advanced Radar Research Center (ARRC) in 2013 and she received the Oklahoma Space Grant Counsortium/

NASA EPSCOR Travel Grant in 2011. She was also selected as a "rising star" who was among the 20 top early career women in civil and environmental engineering by MIT in 2015 (http://ceerisingstars.mit.edu/). Dr. Zhang's current work primarily focuses on flood modeling (e.g., pluvial and fluvial flood modeling), and vulnerability modeling. In addition, her research interests and recent work also includes improving global and regional hydrologic prediction from short-term (hours to days) to seasonal scales with emphasis on the application of remote sensing, numerical weather prediction (NWP), climate forecast system (CFS), land–atmosphere interaction and hydrologic data assimilation, estimating the global terrestrial water budget from multiple sources that range from reanalysis, remotely sensed and hydrologic model products, and hydrologic risk assessment.

Sadiq Ibrahim Khan is currently an associate scientist at the National Oceanic and Atmospheric Agency, National Water Center. Before joining the National Water Center, he was a research scientist in the School of Civil Engineering and Environmental Science at the University of Oklahoma, Norman. Dr. Khan earned his PhD (2011) at the University of Oklahoma. He is the recipient of the NASA Earth and Space Science Fellowship. He received the outstanding paper award at the AGU-Hydrology Section during the Fall 2009 Meeting in San Francisco, California. His main research interests include (1) use of numerical models and spatial analysis techniques to monitor atmospheric and land surface variables and therefore better predict water extremes and (2) satellite remote sensing application in hydrology, particularly optical and microwave remote sensing for water resource monitoring. Dr. Khan has been a member of several research projects supported by NASA, NOAA, the US Department of State, and National Academy of Sciences.

Contributors

Thomas E. Adams III
University Corporation for Atmospheric Research
Boulder, Colorado

Robert F. Adler
NASA Goddard Space Flight Center
Greenbelt, Maryland

and

Earth System Science Interdisciplinary Center
University of Maryland
College Park, Maryland

Emmanouil N. Anagnostou
Department of Civil and Environmental Engineering
University of Connecticut
Storrs, Connecticut

Seleshi Bekele Awulachew
United Nations Department of Economic and Social Affairs
New York, New York

Eun-Chul Chang
Department of Atmospheric Science
Kongju National University
Gongju, Korea

Lu Chen
College of Hydropower and Information Engineering
Huazhong University of Science and Technology
Wuhan, China

Sheng Chen
School of Civil Engineering and Environmental Sciences
University of Oklahoma
Norman, Oklahoma

Zhuoqi Chen
College of Global Change and Earth System Science
Beijing Normal University
Beijing, China

Hongguang Cheng
Green Development Institute
School of Environment
Beijing Normal University
Beijing, China

Robert A. Clark III
Cooperative Institute for Mesoscale Meteorological Studies
University of Oklahoma
and
NOAA National Severe Storms Laboratory
Norman, Oklahoma

Chhimi Dorji
Department of Hydro-Met Services
Ministry of Economic Affairs
Thimphu, Bhutan

Michael Durand
School of Earth Sciences
Ohio State University
Columbus, Ohio

Zachary L. Flamig
Cooperative Institute for Mesoscale
    Meteorological Studies
University of Oklahoma
and
NOAA National Severe Storms
    Laboratory
Norman, Oklahoma

Bing Gao
School of Water Resources and
    Environment
China University of Geosciences
Beijing, China

Huilin Gao
Zachry Department of Civil
    Engineering
Texas A&M University
College Station, Texas

Abebe Sine Gebregiorgis
School of Civil Engineering and
    Environmental Sciences
University of Oklahoma
Norman, Oklahoma

Paola Gonzalez
Environmental Engineering Department
El Bosque University
Bogota, Colombia

Jonathan Gourley
NOAA National Severe Storms
    Laboratory
School of Meteorology at the University
    of Oklahoma
Norman, Oklahoma

Shahid Habib
NASA Goddard Space Flight Center
Greenbelt, Maryland

Fanghua Hao
Green Development Institute
School of Environment
Beijing Normal University
Beijing, China

Zengchao Hao
Green Development Institute
School of Environment
Beijing Normal University
Beijing, China

Xiaogang He
Institute of Industrial Science
University of Tokyo
Tokyo, Japan

Yang Hong
School of Civil Engineering and
    Environmental Sciences
and
School of Meteorology
and
Advanced Radar Research Center
University of Oklahoma
Norman, Oklahoma

George. J. Huffman
NASA Goddard Space Flight Center
Greenbelt, Maryland

Daniel Irwin
NASA Marshall Space Flight Center
Huntsville, Alabama

Maria Jurado
Environmental Engineering
    Department
El Bosque University
Bogota, Colombia

Sadiq Ibrahim Khan
The University Corporation for
    Atmospheric Research
Boulder, Colorado

and

National Oceanic and Atmospheric
    Administration
National Water Center
Tuscaloosa, Alabama

Hyungjun Kim
Institute of Industrial Science
University of Tokyo
Tokyo, Japan

Dalia Kirschbaum
NASA Goddard Space Flight Center
Greenbelt, Maryland

Pierre-Emmanuel Kirstetter
Advanced Radar Research Center
NOAA National Severe Storms
    Laboratory
National Weather Center
Norman, Oklahoma

Tesfaye Korme
African Regional Centre for Mapping
    of Resources for Development
Nairobi, Kenya

Hyongki Lee
Department of Civil and Environmental
    Engineering
University of Houston
Houston, Texas

Bin Li
Key Laboratory of Digital Earth
    Science
Institute of Remote Sensing and
    Digital Earth
Chinese Academy of Sciences
Beijing, China

Li Li
School of Civil Engineering and
    Environmental Sciences
University of Oklahoma
Norman, Oklahoma

Xiaodong Li
State Key Laboratory of Hydraulics and
    Mountain River Engineering
Sichuan University
Chengdu, China

Zhe Li
Key Laboratory of Water Cycle and
    Related Land Surface Processes
Institute of Geographical Sciences and
    Natural Resources Research
Chinese Academy of Sciences
Beijing, China

Zonghu Liao
State Key Laboratory of Petroleum
    Resources and Prospecting
China University of Petroleum
Beijing, China

Ashutosh S. Limaye
NASA Marshall Space Flight Center
Huntsville, Alabama

Di Long
Department of Hydraulic Engineering
Tsinghua University
Beijing, China

and

Bureau of Economic Geology
Jackson School of Geosciences
The University of Texas at Austin
Austin, Texas

Laurent Longuevergne
Geosciences Rennes UMR CNRS 6118
Université de Rennes 1
Rennes, France

Hui Lu
Ministry of Education Key Laboratory
　for Earth System Modelling
Centre for Earth System Science
Tsinghua University
and
Joint Centre for Global Change Studies
Beijing, China

Qinghua Miao
State Key Laboratory of Hydroscience
　and Engineering
Department of Hydraulic Engineering
Tsinghua University
Beijing, China

Semu Ayalew Moges
Civil Engineering Department
Addis Ababa University
Addis Ababa, Ethiopia

Kenneth Ochoa
Environmental Engineering Department
El Bosque University
Bogota, Colombia

Lawrence Okello
African Regional Centre for Mapping
　of Resources for Development
Nairobi, Kenya

Taikan Oki
Institute of Industrial Science
University of Tokyo
Tokyo, Japan

Wei Ouyang
Green Development Institute
School of Environment
Beijing Normal University
Beijing, China

Maritza Paez
Environmental Engineering Department
El Bosque University
Bogota, Colombia

Ming Pan
Department of Civil and Environmental
　Engineering
Princeton University
Princeton, New Jersey

and

State Key Laboratory of Hydraulics and
　Mountain River Engineering
Sichuan University
Chengdu, China

Yang Pan
National Meteorological Information
　Center
Beijing, China

Frederick S. Policelli
NASA Goddard Space Flight Center
Greenbelt, Maryland

Guanghua Qin
State Key Laboratory of Hydraulics and
　Mountain River Engineering
Sichuan University
Chengdu, China

Manabendra Saharia
School of Civil Engineering and
　Environmental Sciences
University of Oklahoma
Norman, Oklahoma

Xinyi Shen
Department of Civil and Environmental
　Engineering
University of Connecticut
Storrs, Connecticut

Yan Shen
National Meteorological Information
　Center
Beijing, China

Yanjun Shen
Center for Agricultural Resources
   Research
Chinese Academy of Sciences
Shijiazhuang, China

Vijay P. Singh
Department of Biological and
   Agricultural Engineering
and
Zachry Department of Civil
   Engineering
Texas A&M University
College Station, Texas

Alexander Sun
Bureau of Economic Geology
Jackson School of Geosciences
The University of Texas at Austin
Austin, Texas

Guoqiang Tang
Department of Hydraulic Engineering
Tsinghua University
Beijing, China

Qiuhong Tang
Institute of Geographical Sciences and
   Natural Resources Research
Key Laboratory of Water Cycle and
   Related Land Surface Processes
Chinese Academy of Sciences
Beijing, China

Humberto Vergara
School of Civil Engineering and
   Environmental Sciences
and
Cooperative Institute for Mesoscale
   Meteorological Studies
University of Oklahoma
Norman, Oklahoma

and

Environmental Engineering Department
El Bosque University
Bogota, Colombia

Zhanming Wan
School of Civil Engineering and
   Environmental Sciences
Hydrometeorology & Remote Sensing
   Laboratory
University of Oklahoma
Norman, Oklahoma

Jiahu Wang
School of Civil Engineering and
   Environmental Sciences
National Weather Center
University of Oklahoma
Norman, Oklahoma

Zhongwang Wei
Department of Natural Environmental
   Studies
and
Atmosphere and Ocean Research
   Institute
University of Tokyo
Tokyo, Japan

Yixin Wen
School of Civil Engineering and
   Environmental Sciences
University of Oklahoma
Norman, Oklahoma

Youlong Xia
Environmental Modeling Center
National Centers for Environmental
   Prediction
I. M. Systems Group
College Park, Maryland

Anyuan Xiong
National Meteorological Information
   Center
Beijing, China

Xianwu Xue
School of Civil Engineering and
    Environmental Sciences
Advanced Radar Research Center
University of Oklahoma
Norman, Oklahoma

Dawen Yang
State Key Laboratory of Hydroscience
    and Engineering
Department of Hydraulic Engineering
Tsinghua University
Beijing, China

Kun Yang
State Key Laboratory of Petroleum
    Resources and Prospecting
China University of Petroleum
Beijing, China

Yuting Yang
National Centre for Groundwater
    Research and Training
and
School of the Environment
Flinders University
Adelaide, Australia

Kei Yoshimura
Department of Climate Variability
    Research
Atmosphere and Ocean Research
    Institute
University of Tokyo
Chiba, Japan

Jingjing Yu
National Meteorological Information
    Center
Beijing, China

Xing Yuan
Key Laboratory of Regional Climate-
    Environment for Temperate East
    Asia
Institute of Atmospheric Physics
Chinese Academy of Sciences
Beijing, China

Ke Zhang
Cooperative Institute for Mesoscale
    Meteorological Studies
University of Oklahoma
Norman, Oklahoma

and

State Key Laboratory of Hydrology-
    Water Resources and Hydraulic
    Engineering
Hohai University
Nanjing, China

Shuai Zhang
Zachry Department of Civil
    Engineering
Texas A&M University
College Station, Texas

Yu Zhang
Department of Civil and Environmental
    Engineering
Princeton University
Princeton, New Jersey

YaoYao Zheng
School of Civil Engineering and
    Environmental Sciences
University of Oklahoma
Norman, Oklahoma

Section I

Remote Sensing Observations and Estimations

1 From Tropical to Global Precipitation Measurement

Initial Validation and Application

Guoqiang Tang, Yixin Wen, YaoYao Zheng,
Di Long, and Yang Hong

CONTENTS

1.1 INTRODUCTION

Precipitation is one of the most important factors that affect the global water and energy balance (Kidd and Huffman, 2011). Researchers can hardly conduct better simulations of the water cycle over regions without accurate precipitation inputs (Xue et al., 2013). Traditionally, there are three ways to measure precipitation, that is, rain gauges, weather radars, and satellite-based sensors. The rain gauge is a conventional way of providing the most straightforward observations of site-based surface precipitation. However, gauge networks are sparse over most of continents and few gauges are located over the ocean (Kidd and Huffman, 2011; Mishra and Coulibaly, 2009). The weather radar can monitor precipitation with relatively higher temporal and spatial resolutions, but is often subject to low data quality in complex terrain mostly due to signal blockage, attenuation by rain, and vertical variability of reflectivity (Dinku et al., 2002; Tian and Peters-Lidard, 2010). Currently, the only

3

practical way to achieve a comprehensive estimate of precipitation on a global basis comes from Earth observation satellites (Hong et al., 2012; Hou et al., 2014). Global satellite-based rainfall products are currently based on passive microwave (PMW), calibrated infrared (IR), and PMW plus IR observations. IR sensors on geostationary Earth orbit (GEO) satellites can provide precipitation estimates at high temporal resolutions, but the accuracy of IR-based estimates is generally not very good due to the indirect linkage between IR signals and precipitation. PMW sensors are more popular in precipitation estimation because their radiative signatures are more directly linked to precipitating particles. However, PMW sensors are only onboard low Earth orbiting (LEO) satellites at present, leading to low temporal sampling. Therefore, combining GEO IR and LEO PMW sensors to improve the accuracy, coverage, and resolution of global precipitation products has been widely recognized and applied (Hong et al., 2012).

Since the launch of Tropical Rainfall Measurement Mission (TRMM) in late 1997, rapid development of precipitation datasets based on PMW, calibrated IR, and PMW plus IR observations has provided a tremendous amount of quasiglobal information for research and applications. To date, a number of satellite precipitation products have been released to the public with various temporal and spatial resolutions, such as TRMM multisatellite precipitation analysis (TMPA) (Huffman et al., 2007), Climate Prediction Center MORPHing (CMORPH) technique (Joyce et al., 2004), Precipitation Estimation from Remotely Sensed Information using Artificial Neural Networks (PERSIANN) (Sorooshian et al., 2000), PERSIANN Cloud Classification System (Hong et al., 2004), Naval Research Laboratory (NRL)-developed blended-satellite precipitation technique (Turk and Miller, 2005), and Global Satellite Mapping of Precipitation (Kubota et al., 2007). The TMPA products used in this study are the real-time 3B42RT (hereafter referred to as 3B42RT) and post-real-time 3B42 V7 (hereafter 3B42V7) products, which have been widely studied and applied in hydrologic simulations and predictions (Habib et al., 2012; Hong et al., 2007; Li et al., 2013; Long et al., 2014; Prakash et al., 2015; Zulkafli et al., 2014). Research and applications of TMPA products have produced great scientific, societal, and economic benefits, such as extreme weather event prediction, disaster forecasts, and water resources planning and management. Despite the great achievement in the TRMM era, TRMM data have some inherent limitations associated with spatiotemporal coverage and uncertainty of solid or light precipitation estimation over higher latitudes/altitudes (Hong et al., 2012; Hou et al., 2014; Huffman et al., 2007; Yong et al., 2014). In addition, the instruments on TRMM were turned off on April 8, 2015, and the spacecraft re-entered the Earth's atmosphere on June 15, 2015, most of which was burnt up in the atmosphere.

Building upon the TRMM heritage, the Global Precipitation Measurement (GPM) mission is an international network of satellites that provides the next-generation global rain and snow products at a spatial resolution of $0.1° \times 0.1°$ with a half-hourly temporal resolution (Hou et al., 2014). As the TRMM successor, the GPM Core Observatory was launched on February 27, 2014, marking the beginning of the GPM era. The GPM constellation consists of this Core Observatory and has approximated 10 partner satellites during the study period. Composition of the constellation will change with the launch and failure of partner satellites. The GPM Core Observatory carries a Dual-frequency Precipitation Radar (DPR; the Ku-band at 13.6 GHz and Ka-band at 35.5 GHz) and a multichannel GPM Microwave Imager (GMI; frequencies range between 10 and 183 GHz), both of which improve on the capabilities of measuring precipitation compared to the predecessor, TRMM instruments (Hou et al., 2014).

As an extension and also upgrade of the highly successful TRMM mission, GPM was anticipated to provide better products. This chapter is generally based on Tang et al. (2015). The objectives are to (1) introduce the GPM core sensor and its constellations, (2) evaluate the GPM-era IMERG product compared with TRMM-era 3B42V7 and 3B42RT estimates statistically, and (3) explore the continuity between IMERG, 3B42V7 and 3B42RT when used in hydrologic calibration and simulation in the Ganjiang River basin, a middle-latitude basin in Southeast China. This study is among the very early attempts to evaluate the newest GPM product, and we expect that the results reported here can shed light on subsequent investigations.

1.2 GPM CORE OBSERVATORY AND CONSTELLATION

1.2.1 THE DUAL-FREQUENCY PRECIPITATION RADAR AND GPM MICROWAVE IMAGER

The GMP core observatory carries a DPR and a multichannel microwave radiometer, which are used as a reference to intercalibrate the partner constellation radiometers, thus providing self-consistent radiometric observations across the constellation (Figure 1.1).

The DRP consists of a Ku-band radar (13.6 GHz) and a Ka-band radar (35.5 GHz). KuPR scan pattern, with 49 footprints, is similar to TRMM and the scan swath is 245 km. The KaPR also has 49 footprints but in two types of scan pattern. The central 25 beams of KaPR are matched to the central 25 beams of KuPR, called matched scan (KaMA). In the second kind of scan, the KaPR is operated in a high-sensitivity mode, called KaHS, which is specifically designed for the detection of light rain and snow. Compared to PR used in TRMM, the main improvement of DPR is that the Ka-band frequency channel can add non-Rayleigh scattering effects at high frequency, add estimates of the phase-transition height in precipitating systems, and add the sensitivity of detection of snow. However, a few uncertainties in DPR precipitation retrieval should be pointed out for better understanding, one comes from conversion from radar reflectivity into rainfall rate, the other comes from variation of raindrop size distribution (Figure 1.2).

The GMI instrument is a multichannel, conical-scanning, microwave radiometer serving an essential role in the near-global-coverage and frequent-revisit-time requirements of GPM. The GMI is characterized by 13 microwave channels ranging in frequency from 10 to 183 GHz. In addition to carrying channels similar to those on the TRMM Microwave Imager (TMI), the GMI carries four high frequency, millimeter-wave, channels about 166 and 183 GHz. Except for the heritage hot load and cold load that are commonly used for linear sensor radiometric calibrations, a hot noise diode and a cold noise diode are implemented in the GMI to determine the nonlinearity and noise levels of the measurements.

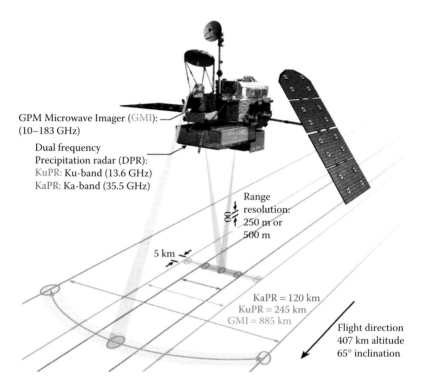

GPM Microwave Imager (GMI):
(10–183 GHz)

   Dual frequency
   Precipitation radar (DPR):
   KuPR: Ku-band (13.6 GHz)
   KaPR: Ka-band (35.5 GHz)

Range
resolution:
250 m or
500 m

5 km

KaPR = 120 km
KuPR = 245 km
GMI = 885 km

Flight direction
407 km altitude
65° inclination

FIGURE 1.1 GPM core observatory and instruments. (From Hou, A. Y. et al., *Bulletin of the American Meteorological Society*, 95, 701–722, 2014.)

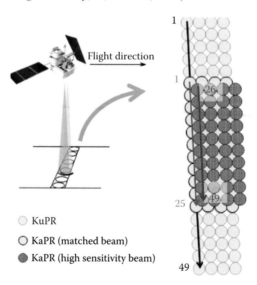

Flight direction

KuPR

KaPR (matched beam)

KaPR (high sensitivity beam)

FIGURE 1.2 DPR instrument. (From Iguchi, T. et al., GPM/DPR level-2 algorithm theoretical basis document, 68 pp, 2015. http://pps.gsfc.nasa.gov/Documents/ATBD_DPR_2015_whole_a.pdf.)

Rotating at 32 rotations per minute, the GMI will gather microwave radiometric brightness measurements over a 140° sector centered about the spacecraft ground track vector. The 140° GMI swath represents a swath of 904 km (562 miles) on the Earth's surface. For comparison, the DPR instrument is characterized by cross-track swath widths of 245 km (152 miles) and 120 km (75 miles), for the Ku- and Ka-band radars, respectively. Only the central portions of the GMI swath will overlap the radar swaths (and with approximately 67 s duration between measurements due to the geometry and spacecraft motion). These measurements within the overlapped swaths are important for improving precipitation retrievals, and in particular, the radiometer-based retrievals.

1.2.2 GPM CONSTELLATION

GPM spacecraft instruments and retrieved precipitation data products using a network/ constellation with GPM core satellite and other partner satellites. The remainder of the GPM constellation is comprised of a number of satellites with GMI-like radiometers or microwave sounding instruments, including the DMSP F19 and F20 (US DoD; imager), GCOM-W1 (JAXA; imager), JPSS-1 (NASA/NOAA; sounder), Megha–Tropiques (CNES/ISRO; imager and sounder), MetOp B and C (EUMETSAT; sounder), NOAA 19 (NOAA; sounder), and NPP (NASA/NOAA; sounder) satellites.

For microwave imager constellation, the imager channels are considered best for low- and mid-latitude use, while the sounding channels maintain some skill in cold and frozen-surface conditions. Research work with the high-frequency channels is beginning to demonstrate that the high-frequency channels on AMSU, GMI, MHS, and SSMIS are also useful at higher latitudes.

1.3 STATISTICAL VALIDATION OF GPM PRECIPITATION PRODUCTS

The GMP data are processed in three levels. The first level is about instrument-independent physical variables like the brightness temperature and radar power. The second level is about derived geophysical parameter at the same resolution and location as level 1. The third level is spatially and time-averaged data. Swath structures are used in level 1 and level 2 while level 3 uses the gridded structure.

The Integrated Multi-satellite Retrievals for GPM (IMERG) algorithm, level 3, is intended to intercalibrate, merge, and interpolate all MW estimates of the GPM constellation, IR estimates, gauge observations, and other potential sensors' data with a $0.1° \times 0.1°$ spatial resolution and 30-min temporal resolution. The Ganjiang River basin, with a drainage area of 81,258 km^2 above the Waizhou hydrologic station, is the seventh largest subcatchment of the Yangtze River, located within 113°30′–116°40′E and 24°29′–29°21′N in Southeast China (Figure 1.3). Precipitation observations of 310 gauges were provided from 2003 to 2009, which were used to calibrate the Coupled Routing and Excess Storage (CREST) model in hydrologic Scenario I as described below. Hu et al. (2014) conducted multiscale evaluation of six high-resolution satellite monthly rainfall estimates over the Ganjiang River basin

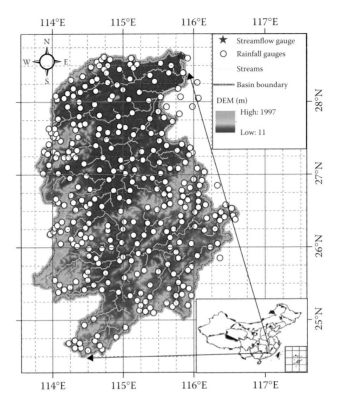

FIGURE 1.3 Map of the Ganjiang River basin, with streams and elevation of the basin, the streamflow gauge station (termed the Waizhou station hereafter), and rain gauges that provide ground reference to evaluate satellite-based precipitation products in this study.

and concluded that TRMM 3B43V6 is the best, and 3B42RTV6 and CMORPH performed better among five other pure satellite-derived products. Therefore, it is meaningful to perform a study to quantitatively evaluate the quality of the latest GPM data and its continuity to TRMM in this area.

Evaluation and comparison were performed for the IMERG and TMPA products both at 39 grid boxes of $0.25° \times 0.25°$, with at least one rain gauge in each grid box, and over the entire basin. Gauge number is less than 310 because during the study time period (from May to September 2014), observed data from 310 gauges in Figure 1.3 were not provided. The $0.1° \times 0.1°$ gridded IMERG estimates were aggregated to $0.25° \times 0.25°$ datasets, the same as TMPA and CGDPA (China Gauge-based Daily Precipitation Analysis). Among all the gauges used in the production of CGDPA, 41 gauges are located in the Ganjiang River basin. So we used the 41 gauges to evaluate the quality of IMERG and TMPA products at the grid scale. In areal comparison, the CGDPA gridded product ($0.25° \times 0.25°$) was used as ground reference. The accuracy of CGDPA over this area is high (Shen and Xiong, 2015), which has also been shown by the excellent performance of CGPDA in a hydrologic simulation in this study.

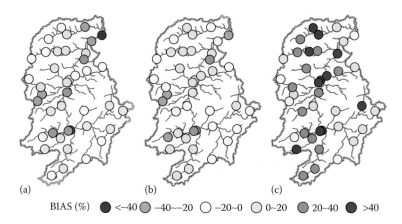

FIGURE 1.4 BIAS distribution for daily precipitation between the satellite precipitation products of (a) IMERG and gauge; (b) 3B42V7 and gauge; and (c) 3B42RT and gauge from May to September 2014.

For the comparison at the grid scale, we derived daily values of satellite products and CGDPA at all grid boxes within the study area to calculate the statistics. In general, IMERG and 3B42V7 match similarly well with gauge observations and both outperform 3B42RT (Figure 1.4a and b). Both post-real-time corrections effectively reduced the biases of IMERG and 3B42V7 to single digits of underestimation (approximately −1% for grid points and −4% over the whole basin) from positive 20+% of 3B42RT. As shown in Figure 1.4, when comparing the precipitation of 41 rain gauges and three satellite precipitation products and calculating the BIAS for each station, we found that 3B42RT overestimates precipitation for 37 rain gauges in all the 41 rain gauges, consistent with the large BIAS at the grid and basin-scale for 3B42RT. But for IMERG and 3B42V7, overestimation and underestimation are close in the number of gauges.

In general, IMERG and 3B42V7 are comparable with little difference in this study, which may be attributed to a number of factors as follows. For example, the latitude of our study area is relatively low and the study timespan was the rainy season with intensive precipitation, which may contribute to the similarity between IMERG and 3B42V7. Overall, all three products demonstrated similarly acceptable (~0.63) and high (0.87) correlation at grid and basin-scales, respectively, but 3B42RT showed much higher overestimation. The limited number of gauges used in this study could be a source of error in the analyses above.

1.4 HYDROLOGIC APPLICATIONS OF GPM

The CREST model (Wang et al., 2011; Shen et al., 2016) is a grid-based distributed hydrologic model developed by the University of Oklahoma (http://hydro.ou.edu) and the NASA-SERVIR Project Team (www.servir.net). Streamflow prediction performance of the three different precipitation products was investigated using two different parameter set-up scenarios for the CREST hydrologic model in this section:

Scenario I (static parameters): Model parameters were first calibrated using observed data of 310 rain gauges from January 2003 to December 2009 and the parameter sets were used in this scenario. Then, the model was run using precipitation from CGDPA, 3B42V7, and 3B42RT in independent validation period 1 (January 1, 2010, to December 31, 2013) with the rain gauge-calibrated model parameters. The three precipitation products afore-mentioned and the IMERG product were used to force model in validation period 2 (May 1, 2014, to September 30, 2014) with the same parameter set.

Scenario II (dynamic parameters): Model parameters were dynamically recali-brated according to precipitation inputs respectively, for example, 3B42V7 and 3B42RT, from January 2008 to December 2010, respectively. The two calibrations of the model were subsequently validated using 3B42V7 and 3B42RT in validation period 1 (January 1, 2011, to December 31, 2013), respectively. Then, we used the product-specific parameter sets to simulate streamflow based on the IMERG precipitation input to compare the perfor-mance of IMERG, 3B42V7, and 3B42RT over validation period 2 (May 1, 2014, to September 30, 2014).

1.4.1 SCENARIO I: STATIC PARAMETERS

The simulated streamflow based on gauge precipitation agrees well with the observed streamflow in the calibration period. Metrics of the calibration period show the great utility of gauge-calibrated model parameters with very high NSCE (0.90) and CC (0.95), as well as small BIAS (−6.89%) and the least RMSE (534.81 m³/s).

In validation period 2 (May 1, 2014, to September 30, 2014), the CREST model was forced by CGDPA, 3B42V7, 3B42RT, and IMERG precipitation data, respec-tively (Figure 1.5). The CGDPA reference data have the best skill scores as expected, closely followed by the IMERG metrics (IMERG/CGDPA: NSCE = 0.77/0.86, CC = 0.91/0.94, BIAS = −14.09%/−8.76%, and RMSE = 1080.87/822.73 m³/s). The hydrograph of IMERG is remarkably similar to that of CGDPA (Figure 1.5a and b). Among the three satellite precipitation products, IMERG performed the best and matched well with the observed streamflow, especially for the second flow peak compared with 3B42V7 and 3B42RT (Figure 1.4b and c). The 3B42RT product came at the bottom as its NSCE declined to 0.46 and RMSE increased to 1637.53 m³/s.

We speculate that three factors led to the better performance of IMERG. First, IMERG uses the Goddard Profiling Algorithm 2014 (GPROF 2014) to retrieve rain-fall and vertical structure information from satellite-based passive MW observa-tions, whereas 3B42V7 uses GPROF 2010. Second, the IMERG product involved more sensors than the 3B42V7 product, which could improve the quality of IMERG. Third, in the statistical comparison, IMERG and 3B42V7 show comparable per-formance, but IMERG is slightly better at the basin-scale. As such, better spatial representation of precipitation systems could most likely reduce error propagations

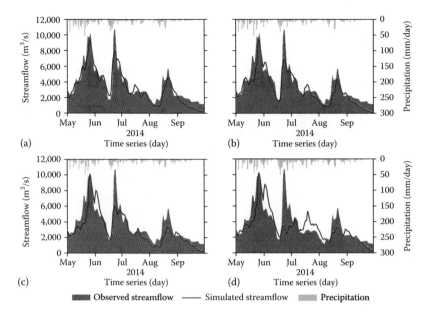

FIGURE 1.5 Comparison of CREST-simulated streamflow with gauge-calibrated parameters and observed streamflow validation period 2 (May 1, 2014, to September 30, 2014). (a) Daily data from CGDPA; (b) daily data from IMERG; (c) daily data from 3B42V7; and (d) daily data from 3B42RT.

from forcing data to hydrologic prediction at the basin outlet because CREST is a distributed hydrologic model that takes into account not only the intensity but also spatial variability in rainfall data.

In summary, IMERG performs very well, showing comparable hydrologic utility with gauge-based CGDPA during the extended rainy season in the Ganjiang River basin with the calibrated static parameter set during the baseline period, followed by the acceptable 3B42V7 and the less-reliable 3B42RT products (due to its large uncertainty). But the difference between IMERG and 3B42V7 is modest from statistics.

1.4.2 Scenario II: Dynamic Parameters

In Scenario II, 3B42V7 and 3B42RT were separately used to recalibrate the CREST model from January 2008 to December 2010, to further investigate the applicability of the IMERG product in the TMPA product-specific calibrated model. Then the CREST model was validated for two periods with IMERG and TMPA products, based on the corresponding dynamic parameter sets.

In validation period 2, the 3B42V7-forced simulation curve fits well with the curve of observation and captures the main flow peaks, whereas it underestimates the third main flow peak and overestimates the streamflow trend after the first flow peak (Figure 1.6a). The IMERG simulation based on the 3B42V7-specific parameter set shows a similar curve with 3B42V7, with slightly higher NSCE (Figure 1.6a). As for 3B42RT, deviation from observations is obvious after the first and second flow

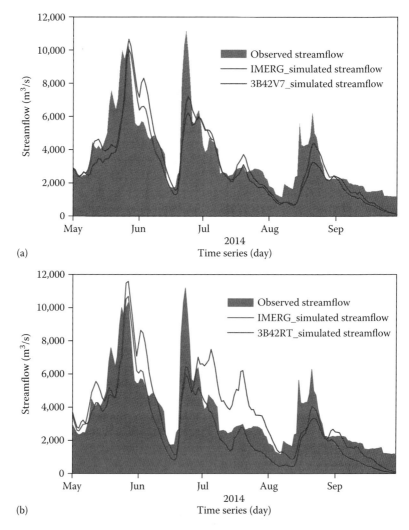

FIGURE 1.6 Comparison of CREST-simulated streamflow with parameters recalibrated using (a) 3B42V7 and (b) 3B42RT, respectively, in validation period 2 (May 1, 2014, to September 30, 2014). The red and blue lines represent streamflow simulated using IMERG and corresponding TMPA products.

peaks, while the deviation is reduced when forcing the model using IMERG based on the 3B42RT-specific parameter set (Figure 1.6b). All metrics of IMERG show overall very desirable hydrologic continuity from TRMM-era calibrated 3B42V7-specific and 3B42RT-specific parameter sets in previous period, except the BIAS of IMERG that shows slight deterioration from 3B42RT.

The comparison results under both gauge reference-based Scenario I and forcing-specific Scenario II above all showed that IMERG has high hydrologic utility and continuity (i.e., a good and robust successor) from TRMM-era standard products whether the hydrologic model is calibrated using rain gauge data or TMPA-specific products. If the hydrologic model is calibrated using rain gauge data, the IMERG product can be used directly in the model and the results could be comparable to that using rain gauge data. If the model is calibrated using TMPA 3B42 products, it is anticipated that consistent results can be obtained compared with 3B42V7 and 3B42RT, and some deviations from observed streamflow could even be reduced, especially for the 3B32RT product with higher uncertainty in this study. The conclusion from our basin specific study may not apply to other basins especially in different climates.

1.5 CONCLUSIVE REMARKS AND OUTLOOKS

After 17-year success of quasiglobal precipitation measurement, TRMM was decommissioned and GPM is providing the next generation of global precipitation products. This chapter first overviews the GPM and its products; then focuses on the statistical evaluation of TRMM-era standard products 3B42V7/3B42RT and the GPM-era IMERG research product; and finally, hydrologically evaluates their streamflow prediction utility using the CREST hydrologic model in the Ganjiang River basin, a midlatitude basin in China. Because of the delayed IMERG data availability and also the reliable ground gauge network, this study is limited to the extended rainy season of May–September 2014. Results from the above analyses may be specific to the Ganjiang River basin but are likely to be more generally applicable to basins situated in the subtropics.

The main conclusions from the statistical comparison of IMERG, 3B42V7, and 3B42RT, conducted at both grid and basin scales from May to September 2014, are summarized as follows:

1. In general, both IMERG and 3B42V7 fit similarly well with gauge observations and outperform 3B42RT. The post-real-time monthly gauge analysis-based corrections used in both datasets effectively reduce the biases of IMERG and 3B42V7 to single digits of underestimation, approximately −1% for grid points and −4% over the whole basin, compared to the positive 20+% of 3B42RT.
2. All three products show acceptable (~0.63) and high (0.87) correlation at grid and basin-scales, respectively, indicating a similar capability of capturing trends in precipitation, but 3B42RT shows significant higher overestimation.

In terms of the hydrologic evaluation, two scenarios of hydrologic parameter sets are designed, with Scenario I benchmarked by in situ gauges and Scenario II calibrated with TMPA-based specific products. Scenario I is conventionally used over gauged basins where the hydrologic model can be tuned up using ground observations. In Scenario II, input-specific recalibration can be used as an alternative over not or sparsely gauged regions where remotely sensed data are acquired as substitutes to drive the models. The main conclusions are as follows.

1. In Scenario I, IMERG performs well, showing comparable hydrologic utility with gauge-based CGDPA during the extended rainy season in the Ganjiang River basin with the static parameter set calibrated during the baseline period (2003–2009), followed by the acceptable performance of 3B42V7 and less reliable skill of 3B42RT due to its large uncertainty.

2. In Scenario II, the CREST model was recalibrated with two best parameter sets using 3B42V7 and 3B42RT, respectively, and then validated with the three satellite products over later periods. As anticipated, 3B42V7 shows consistent hydrologic utility over the two periods, whereas the NSCE for 3B42RT reduces from 0.86 to 0.42, indicating that the recalibration for 3B42RT only improves the performance in the calibration period and its benchmarked hydrologic skill fails to carry over to later validation periods. Metrics of IMERG show overall very high hydrologic continuity from TRMM-era standard precipitation-specific parameter sets calibrated in the previous period, with only slight deterioration with 3B42RT-specfic calibration.

In conclusion, IMERG performs comparably to reference data and in many cases outperforms TMPA standard products in this study, indicating a promising prospect of hydrologic utility and also a desirable hydrologic continuity from TRMM-era product heritages to the GPM-era IMERG product even with its limited data availability to date in this well-gauged and midlatitude basin. It is reasonably anticipated that IMERG data would particularly outdistance TMPA products when it comes to high altitudes and/or high latitudes, given the capability of the GPM Core Observatory to detect light and solid precipitation (Hou et al., 2014). As more IMERG data and even retrospectively processed TRMM/GPM-era IMERG data sets are released, more studies to explore the potential of IMERG and other GPM-era products in water, weather, and climate studies are needed in the future.

REFERENCES

Dinku, T., E. N. Anagnostou, and M. Borga, 2002: Improving radar-based estimation of rainfall over complex terrain. *Journal of Applied Meteorology*, 41, 1163–1178.

Habib, E., M. ElSaadani, and A. T. Haile, 2012: Climatology-focused evaluation of CMORPH and TMPA satellite rainfall products over the Nile Basin. *Journal of Applied Meteorology and Climatology*, 51, 2105–2121.

Hong, Y., K.-L. Hsu, S. Sorooshian, and X. Gao, 2004: Precipitation estimation from remotely sensed imagery using an artificial neural network cloud classification system. *Journal of Applied Meteorology*, 43, 1834–1853.

Hong, Y., S. Chen, X. Xue, and G. Hodges, 2012: Global precipitation estimation and applications. In *Multiscale Hydrologic Remote Sensing: Perspectives and Applications*, N. -B. Chang (ed.), CRC Press, Boca Raton, FL, pp. 371–386.

Hong, Y., R. F. Adler, F. Hossain, S. Curtis, and G. J. Huffman, 2007: A first approach to global runoff simulation using satellite rainfall estimation. *Water Resources Research*, 43, W08502.

Hou, A. Y. and Coauthors, 2014: The global precipitation measurement mission. *Bulletin of the American Meteorological Society*, 95, 701–722.

Hu, Q., D. Yang, Z. Li, A. K. Mishra, Y. Wang, and H. Yang, 2014: Multi-scale evaluation of six high-resolution satellite monthly rainfall estimates over a humid region in China with dense rain gauges. *International Journal of Remote Sensing*, 35, 1272–1294.

Huffman, G. J. and Coauthors, 2007: The TRMM multisatellite precipitation analysis (TMPA): Quasi-global, multiyear, combined-sensor precipitation estimates at fine scales. *Journal of Hydrometeorology*, 8, 38–55.

Iguchi, T., S. Seto, R. Meneghini, N. Yoshida, J. Awaka, and T. Kubota, 2015: GPM/DPR level-2 algorithm theoretical basis document, 68 pp. http://pps.gsfc.nasa.gov/Documents/ATBD_DPR_2015_whole_a.pdf.

Joyce, R. J., J. E. Janowiak, P. A. Arkin, and P. Xie, 2004: CMORPH: A method that produces global precipitation estimates from passive microwave and infrared data at high spatial and temporal resolution. *Journal of Hydrometeorology*, 5, 487–503.

Kidd, C. and G. Huffman, 2011: Global precipitation measurement. *Meteorological Applications*, 18, 334–353.

Kubota, T. and Coauthors, 2007: Global precipitation map using satellite-borne microwave radiometers by the GSMaP project: Production and validation. *IEEE Transactions on Geoscience and Remote Sensing*, 45, 2259–2275.

Li, Z., D. Yang, and Y. Hong, 2013: Multi-scale evaluation of high-resolution multi-sensor blended global precipitation products over the Yangtze River. *Journal of Hydrology*, 500, 157–169.

Long, D. and Coauthors, 2014: Drought and flood monitoring for a large karst plateau in Southwest China using extended GRACE data. *Remote Sensing of Environment*, 155, 145–160.

Mishra, A. K. and P. Coulibaly, 2009: Developments in hydrometric network design: A review. *Reviews of Geophysics*, 47, RG2001.

Prakash, S., A. K. Mitra, I. M. Momin, D. S. Pai, E. N. Rajagopal, and S. Basu, 2015: Comparison of TMPA-3B42 versions 6 and 7 precipitation products with gauge-based data over India for the Southwest monsoon period. *Journal of Hydrometeorology*, 16, 346–362.

Shen, X., Y. Hong, K. Zhang, and Z. Hao, 2016: Refine a distributed reservoir routing method to improve performance of the CREST model (accepted). *Journal of Hydrologic Engineering*.

Shen, Y. and A. Xiong, 2015: Validation and comparison of a new gauge-based precipitation analysis over mainland China. *International Journal of Climatology*, 36, 252–265.

Sorooshian, S., K.-L. Hsu, X. Gao, H. V. Gupta, B. Imam, and D. Braithwaite, 2000: Evaluation of PERSIANN system satellite-based estimates of tropical rainfall. *Bulletin of the American Meteorological Society*, 81, 2035–2046.

Tang, G., Z. Zeng, D. Long, X. Guo, B. Yong, and W. Zhang, 2015: Statistical and hydrological comparisons between TRMM and GPM level-3 products over a mid-latitude basin: Is day-1 IMERG a good successor for TMPA 3B42V7? *Journal of Hydrometeorology*, 17(1), 121–137.

Tian, Y. and C. D. Peters-Lidard, 2010: A global map of uncertainties in satellite-based precipitation measurements. *Geophysical Research Letters*, 37, L24407.

Turk, F. J. and S. D. Miller, 2005: Toward improved characterization of remotely sensed precipitation regimes with MODIS/AMSR-E blended data techniques. *IEEE Transactions on Geoscience and Remote Sensing*, 43, 1059–1069.

Wang, J. and Coauthors, 2011: The coupled routing and excess storage (CREST) distributed hydrological model. *Hydrological Sciences Journal*, 56, 84–98.

Xue, X. and Coauthors, 2013: Statistical and hydrological evaluation of TRMM-based multi-satellite precipitation analysis over the wangchu basin of bhutan: Are the latest satellite precipitation products 3B42V7 ready for use in ungauged basins? *Journal of Hydrology*, 499, 91–99.

Yong, B., D. Liu, J. J. Gourley, Y. Tian, G. J. Huffman, L. Ren, and Y. Hong, 2014: Global view of real-time TRMM multi-satellite precipitation analysis: Implication to its successor global precipitation measurement mission. *Bulletin of the American Meteorological Society*, 96, 283–296.

Zulkafli, Z., W. Buytaert, C. Onof, B. Manz, E. Tarnavsky, W. Lavado, and J.-L. Guyot, 2014: A comparative performance analysis of TRMM 3B42 (TMPA) versions 6 and 7 for hydrological applications over Andean–Amazon river basins. *Journal of Hydrometeorology*, 15, 581–592.

2 Evapotranspiration Mapping Utilizing Remote Sensing Data

Ke Zhang and Yang Hong

CONTENTS

2.1 INTRODUCTION

The net solar radiation over the land surface is partitioned into sensible, latent, and ground heat fluxes. The latent heat is absorbed by a body, such as plants, or a thermodynamic system to convert liquid/solid water into water vapor. The total evaporated water is called evapotranspiration (ET) as a sum of soil evaporation, vegetation evaporation, and vegetation transpiration; the latter is a process that couples with carbon uptake through photosynthesis. Therefore, land surface energy, water, and carbon fluxes are interconnected through the coupled land surface processes, in particular the ET process. Accurate measurements and estimates of these energy–water fluxes, that is, the sensible, latent, and ground heat fluxes, are critical to quantify the surface energy and water budgets (Diak et al., 2004), predict short-term weather and longer-term climate (Pielke et al., 1998; Schumacher et al., 2004), diagnose climate change (IPCC, 2013), and evaluate and improve physics of climate models (Bony et al., 2006).

The world has experienced persistent climatic warming attributed largely to human activities over the past century (IPCC, 2013), and the warming is projected to continue (IPCC, 2013). Recent climatic changes have altered the global water cycle and surface energy budget (Wentz et al., 2007; Huntington, 2010). To better understand these regional and global water balance changes, each term in the terrestrial water balance equation, $\Delta s = P - \text{ET} - R$, must be accurately measured or quantified. Precipitation (P) and runoff (R) can be directly measured by *in situ* weather stations and streamflow gauge networks. However, ET is inherently difficult to measure and predict especially at large spatial scales. Recent advances in retrieval algorithms and satellite remote sensing technology now enable large-scale mapping and monitoring of ET (Cleugh et al., 2007; Mu et al., 2007, 2011; Fisher et al., 2008; Zhang et al., 2009, 2010), P (Hsu et al., 1997; Sorooshian et al., 2000; Kummerow et al., 2001; Joyce et al., 2004; Hong et al., 2005; Amitai et al., 2006; Huffman et al., 2007), and water storage (Δs) (Tapley et al., 2004; Landerer and Swenson, 2012). However, basin-scale water budget closure is rare, if ever achieved due to large and variable uncertainties and inconsistencies among the water budget terms and associated products (Pan and Wood, 2006; Sheffield et al., 2009; Gao et al., 2010; Rodell et al., 2015). Improved accuracy in quantifying the magnitude and variability of regional and global water and energy fluxes, closing the water budget worldwide, and hence improving weather forecasting, climate and water availability assessments are the ultimate goals of current water and energy cycle research.

Remote sensing (RS), especially from polar-orbiting satellites, provides relatively frequent and spatially contiguous measurements for global monitoring of surface biophysical variables affecting ET, including albedo, vegetation type, and density. RS-based mapping of ET is a cost-effective way to estimate and monitor this flux. Currently, satellites cannot provide temporally continuous observations of spectral reflectance of Earth's surface. Therefore, the episodic satellite observations must rely on models to be extrapolated or scaled to produce averaged estimates of this flux. As a result, the current satellite observations of this flux are usually available as instantaneous values or at daily, weekly, and even monthly resolution. However, with the rapid development of RS technologies and the Earth Observing System, many studies have been conducted in the past decade to develop RS-based ET mapping methods. A brief overview of the past ET mapping studies is described in the following section. The RS-based ET records have been widely used in numerous studies and serve for both scientific community and nongovernmental organizations.

2.2 OVERVIEW OF REMOTE SENSING-BASED ET MAPPING METHODS

Energy budget at Earth's land surface has three heat flux components, latent (λE), sensible (H), and ground heat (G) flux, satisfying the surface energy balance equation,

$$R_n = \lambda E + H + G, \tag{2.1}$$

where R_n is the net radiation as the sum of difference between downward and upward short- and long-wave radiative fluxes. Each of the three components on the right-hand side of Equation 2.1 can be computed separately or estimated as a residual of the energy balance equation. Satellite-based estimation methods of land surface ET or latent heat (λE) can be classified into three general approaches in terms of the way in ET/latent heat is computed as a direct or indirect quantity. The first approach estimates ET/λE first so that sensible heat flux is or can be obtained as a residual in the energy balance equation. Methods of this type include modified Penman–Monteith (PM)-type models with parameterized canopy conductance and water stress (Cleugh et al., 2007; Mu et al., 2007, 2011; Leuning et al., 2008; Zhang et al., 2008, 2009, 2010), modified Priestly–Taylor method (Fisher et al., 2008), vegetation index–land surface temperature empirical relationship methods (Gillies et al., 1997; Nishida et al., 2003; Tang et al., 2009), and other empirical methods (Wang and Liang, 2008; Jung et al., 2010). The second approach, referred to as the "surface energy balance" approach, computes sensible heat flux directly and estimates latent heat as a residual of the surface energy balance equation. Methods of the second type include the surface energy balance algorithm for land algorithm (Bastiaanssen et al., 1998, 2003), the surface energy balance system (Su, 2002), and the two-source energy balance model (Norman et al., 1995). The third approach partitions the net radiation into sensible and latent heat (and ground) fluxes directly. A model of this type is the maximum entropy production (MEP) model (Wang and Bras, 2009, 2011). Additional review of retrieval and estimation methods of land ET or latent heat can be found in previous publications (Parlange et al., 1995; Glenn et al., 2007; Kalma et al., 2008; Petropoulos et al., 2009; Wang and Dickinson, 2012).

Although there are many existing algorithms that can be used to estimate one or all components of the surface energy–water fluxes, their applications are limited by the availability of input data and model parameters, and/or inherent model assumptions. Table 2.1 summarizes the assumptions, requirements, and limitations of these methods. In summary, the global implementation of these existing algorithms/models is restricted by one or more of the following limitations/challenges: (1) high requirement of meteorological forcings resulting in potential errors propagated from these input data, (2) susceptibility to bad weather conditions such as cloudy sky, and (3) requirement of scaling instantaneous values to time-average values with assumptions that are not always valid.

Among a variety of RS-based ET mapping methods, the PM-based ET mapping methods have become popular for several reasons. First, it is a biophysically sound and robust framework for estimating ET at regional to global scales using RS data. Second, it can be easily applied at multiple timescales and can produce continuous ET mapping without extrapolation. Finally, this method has been tested and evaluated at regional to global applications and has been proven to be a valid method. In the following section, we describe the PM-based ET mapping methods in detail.

2.3 PENMAN–MONTEITH ET MAPPING METHODS

2.3.1 Surface Energy Balance and Penman–Monteith Model

Almost every PM-based ET mapping method can start with the energy balance equation shown in Equation 2.1. A general flow of mapping ET based on the PM method using satellite data can be summarized in the following steps, as shown in Figure 2.1: (1) determine whether a grid cell is a vegetated pixel; (2) compute net radiation and

TABLE 2.1

Summary of Existing Major Satellite-Based ET Mapping Methods

Model	Advantages	Assumptions/Limitations	ET Partition
Penman–Monteith methods	Process-based, temporally continuous coverage, flexible for time step, no or low requirements for surface temperature	High requirement for meteorological forcing; require parameterization of canopy conductance	Soil, vegetation, and/or open water components
Surface energy balance methods	Low requirement for meteorological forcing; potentially completely driven by satellite data	Only available for clear-sky conditions; high sensitivity to errors of surface temperature; assuming constant evaporative fraction to extrapolate the instantaneous values	Soil and vegetation components
Priestly–Taylor method	Simple; no calibration required; moderate requirement for meteorological forcing	Many simplifications of the physical processes; requires ground heat flux as an input or assumes that it is negligible; it was applied on the monthly timescale	Soil and vegetation components
MEP model	Low requirement for meteorological forcing	Requires continuous land surface temperature to produce continuous record	Soil and vegetation components
Other empirical models	Simple; most models have low requirement for forcing data	Requires calibration; oversimplification of physical processes; subject to weather condition if land surface temperature is required	Usually do not partition ET into soil and vegetation components

partition it into different components for soil evaporation, plant ET, and/or water evaporation based on energy balance and radiative transfer equations; (3) estimate soil evaporation, plant ET, and water evaporation using the PM or PM-alike methods; and (4) mapping the total ET/latent heat at this grid cell by summing up all components. When the PM method is applied to estimate canopy ET/latent heat, a canopy conductance model is usually required to couple with the PM method.

Net radiation (R_n) can be calculated using the following equation or any similar one:

$$R_n = R_{ns} - R_{nl} = (1 - \alpha) R_{s\downarrow} - R_{nl} \tag{2.2}$$

where:

R_{ns} is net shortwave (i.e., solar) radiation

$R_{s\downarrow}$ is incoming shortwave radiation

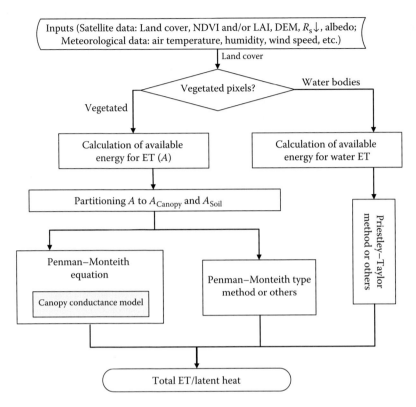

FIGURE 2.1 A common flow for ET mapping using Penman–Monteith method driven by satellite data.

α is surface albedo

R_{nl} is outgoing net longwave radiation

R_{nl} can be calculated using the method of Allen et al. (1998):

$$R_{nl} = \sigma \left[\frac{T_{max,K}^4 + T_{min,K}^4}{2} \right] \left(0.34 - 0.14\sqrt{e_a} \right) \left(1.35 \frac{R_{s\downarrow}}{R_{so}} - 0.35 \right) \qquad (2.3)$$

where:

σ is the Stefan–Boltzmann constant (4.903×10^9 MJ K^{-4} m^{-2})

$T_{max,K}$ and $T_{min,K}$ are the daily maximum and minimum air temperature in kelvin

e_a (Pa) is the actual daily air water vapor pressure

R_{so} (W m^{-2}) is clear-sky incoming shortwave radiation

ET for vegetated areas is usually partitioned into soil evaporation and canopy transpiration by partitioning available energy for ET using the fractional vegetation cover (f_c) derived either from satellite observed vegetation index (VI), such as normalized difference vegetation index (NDVI) and enhanced vegetation index (EVI), or from satellite observed leaf area index (LAI). For example, f_c can be estimated using VI:

$$f_c = \left(\frac{VI - VI_{min}}{VI_{max} - VI_{min}} \right)^2 \tag{2.4}$$

where VI_{min} and VI_{max} are the VI values of bare ground and closed canopy, respectively. The available energy for ET (A: W m^{-2}) is determined as the difference between R_n and G. For vegetated areas, G is calculated as a function of R_n and f_c according to Su et al. (2001):

$$G = R_n \times \left[\Gamma_c + (1 - f_c) \times (\Gamma_s - \Gamma_c) \right] \tag{2.5}$$

where Γ_c and Γ_s are the ratios of G to R_n for full vegetation canopy and bare soil, respectively. Su et al. (2001) assumed Γ_c and Γ_s as global constants, while Zhang et al. (2010) regarded Γ_c and Γ_s as biome-specific constants. The A term is then linearly partitioned into available energy components for the canopy (A_{Canopy}: W m^{-2}) and soil surface (A_{Soil}: W m^{-2}) using f_c such that

$$A_{Canopy} = A \times f_c \tag{2.6}$$

$$A_{Soil} = A \times (1 - f_c) \tag{2.7}$$

The PM equation is used to calculate vegetation transpiration as

$$\lambda E_{Canopy} = \frac{\Delta A_{Canopy} + \rho C_P (e_{sat} - e_a) g_a}{\Delta + \gamma (1 + g_a / g_c)} \tag{2.8}$$

where:

λE_{Canopy} (W m^{-2}) is the latent heat flux of the canopy (i.e., LE_{Canopy})

λ (J kg^{-1}) is the latent heat of vaporization

$\Delta = d(e_{sat})/dT$(Pa K^{-1}) is the slope of the curve relating saturated water vapor pressure (e_{sat}: Pa) to air temperature (T: K)

$e_{sat} - e$ is equal to the vapor pressure deficit (VPD: Pa)

ρ (kg m^{-3}) is the air density

C_P (J kg^{-1} K^{-1}) is the specific heat capacity of air

g_a (m s^{-1}) is the aerodynamic conductance

The psychrometric constant is given by $\gamma = (M_a / M_w)(C_P P_{air} / \lambda)$, where M_a (kg mol^{-1}), M_w (kg mol^{-1}), and P_{air} (Pa) are the molecular mass of dry air, the molecular mass of wet air, and the air pressure, respectively. The g_c (s m^{-1}) term is the canopy conductance (g_c), which is usually estimated by a canopy conductance model.

2.3.2 Canopy Conductance Models

2.3.2.1 Vegetation Index-Based Canopy Conductance Model

Zhang et al. (2009, 2010) developed an empirical biome-specific NDVI-based Jarvis–Stewart-type canopy conductance model by deriving canopy conductance values through the PM equation using an inverse method. This model can be expressed as

$$g_c = g_0(\text{NDVI}) \cdot m(T_{\text{day}}) \cdot m(\text{VPD}) \qquad (2.9)$$

where:

$g_0(\text{NDVI})$ is the biome-dependent potential (i.e., maximum) value of g_c, which is a function of NDVI

T_{day} (°C) is the daylight average air temperature

$m(T_{\text{day}})$ is a temperature stress factor and function of T_{day}

$m(\text{VPD})$ is a water/moisture stress factor and function of VPD

The temperature stress factor $m(T_{\text{day}})$ follows the equation detailed by June et al. (2004) with an optimum temperature, T_{opt}:

$$m(T_{\text{day}}) = \begin{cases} 0.01 & T_{\text{day}} \leq T_{\text{close\_min}} \\ \exp\left(-\left(\dfrac{T_{\text{day}} - T_{\text{opt}}}{\beta}\right)^2\right) & T_{\text{close\_min}} < T_{\text{day}} < T_{\text{close\_max}} \\ 0.01 & T_{\text{day}} \geq T_{\text{close\_max}} \end{cases} \qquad (2.10)$$

where:

T_{opt} (°C) is a biome-specific optimal air temperature for photosynthesis

$T_{\text{close\_min}}$ (°C) and $T_{\text{close\_max}}$ (°C) are respectively the biome-specific minimum and maximum critical temperatures for stomatal closure and the effective cessation of plant photosynthesis

β (°C) is a biome-specific parameter and is the difference in temperature from T_{opt} at which temperature stress factor falls to 0.37 (i.e., e^{-1})

The $m(\text{VPD})$ term is calculated as

$$m(\text{VPD}) = \begin{cases} 1.0 & \text{VPD} \leq \text{VPD}_{\text{open}} \\ \dfrac{\text{VPD}_{\text{close}} - \text{VPD}}{\text{VPD}_{\text{close}} - \text{VPD}_{\text{open}}} & \text{VPD}_{\text{open}} < \text{VPD} < \text{VPD}_{\text{close}} \\ 0.1 & \text{VPD} \geq \text{VPD}_{\text{close}} \end{cases} \qquad (2.11)$$

where:

VPD_{open} (Pa) is the biome-specific critical value of VPD at which the canopy stomata are completely open

$\text{VPD}_{\text{close}}$ (Pa) is the biome-specific critical value of VPD at which canopy stomata are completely closed

Zhang et al. (2010) applied a series of sigmoid functions to fit the hypothetical biome-specific $g_0(\text{NDVI})$ relationships using surface heat flux observations from

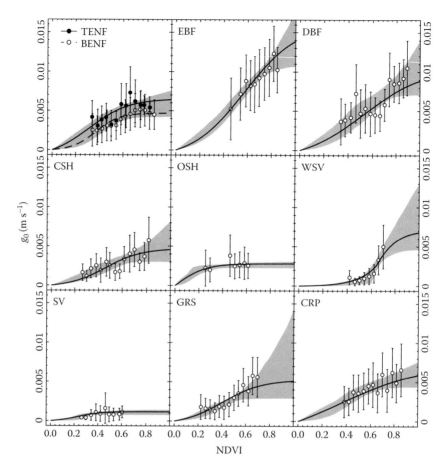

FIGURE 2.2 Calculated biome-specific maximum canopy conductance (g_0) versus NDVI curves and corresponding fitting lines. Empirical sigmoid functions are used to fit the relationship between g_0 and NDVI for temperate evergreen needleleaf forest (TENF), boreal evergreen needleleaf forest (BENF), evergreen broadleaf forest (EBF), deciduous broadleaf forest (DBF), closed shrubland (CSH), open shrubland (OSH), woody savannah (WSV), savannah (SV), grassland (GRS), and cropland (CRP) vegetation types. Error bars denote the standard deviations of g_0. Gray areas correspond to the 99% posterior limits of the fitting model uncertainty derived from an adaptive MCMC method. (From Zhang, K. et al., *Water Resour. Res.*, 46, W09522, 2010. With permission.)

a large number of flux towers across the globe and corresponding satellite-observed NDVI values (Figure 2.2). To analyze the uncertainty in the fitted relationship of g_0 versus NDVI, Zhang et al. (2010) applied an adaptive Markov chain Monte Carlo (MCMC) method (Haario et al., 2006) with a chain of length 6000 to produce the 99% posterior distribution of the fitted relationship of g_0 versus NDVI for each biome type. As shown in Figure 2.2, g_0 derived from flux and meteorology measurements generally follows a sigmoid response curve with increasing NDVI that gradually levels off at higher NDVI values. The g_0 values generally vary between 0.001 and

0.012 ms^{-1} and are biome-specific. The NDVI is an effective surrogate for canopy density. The reduced slope of this relationship at higher NDVI levels reflects increasing shading of individual leaves and leaf boundary layer adjustments with increasing canopy density. The derived g_0 versus NDVI relationships for the 10 biome types show clear differences among each other despite having similar functional shapes (Figure 2.2); these differences reflect variations in leaf traits and physiologies among the different biome types. For the same values of NDVI, g_0 values are highest for the evergreen broadleaf forest (EBF) and deciduous broadleaf forest (DBF) biome types followed by the cropland (CRP), grassland (GRS), temperate evergreen needleleaf forest (TENF), and boreal evergreen needleleaf forest (BENF) types, while the closed shrubland (CSH), open shrubland (OSH), woody savannah (WS), and savannah (SV) types have the lowest g_0 values. The fitted curves for the 10 biome types show generally favorable agreement with the tower observations, while the derived 99% posterior distributions of the g_0 versus NDVI relationships are generally narrow for each biome type (Figure 2.2).

2.3.2.2 Leaf Area-Based and Other Canopy Conductance Models

The leaf area-based canopy conductance model usually scales leaf-level stomatal conductance (g_L) to canopy-level conductance using the LAI. For example, Mu et al. (2007) developed a simple leaf area-based canopy conductance model:

$$g_c = g_L \cdot m(T_{day}) \cdot m(VPD) \cdot LAI \qquad (2.12)$$

where $m(T_{day})$ and $m(VPD)$ are two multipliers the same as or similar to those in Equations 2.10 and 2.11. LAI is the LAI value of a given grid cell.

Kelliher et al. (1995) developed an expression for canopy conductance in terms of maximum stomatal conductance (g_{sx}) of leaves at the top of canopy, LAI, and a hyperbolic response to absorbed shortwave radiation:

$$g_c = \frac{g_{sx}}{k_Q} \ln\left[\frac{Q_h + Q_{50}}{Q_h \exp(-k_Q LAI) + Q_{50}}\right] \qquad (2.13)$$

where:

Q_h is the flux density of visible radiation at the top of the canopy (about half of incoming solar radiation)

k_Q is the extinction coefficient for shortwave radiation

Q_{50} is the visible radiation flux when stomatal conductance is half its maximum value

Leuning et al. (2008) further modified this expression to include the response of stomatal conductance to humidity deficit as developed by Leuning (1995) to give

$$g_c = \frac{g_{sx}}{k_Q} \ln\left[\frac{Q_h + Q_{50}}{Q_h \exp(-k_Q LAI) + Q_{50}}\right]\left[\frac{1}{1 + (D_a / D_{50})}\right] \qquad (2.14)$$

where:

D is the humidity deficit

D_{50} is the humidity deficit at which stomatal conductance is half its maximum value

2.3.3 Soil Evaporation

Soil evaporation can be calculated using a PM-alike model (Mu et al., 2007; Zhang et al., 2009), which is a combination of an adjusted PM equation and the complementary relationship hypothesis (Bouchet, 1963; Fisher et al., 2008). The soil evaporation equation and its auxiliary equations include

$$\lambda E_{\text{Soil}} = RH^{(\text{VPD}/k)} \frac{\Delta A_{\text{Soil}} + \rho C_{\text{P}} \text{VPD}/r_{\text{a}}}{\Delta + \gamma \cdot r_{\text{totc}}/r_{\text{a}}} \tag{2.15}$$

$$r_{\text{a}} = r_{\text{c}} r_{\text{r}} / \left(r_{\text{c}} + r_{\text{r}} \right) \tag{2.16}$$

$$r_{\text{r}} = \rho C_{\text{P}} / \left(4.0 \times \sigma \times T_{\text{day}}^3 \right) \tag{2.17}$$

$$r_{\text{totc}} = r_{\text{tot}} \times R_{\text{corr}} \tag{2.18}$$

$$R_{\text{corr}} = \frac{1.0}{[(273.15 + T_{\text{day}})/293.15] \times (101,300/P_{\text{air}})} \tag{2.19}$$

where RH is the relative humidity of air with values between 0 and 1; $RH^{(\text{VPD}/k)}$ is a moisture constraint on soil evaporation (Fisher et al., 2008), which is an index of soil water deficit based on the complementary relationship of Bouchet (1963) whereby surface moisture status is linked to and reflects the evaporative demand of the atmosphere. The assumption is that soil moisture is reflected in the adjacent atmospheric moisture. k (Pa) is a parameter to fit the complementary relationship and reflects the relative sensitivity to VPD (Fisher et al., 2008). Considering the possible impacts of different vegetation morphology and root zone structure among different biomes on this complementary relationship, Zhang et al. (2010) empirically adjusted the k parameter for different vegetation types. The r_{r} (s m^{-1}) term is the resistance to radiative heat transfer and is calculated using Equation 2.16 (Choudhury and DiGirolamo, 1998). The r_{c} (s m^{-1}) term is the resistance to convective heat transfer and is assumed to be equal to the boundary layer resistance, and assigned as a biome-specific constant (Mu et al., 2007; Zhang et al., 2010). The r_{tot} (s m^{-1}) term is the total aerodynamic resistance to vapor transport and the sum of surface and aerodynamic resistance components. The r_{totc} (s m^{-1}) term is the corrected value of r_{tot} from the standard conditions for temperature and pressure using the correction coefficient (R_{corr}) (Jones, 1992).

2.3.4 EVAPORATION OVER WATER BODIES

For water bodies, G is calculated as a function of air temperature and effective water depth (ΔZ: m) for heat exchange, based on the premise that water surface temperature generally follows air temperature (Pilgrim et al., 1998; Livingstone and Dokulil, 2001; Morrill et al., 2005):

$$G = \rho_W \cdot c_W \cdot K \cdot (T_{avg,i} - T_{avg,i-1}) \cdot \Delta Z \tag{2.20}$$

where:

ρ_W ($1.0 \times 10^3 \, kg \, m^{-3}$) is the water density

c_W ($4.186 \, J \, g^{-1} \, {}^{\circ}C^{-1}$) is the specific heat of water

$T_{avg,i}$ and $T_{avg,i-1}$ are the respective daily average air temperatures for the current day and previous day

K is the slope of the simple linear regression of water surface temperature on air temperature and represents the ratio of water temperature change to surface air temperature change

Zhang et al. (2010) set K to a constant, that is, the mean (0.7) of previously reported values (Pilgrim et al., 1998, Morrill et al., 2005). The effective water depth is the uppermost well-mixed layer of the epilimnion and depends on morphology of open water bodies and other climatic factors.

The evaporation for open water pixels can then be calculated using the Priestley–Taylor (PT) model (Priestley and Taylor, 1972):

$$\lambda E_{Water} = a \frac{\Delta A}{\Delta + \gamma} \tag{2.21}$$

where the PT coefficient a accounts for evaporation arising from the atmospheric vapor pressure deficit in addition to the equilibrium term and is set to 1.26 following Priestley and Taylor (1972). The PT coefficient of 1.26 is generally valid for the saturated surface (Priestley and Taylor, 1972) and is even valid for wet meadow (Stewart and Rouse, 1977) and well-watered grass (Lhomme, 1997).

2.4 GLOBAL IMPLEMENTATION OF PM ET MAPPING

The PM ET mapping methods have been implemented for estimating global ET. Mu et al. (2007) has applied a PM ET mapping method and its improved version (Mu et al., 2011) to estimate global land 8-day composite ET for the MODIS-era. Vinukollu et al. (2011) used the PM ET mapping model of Mu et al. (2007) with two other models to estimate global ET using satellite forcings and conducted intercomparison between these models. On the basis of the PM model, NDVI-based conductance model, soil evaporation model, and water evaporation model introduced in Section 1.3, Zhang et al. (2010) developed an ET mapping algorithm called process-based land surface ET/heat fluxes (P-LSH) and applied it for global daily land ET mapping for the satellite era (Zhang et al., 2015).

The estimated ET by the P-LSH mapping method has gone through extensive validation by comparing with site-level observations such as eddy covariance measurements and regional-level ET reconstruction through a water-balance approach. Figure 2.3 shows the daily time series of LE fluxes modeled by P-LSH with corresponding tower LE measurements at 10 tower sites. Each of the 10 towers represents one biome type. It is clear that the modeled LE fluxes generally match well with the

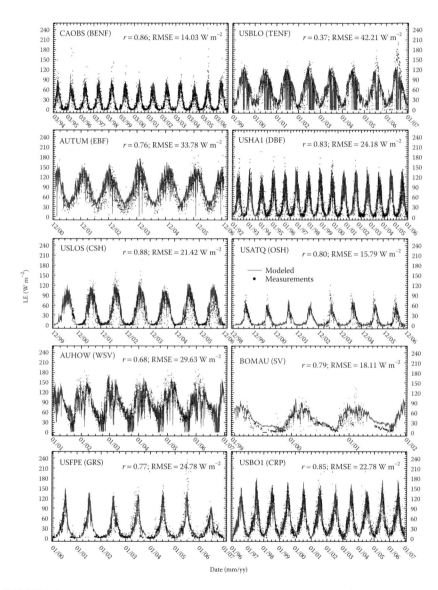

FIGURE 2.3 Comparison of daily measured latent heat fluxes (LE: w m^{-2}) and modeled LE using the P-LSH algorithm driven by AVHRR, NDVI, and NCEP reanalysis meteorology at 10 representative tower sites.

observations in terms of the statistical metrics, simple linear correlation coefficient (r), and root-mean-square error (RMSE) (Figure 2.3).

The monthly ET modeled by P-LSH were also evaluated at 82 flux tower sites distributed across the world and representing all major biome types in the world. The longest measurement record of these sites is about 20 years, while the shortest record is larger than 3 years (Figure 2.4a). The modeled ET fluxes account for approximately 80% of the observed variation in monthly ET measurements with respective RMSE and mean bias (MB) of 14.3 and −2.0 mm month^{-1} (Figure 2.4b). The high coefficients of determination (i.e., R^2), low RMSE and MB differences for the monthly ET results indicate that the P-LSH algorithm generally captures observed seasonal and interannual variations and site-to-site differences in ET.

The P-LSH ET results (ET$_{RS}$) were also evaluated against inferred basin-scale average ET (ET$_{Inferred}$) derived from observed discharge and gauge-based (GPCC)

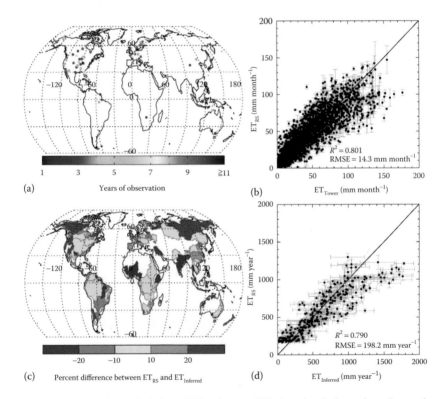

(a) Years of observation

(b)

(c) Percent difference between ET$_{RS}$ and ET$_{Inferred}$

(d)

FIGURE 2.4 Validation of RS-based ET estimates (ET$_{RS}$) against independent observations at the site and basin levels. (a) Locations of 82 FLUXNET sites used for the ET validation; the color code refers to the number of years of observation at each site. (b) Comparison of monthly ET$_{RS}$ against 1-km tower footprints and associated ET measurements at the 82 FLUXNET sites for the period of record 1992–2006. (c) Percent difference between multiyear (1983–2006) average ET$_{RS}$ and inferred ET (ET$_{Inferred}$) from basin-scale multiyear water balance calculations for 284 major global basins. (d) Scatter plot of ET$_{RS}$ versus ET$_{Inferred}$ ($n = 284$); error bars denote the min–max range of ET$_{Inferred}$ resulted from differences in the global precipitation datasets.

precipitation records for 261 global basins (Figure 2.4c). Figure 2.4c shows the global distribution of selected basins and the relative difference (%) between ET_{RS} and $ET_{Inferred}$ defined as $\left(ET_{RS} - ET_{Inferred}\right) \times 100 / ET_{Inferred}$. Figure 2.4d shows the scatter plot of the relationship between ET_{RS} and $ET_{Inferred}$; these results indicate that ET_{RS} and $ET_{Inferred}$ are similar for most basins (RMSE = 198.2 mm year^{-1}; $R^2 = 0.79$). The relative difference between ET_{RS} and $ET_{Inferred}$ falls within $\pm 50\%$, $\pm 20\%$, and $\pm 10\%$ for 95%, 68%, and 50% of the area covered by the 284 basins, respectively. The largest ET_{RS} and $ET_{Inferred}$ difference occurs in some northern high-latitude, subtropical and tropical basins (Figure 2.4c). ET_{RS} is much higher than $ET_{Inferred}$ in some northern high-latitude basins, including the Yukon, Mackenzie, Yenisei, Lena, and Yana basins (Figure 2.4c). The mean ET_{RS} and $ET_{Inferred}$ difference in these basins is approximately 100 mm year^{-1}. This systematic difference is at least partially attributable to the substantial underestimation of GPCC precipitation from snow- and wind-related biases of gauge observations, and the sparse weather station network density in the northern high latitudes (Yang et al., 2005). Zhang et al. (2009) showed that the GPCC product underestimates precipitation by 7.15 mm month^{-1} in relation to bias-corrected observations in these regions (Yang et al., 2005). The GPCC precipitation bias can contribute to an underestimation of 90 mm year^{-1} in $ET_{Inferred}$, which approximates the average difference between ET_{RS} and $ET_{Inferred}$ in these basins. ET_{RS} is lower than $ET_{Inferred}$ in Western Africa and Indian subcontinent basins, indicating that the RS model may underestimate actual ET in these regions. However, uncertainty in the coarse GPCC precipitation and discharge measurements may also contribute to ET_{RS} and $ET_{Inferred}$ differences in these regions. Although there are large differences between ET_{RS} and $ET_{Inferred}$ in some basins, the generally favorable agreement in these results for most areas indicates that the RS-based ET product is relatively accurate on a global basis.

The multiyear (1982–2013) average annual ET by P-LSH is shown in Figure 2.5 and shows strong regional variations and latitudinal gradients corresponding to global

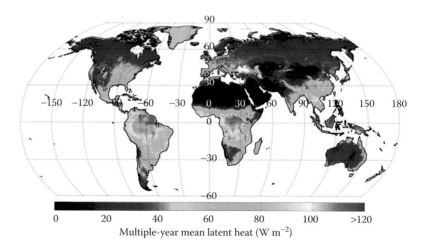

FIGURE 2.5 Global map of multiyear (1982–2013) mean annual ET by P-LSH. Permanent ice and snow covered areas (in gray), and ocean (in white) were excluded from the model calculations.

climate patterns. Tropical rainforests in South America, Africa, and Southeast Asia have the highest annual ET, while drier areas within temperate and subtropical regions and the arctic have the lower annual ET. Not surprisingly, the deserts have the lowest ET values. Annual ET values for temperate and boreal forests are generally intermediate between these two extremes. The estimated ET over water bodies is generally much larger than for adjacent vegetated areas within the same climate zone due to lower surface resistance to evaporation over water relative to land. The global terrestrial average annual ET weighted by area is 539.3 ± 9.1 mm year^{-1}, which is about 0.63 ± 0.02 of the global average annual GPCC precipitation. The estimated global average ET to P ratio is similar to values reported from previous studies (Oki and Kanae, 2006; Jung et al., 2010; Mueller et al., 2011; Rodell et al., 2015).

2.5 SUMMARY AND PROSPECT

In this chapter, we briefly reviewed the existing RS-based ET mapping methods. In particular, we reviewed and summarized the PM-type approaches by describing the detailed theories behind this type of methods and providing the detailed mathematical equations. The PM-based ET mapping has become a popular method for regional to global applications due to its sound theories and easy applicability. Several global ET records have been produced through the PM-type approaches. As one of many researchers, the author has developed a PM-type ET mapping method named P-LSH and successfully applied it to estimate global long-term land ET. The results from these studies have benefited many other studies (Jung et al., 2010; Cai et al., 2011; Cheng et al., 2011) and provide a useful record of global ET climatologies.

Although the PM methods have been widely tested and used, there are a few limitations on these methods. The application of the PM methods requires meteorological forcings, such as air temperature and humidity, which are usually not directly observed by the satellites. Therefore, this limitation makes the PM methods a semisatellite approach and may limit these methods for real-time estimates of ET. However, recent studies have made it possible to retrieve air temperatures and humidity from satellite observations (Jackson et al., 2006; Jones et al., 2010; Jin et al., 2015). By coupling with these retrieval methods or results, it is possible to make the PM methods a completely RS-based ET mapping method. In addition, the PM model needs to be coupled with a canopy conductance model to estimate ET/latent heat. More attention should be drawn on investigating the sensitivity of the PM model on the canopy conductance and further validation and development of this canopy conductance model.

REFERENCES

Allen, R. G., L. S. Pereira, D. Raes, and M. Smith. 1998. *Crop Evapotranspiration: Guidelines for Computing Crop Requirements, FAO Irrigation and Drainage Paper 56*. Rome, Italy: Food and Agricultural Organization of the U.N.

Amitai, E., X. Llort, and D. Sempere-Torres. 2006. Opportunities and challenges for evaluating precipitation estimates during GPM mission. *Meteorologische Zeitschrift* 15(5):551–557. doi:10.1127/0941–2948/2006/0157.

Bastiaanssen, W. G. M., M. Menenti, R. A. Feddes, and A. A. M. Holtslag. 1998. A remote sensing surface energy balance algorithm for land (SEBAL): 1. Formulation. *Journal of Hydrology* 212–213:198–212.

Bastiaanssen, W. G. M., E. J. M. Noordman, H. Pelgrum, G. Davids, B. P. Thoreson, and R. G. Allen. 2003. SEBAL model with remotely sensed data to improve water-resources management under actual field conditions. *Journal of Irrigation and Drainage Engineering* 131(1):85–93.

Bony, S., R. Colman, V. M. Kattsov, R. P. Allan, C. S. Bretherton, J. L. Dufresne, A. Hall et al. 2006. How well do we understand and evaluate climate change feedback processes? *Journal of Climate* 19(15):3445–3482.

Bouchet, R. J. 1963. Evapotranspiration réelle evapotranspiration potentielle, signification climatique. In *General Assembly of Berkeley, Transactions, vol. 2, Evaporation, International Association of Scientific Hydrology,* Berkeley, CA, pp. 134–142.

Cai, X. M., Y. C. E. Yang, C. Ringler, J. S. Zhao, and L. Z. You. 2011. Agricultural water productivity assessment for the Yellow River Basin. *Agricultural Water Management* 98(8):1297–1306.

Cheng, L., Z. X. Xu, D. B. Wang, and X. M. Cai. 2011. Assessing interannual variability of evapotranspiration at the catchment scale using satellite-based evapotranspiration data sets. *Water Resources Research* 47:W09509.

Choudhury, B. J. and N. E. DiGirolamo. 1998. A biophysical process-based estimate of global land surface evaporation using satellite and ancillary data I. Model description and comparison with observations. *Journal of Hydrology* 205:164–185.

Cleugh, H. A., R. Leuning, Q. Mu, and S. W. Running. 2007. Regional evaporation estimates from flux tower and MODIS satellite data. *Remote Sensing of Environment* 106:285–304.

Diak, G. R., J. R. Mecikalski, M. C. Anderson, J. M. Norman, W. P. Kustas, R. D. Torn, and R. L. DeWolf. 2004. Estimating land surface energy budgets from space–Review and current efforts at the University of Wisconsin-Madison and USDA-ARS. *Bulletin of the American Meteorological Society* 85(1):65–78.

Fisher, J. B., K. P. Tu, and D. D. Baldocchi. 2008. Global estimates of the land-atmosphere water flux based on monthly AVHRR and ISLSCP-II data, validated at 16 FLUXNET sites. *Remote Sensing of Environment* 112:901–919.

Gao, H. L., Q. H. Tang, C. R. Ferguson, E. F. Wood, and D. P. Lettenmaier. 2010. Estimating the water budget of major US river basins via remote sensing. *International Journal of Remote Sensing* 31(14):3955–3978. doi:10.1080/01431161.2010.483488.

Gillies, R. R., T. N. Carlson, J. Cui, W. P. Kustas, and K. S. Humes. 1997. A verification of the 'triangle' method for obtaining surface soil water content and energy fluxes from remote measurements of the Normalized Difference Vegetation Index (NDVI) and surface radiant temperature. *International Journal of Remote Sensing* 18(15):3145–3166.

Glenn, E. P., A. R. Huete, P. L. Nagler, K. K. Hirschboeck, and P. Brown. 2007. Integrating remote sensing and ground methods to estimate evapotranspiration. *Critical Reviews in Plant Sciences* 26(3):139–168.

Haario, H., M. Laine, A. Mira, and E. Saksman. 2006. DRAM: Efficient adaptive MCMC. *Statistics and Computing* 16(4):339–354.

Hong, Y., K. L. Hsu, S. Sorooshian, and X. G. Gao. 2005. Improved representation of diurnal variability of rainfall retrieved from the tropical rainfall measurement mission microwave imager adjusted precipitation estimation from remotely sensed information using artificial neural networks (PERSIANN) system. *Journal of Geophysical Research-Atmospheres* 110(110):D06102. doi:10.1029/2004JD005301.

Hsu, K. L., X. G. Gao, S. Sorooshian, and H. V. Gupta. 1997. Precipitation estimation from remotely sensed information using artificial neural networks. *Journal of Applied Meteorology* 36(9):1176–1190.

Huffman, G. J., R. F. Adler, D. T. Bolvin, G. J. Gu, E. J. Nelkin, K. P. Bowman, Y. Hong, E. F. Stocker, and D. B. Wolff. 2007. The TRMM multisatellite precipitation analysis (TMPA): Quasi-global, multiyear, combined-sensor precipitation estimates at fine scales. *Journal of Hydrometeorology* 8(1):38–55.

Huntington, T. G. 2010. Climate warming-induced intensification of the hydrologic cycle: An assessment of the published record and potential impacts on agriculture. *Advances in Agronomy* 109:1–53. doi:10.1016/S0065-2113(10)09001-2.

IPCC. 2013. Climate change 2013: The physical science basis. In *Contribution of Working Group I to the Fifth Assessment Report of the Intergovernmental Panel on Climate Change*. Edited by T. F. Stocker, D. Qin, G.-K. Plattner, M. Tignor, S. K. Allen, J. Boschung, A. Nauels, Y. Xia, V. Bex, and P. M. Midgley. Cambridge, UK: Cambridge University Press.

Jackson, D. L., G. A. Wick, and J. J. Bates. 2006. Near-surface retrieval of air temperature and specific humidity using multisensor microwave satellite observations. *Journal of Geophysical Research-Atmospheres* 111(D10). doi:10.1029/2005jd006431.

Jin, X. Z., L. S. Yu, D. L. Jackson, and G. A. Wick. 2015. An improved near-surface specific humidity and air temperature climatology for the SSM/I satellite period. *Journal of Atmospheric and Oceanic Technology* 32(3):412–433. doi:10.1175/Jtech-D-14-00080.1.

Jones, H. G. 1992. *Plants and Microclimate: A Quantitative Approach to Environmental Plant Physiology*. 2nd ed. Cambridge: Cambridge University Press.

Jones, L. A., C. R. Ferguson, J. S. Kimball, K. Zhang, S. T. K. Chan, K. C. McDonald, E. G. Njoku, and E. F. Wood. 2010. Satellite microwave remote sensing of daily land surface air temperature minima and maxima from AMSR-E. *IEEE Journal of Selected Topics in Applied Earth Observations and Remote Sensing* 3(1):111–123. doi:10.1109/Jstars.2010.2041530.

Joyce, R. J., J. E. Janowiak, P. A. Arkin, and P. P. Xie. 2004. CMORPH: A method that produces global precipitation estimates from passive microwave and infrared data at high spatial and temporal resolution. *Journal of Hydrometeorology* 5(3):487–503.

June, T., J. R. Evans, and G. D. Farquhar. 2004. A simple new equation for the reversible temperature dependence of photosynthetic electron transport: A study on soybean leaf. *Functional Plant Biology* 31:275–283.

Jung, M., M. Reichstein, P. Ciais, S. I. Seneviratne, J. Sheffield, M. L. Goulden, G. Bonan et al. 2010. Recent decline in the global land evapotranspiration trend due to limited moisture supply. *Nature* 467(7318):951–954. doi:10.1038/Nature09396.

Kalma, J. D., T. R. McVicar, and M. F. McCabe. 2008. Estimating land surface evaporation: A review of methods using remotely sensed surface temperature data. *Surveys in Geophysics* 29(4–5):421–469.

Kelliher, F. M., R. Leuning, M. R. Raupach, and E. D. Schulze. 1995. Maximum conductances for evaporation from global vegetation types. *Agricultural and Forest Meteorology* 73(1–2):1–16. doi:10.1016/0168-1923(94)02178-M.

Kummerow, C., Y. Hong, W. S. Olson, S. Yang, R. F. Adler, J. McCollum, R. Ferraro, G. Petty, D. B. Shin, and T. T. Wilheit. 2001. The evolution of the Goddard profiling algorithm (GPROF) for rainfall estimation from passive microwave sensors. *Journal of Applied Meteorology* 40(11):1801–1820.

Landerer, F. W. and S. C. Swenson. 2012. Accuracy of scaled GRACE terrestrial water storage estimates. *Water Resources Research* 48:W04531. doi: 10.1029/2011WR011453.

Leuning, R. 1995. A critical-appraisal of a combined stomatal-photosynthesis model for C-3 plants. *Plant Cell and Environment* 18(4):339–355. doi:10.1111/j.1365-3040.1995.tb00370.x.

Leuning, R., Y. Q. Zhang, A. Rajaud, H. Cleugh, and K. Tu. 2008. A simple surface conductance model to estimate regional evaporation using MODIS leaf area index and the Penman-Monteith equation. *Water Resources Research* 44:W10419. doi:10.1029/2007WR006562.

Lhomme, J. P. 1997. A theoretical basis for the Priestley-Taylor coefficient. *Boundary-Layer Meteorology* 82(2):179–191.

Livingstone, D. M. and M. T. Dokulil. 2001. Eighty years of spatially coherent Austrian lake surface temperatures and their relationship to regional air temperature and the North Atlantic Oscillation. *Limnology and Oceanography* 46(5):1220–1227.

Morrill, J. C., R. C. Bales, and M. H. Conklin. 2005. Estimating stream temperature from air temperature: Implications for future water quality. *Journal of Environmental Engineering-ASCE* 131(1):139–146.

Mu, Q., F. A. Heinsch, M. Zhao, and S. W. Running. 2007. Development of a global evapotranspiration algorithm based on MODIS and global meteorology data. *Remote Sensing of Environment* 111(4):519–536.

Mu, Q. Z., M. S. Zhao, and S. W. Running. 2011. Improvements to a MODIS global terrestrial evapotranspiration algorithm. *Remote Sensing of Environment* 115(8):1781–1800.

Mueller, B., S. I. Seneviratne, C. Jimenez, T. Corti, M. Hirschi, G. Balsamo, P. Ciais et al. 2011. Evaluation of global observations-based evapotranspiration datasets and IPCC AR4 simulations. *Geophysical Research Letters* 38:L06402. doi:10.1029 / 2010gl046230.

Nishida, K., R. R. Nemani, J. M. Glassy, and S. W. Running. 2003. Development of an evapotranspiration index from aqua/MODIS for monitoring surface moisture status. *IEEE Transactions on Geoscience and Remote Sensing* 41(2):493–501.

Norman, J. M., W. P. Kustas, and K. S. Humes. 1995. A two-source approach for estimating soil and vegetation energy fluxes from observations of directional radiometric surface temperature. *Agricultural and Forest Meteorology* 77:263–293.

Oki, T. and S. Kanae. 2006. Global hydrological cycles and world water resources. *Science* 313(5790):1068–1072.

Pan, M. and E. F. Wood. 2006. Data assimilation for estimating the terrestrial water budget using a constrained ensemble Kalman filter. *Journal of Hydrometeorology* 7(3):534–547. doi:10.1175/Jhm495.1.

Parlange, M. B., W. E. Eichinger, and J. D. Albertson. 1995. Regional-scale evaporation and the atmospheric boundary-layer. *Reviews of Geophysics* 33(1):99–124.

Petropoulos, G., T. N. Carlson, M. J. Wooster, and S. Islam. 2009. A review of T-s/VI remote sensing based methods for the retrieval of land surface energy fluxes and soil surface moisture. *Progress in Physical Geography* 33(2):224–250.

Pielke, R. A., R. Avissar, M. Raupach, A. J. Dolman, X. B. Zeng, and A. S. Denning. 1998. Interactions between the atmosphere and terrestrial ecosystems: Influence on weather and climate. *Global Change Biology* 4(5):461–475.

Pilgrim, J. M., X. Fang, and H. G. Stefan. 1998. Stream temperature correlations with air temperatures in Minnesota: Implications for climate warming. *Journal of the American Water Resources Association* 34(5):1109–1121.

Priestley, C. H. B. and R. J. Taylor. 1972. On the assessment of surface heat flux and evaporation using large-scale parameters. *Monthly Weather Review* 100(2):81–92.

Rodell, M., H. K. Beaudoing, T. S. L'Ecuyer, W. S. Olson, J. S. Famiglietti, P. R. Houser, R. Adler et al. 2015. The observed state of the water cycle in the early twenty-first century. *Journal of Climate* 28:8289–8318.

Schumacher, C., R. A. Houze, and I. Kraucunas. 2004. The tropical dynamical response to latent heating estimates derived from the TRMM precipitation radar. *Journal of the Atmospheric Sciences* 61(12):1341–1358.

Sheffield, J., C. R. Ferguson, T. J. Troy, E. F. Wood, and M. F. McCabe. 2009. Closing the terrestrial water budget from satellite remote sensing. *Geophysical Research Letters* 36(36):L07403. doi:10.1029/2009GL037338.

Sorooshian, S., K. L. Hsu, X. Gao, H. V. Gupta, B. Imam, and D. Braithwaite. 2000. Evaluation of PERSIANN system satellite-based estimates of tropical rainfall. *Bulletin of the American Meteorological Society* 81(9):2035–2046.

Stewart, R. B. and W. R. Rouse. 1977. Substantiation of Priestley and Taylor parameter a = 1.26 for potential evaporation in high latitudes. *Journal of Applied Meteorology* 16(6):649–650.

Su, Z. 2002. The surface energy balance system (SEBS) for estimation of turbulent heat fluxes. *Hydrology and Earth System Sciences* 6(1):85–99.

Su, Z., T. Schmugge, W. P. Kustas, and W. J. Massman. 2001. An evaluation of two models for estimation of the roughness height for heat transfer between the land surface and the atmosphere. *Journal of Applied Meteorology* 40(11):1933–1951.

Tang, Q. H., S. Peterson, R. H. Cuenca, Y. Hagimoto, and D. P. Lettenmaier. 2009. Satellite-based near-real-time estimation of irrigated crop water consumption. *Journal of Geophysical Research-Atmospheres* 114(114):D05114. doi:10.1029/2008JD010854.

Tapley, B. D., S. Bettadpur, M. Watkins, and C. Reigber. 2004. The gravity recovery and climate experiment: Mission overview and early results. *Geophysical Research Letters* 31(31):L09607. doi:10.1029/2004GL019920.

Vinukollu, R. K., E. F. Wood, C. R. Ferguson, and J. B. Fisher. 2011. Global estimates of evapotranspiration for climate studies using multi-sensor remote sensing data: Evaluation of three process-based approaches. *Remote Sensing of Environment* 115(3):801–823. doi: 10.1016/j.rse.2010.11.006.

Wang, J. and R. L. Bras. 2009. A model of surface heat fluxes based on the theory of maximum entropy production. *Water Resources Research* 45:W11422. doi:10.1029/2009wr007900.

Wang, J. F. and R. L. Bras. 2011. A model of evapotranspiration based on the theory of maximum entropy production. *Water Resources Research* 47:W03521. doi: 10.1029/2010wr009392.

Wang, K. C. and R. E. Dickinson. 2012. A review of global terrestrial evapotranspiration: Observation, modeling, climatology, and climatic variability. *Reviews of Geophysics* 50:RG2005. doi:10.1029/2011RG000373.

Wang, K. C. and S. L. Liang. 2008. An improved method for estimating global evapotranspiration based on satellite determination of surface net radiation, vegetation index, temperature, and soil moisture. *Journal of Hydrometeorology* 9(4):712–727.

Wentz, F. J., L. Ricciardulli, K. Hilburn, and C. Mears. 2007. How much more rain will global warming bring? *Science* 317(5835):233–235.

Yang, D., D. Kane, Z. Zhang, D. Legates, and B. Goodison. 2005. Bias corrections of long-term (1973–2004) daily precipitation data over the northern regions. *Geophysical Research Letters* 32:L19501. doi:10.1029/2005GL024057.

Zhang, K., J. S. Kimball, Q. Mu, L. A. Jones, S. J. Goetz, and S. W. Running. 2009. Satellite based analysis of northern ET trends and associated changes in the regional water balance from 1983 to 2005. *Journal of Hydrology* 379(1–2):92–110.

Zhang, K., J. S. Kimball, R. R. Nemani, and S. W. Running. 2010. A continuous satellite-derived global record of land surface evapotranspiration from 1983 to 2006. *Water Resources Research* 46:W09522. doi:10.1029/2009WR008800.

Zhang, K., J. S. Kimball, R. Nemani, S. Running, Y. Hong, J. J. Gourley, and Z. Yu. 2015. Vegetation greening and climate change promote multidecadal rises of global land evapotranspiration. *Scientific Reports* 5:15956. doi:10.1038/srep15956.

Zhang, Y. Q., F. H. S. Chiew, L. Zhang, R. Leuning, and H. A. Cleugh. 2008. Estimating catchment evaporation and runoff using MODIS leaf area index and the Penman-Monteith equation. *Water Resources Research* 44(44):W10420. doi:10.1029/2007WR006563.

3 Soil Moisture Estimation Using Active and Passive Remote Sensing Techniques

Ming Pan, Xing Yuan, Hui Lu,
Xiaodong Li, and Guanghua Qin

CONTENTS

3.1   INTRODUCTION

Soil moisture is a key land surface state variable for its important role in regulating the energy and moisture fluxes between the atmosphere and land surface. It is an important boundary condition for the atmosphere and imposes a strong control on the water and energy exchanges between the land and atmosphere through regulating the evapotranspiration flux (Koster and Suarez, 1992). Better knowledge of soil moisture greatly contributes to our understanding of land surface processes (Albertson and Parlange, 1999; Cahill et al., 1999). Because the land surface soil moisture has a longer memory than the dynamic processes in the atmosphere, better quantification of land surface soil moisture status has the potential to improve numerical weather forecasts at both short-term and seasonal scales (Koster and Suarez, 2001), especially in the warm season when land–atmosphere interactions are greatest. Operational large-scale soil moisture observational products would likely enhance the accuracy of numerical weather prediction products, hydrologic flood forecasting, agricultural drought monitoring as well as water cycle research related to climate studies. Soil moisture information is also of great value in agriculture and water resources management, especially for drought management (Sheffield and Wood, 2007; Mo, 2008).

Therefore, efforts have been made in developing soil moisture observational networks (Robock et al., 2000) and evaluating soil moisture as modeled by land surface schemes (Schaake et al., 2004) and climate models (Luo and Wood, 2008). However, soil moisture is highly variable across a range of temporal and spatial scales (Crow and Wood, 1999; Pan et al., 2009), making it difficult to either model or measure. *In situ* measurement of soil moisture is carried out by networks of ground sensors, for example, the Natural Resource Conservation Service's (NRCS) Soil Climate Analysis Network (SCAN) and the National Oceanic and Atmospheric Administration's (NOAA) United States Climate Reference Network (USCRN) (Bell et al., 2013; Diamond et al., 2013), both of which use a frequency domain reflectometer (Seyfried et al., 2005) to measure soil moisture content at different depths. More recently, more coherent efforts have been made to implement *in situ* networks across the conterminous United States (Jackson et al., 2012b). Still, ground networks are generally very expensive and cannot offer large-scale coverage with reasonable density (Robock et al., 2000). Therefore, space-borne remote sensing provides an important global alternative to the limited *in situ* observations of soil moisture (Owe et al., 2008).

3.2   ESTIMATING SOIL MOISTURE CONTENT FROM MICROWAVE SENSORS

The underpinnings of space-borne soil moisture remote sensing come from the fact that changes in surface soil moisture content lead to changes in the surface emissivity and backscattering properties in microwave frequencies. Therefore, passive satellite sensors (radiometers) can detect soil moisture variations by measuring the brightness temperature (Njoku, 1977) and active sensors (radars) can detect soil moisture variations by measuring the backscatter returned to a satellite sensor (Dobson and Ulaby, 1986). Simply speaking, wetter soil looks cooler in the eyes of the radiometer relative to its physical temperature and appears brighter on a radar scan. However, vegetation also emits and backscatters microwave signals, and dense vegetation (e.g., forests) can significantly attenuate or overwhelm the soil moisture signals especially at short wavelengths, making it hard or impossible to retrieve soil moisture information. Heavy rain clouds also add noise to the retrieval process and retrievals over active raining areas are less reliable. Also, microwave signals can only penetrate a thin layer of soil, about 1/10 to half of the wavelength, so the remotely sensed soil moisture is representative of a shallow surface layer. Longer wavelengths result in a deeper penetration and so far 1.4 GHz (L-band; 1–5 cm penetration) is the longest wavelength in use due to technical limitations and tradeoffs in antenna size, orbital geometry (including the implications for revisit time), and other factors.

Nevertheless, the global coverage and long-term availability of remote sensing products makes them an extremely valuable source of observations for large-scale analysis in applications such as model validation, drought analysis, and climate studies.

3.2.1 RADIATIVE TRANSFER PROCESS FROM GROUND TO SPACE AND PASSIVE REMOTE SENSING

Passive measurements in microwave frequencies can be used for monitoring surface moisture conditions for a number of reasons. First, as the soil water content increases, the dielectric conductivity of the wet soil decreases, resulting in reduced surface emission (or brightness temperature). The lower the frequency, the higher is the sensitivity to soil moisture. Second, atmospheric emissions are minimal at many microwave frequencies such that land surface emission can penetrate through the atmosphere and thin clouds with little atmospheric attenuation (Ulaby et al., 1982). Research has been done to establish the physical relationship between soil moisture and other surface conditions and the brightness temperature measured in space. Such a physical relationship is usually parameterized into a radiative transfer model (RTM). An RTM calculates the brightness temperature measured at the top of atmosphere (TOA) given the conditions of soil (temperature, wetness, texture, etc.), vegetation (thickness, structure, temperature, etc.), and atmosphere (temperature, pressure, humidity, etc.). Usually, the RTM is the basis for retrieving surface soil moisture from satellite measurements.

Figure 3.1 shows the radiative transfer process over a vegetated soil surface. T_s and T_v are the physical temperature of the soil and vegetation, $T_{b,a\uparrow}$ and $T_{b,a\downarrow}$ are the up- and downward brightness temperature of the atmosphere, $\Gamma_v = \exp(-\tau_v)$ and

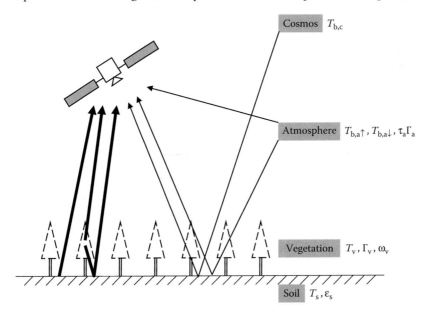

FIGURE 3.1   Radiative transfer over a vegetated soil surface.

$\Gamma_a = \exp(-\tau_a)$ are the transmissivity of the vegetation layer and atmosphere (τ_v and τ_a are the optical depth of the vegetation and atmosphere), ε_s and ω_v are the emissivity of the wet soil and single scattering albedo of the vegetation layer, and $T_{b,c}$ is the brightness temperature of the cosmic background.

The figure shows that the emission reaching TOA consists of six components: (1) soil emission, (2) direct vegetation emission, (3) reflected vegetation emission, (4) reflected cosmic emission, (5) reflected atmospheric emission, and (6) upward atmospheric emission. Not all these six components are important and almost all RTM schemes will do some simplifications, for example, the cosmic and atmospheric contributions are constants and these constants are usually determined through calibrations (Drusch et al., 2001) and only the soil and vegetation emissions need to be parameterized. And the brightness temperature at TOA is

$$T_b = \varepsilon_s T_s \Gamma_v \Gamma_a + T_v (1 - \omega_v)(1 - \Gamma_v)\big(1 + (1 - \varepsilon_s)\Gamma_v\big)\Gamma_a + \text{constant}$$

An important goal of an RTM is to parameterize the relationship between moisture content θ and wet soil emissivity ε_s, together with other soil and vegetation parameters, such that once the surface moisture/temperature is known, the brightness temperature at TOA can be calculated. To this end, some researchers use very detailed physical models that include dielectric property of wet soil (Wang and Schmugge, 1980; Dobson et al., 1985) and soil polarization mixing (Choudhury et al., 1979; Wang and Choudhury, 1981), while others simply lump everything (emissivity, soil roughness, etc.) into one effective emissivity parameter (de Jeu et al., 2008; Pan et al., 2014). For the vegetation properties, they can be treated as a simple function of the leaf area index and vegetation type (Kirdyashev et al., 1979), or in most cases, estimated simultaneously with soil moisture from measurements of multiple polarizations and/or multiple frequencies.

The bare soil is considered a special case of vegetation surface where the vegetation is completely transparent; however, the water surface needs to be treated differently as the water body has a dramatically lower emissivity.

Once the radiative transfer process is parameterized, the retrieval of soil moisture from satellite measurements is simply a reverse estimation problem. This could be done by inverting the RTM using a root-finding algorithm, for example, the land surface microwave emission model (LSMEM) approach (Drusch et al., 2001; Gao et al., 2004, 2006), the retrieval products by the United States Department of Agriculture (USDA) (Jackson, 1993; Jackson et al., 1999; Jackson and Hsu, 2001) and the L-band Microwave Emission of the Biosphere (L-MEB) model (Wigneron et al., 2007) used for SMOS soil moisture retrievals at the European Centre for Medium Range Weather Forecasts. This approach usually solves for one unknown (soil moisture) given the brightness temperature from a single channel/polarization and the inversion is algorithmically simple. But it requires many physical parameters, some of which we lack any accurate knowledge of, for example, the emission/optical properties of the soil surface and vegetation cover in target frequency.

The alternative to parameter-intensive approaches is to solve for more parameters altogether. The initial production of the official National Aeronautics and Space

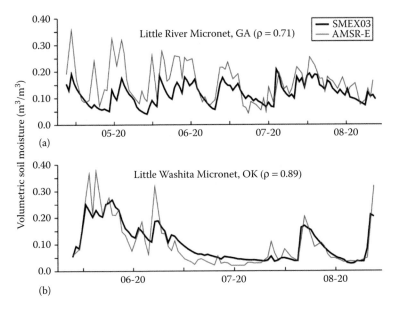

FIGURE 3.2   Soil moisture retrieval (LSMEM) time series versus ground observations from the SMEX03 campaign over two areas in (a) Little River, GA, and (b) Little Washita, OK.

Administration (NASA) Advanced Microwave Scanning Radiometer (AMSR-E) soil moisture product (Njoku et al., 2003) takes a multichannel and multipolarization approach (Njoku and Li, 1999) where the vegetation properties and soil moisture (as well as the surface temperature) are solved simultaneously. Later this approach was revised such that the combined vegetation-roughness effect is modeled as a function of the microwave polarization difference index (MPDI) and the soil moisture is estimated as a deviation relative to a reference dry moisture condition (Njoku and Chan, 2006). The land parameter retrieval model (LPRM) (Owe et al., 2001) provides another alternative based on the concept of the MPDI and it solves for the soil moisture and other parameters through an iterative optimization procedure (de Jeu et al., 2008). The improved LSMEM approach (Pan et al., 2014) also falls into this category. Figure 3.2 gives an example of soil moisture retrievals (Pan et al., 2014) and the comparison against *in situ* observations over two sites in the United States. These time series clearly show that the passive microwave estimates can very well capture the dynamics of daily soil moisture variations.

3.2.2   RADAR BACKSCATTER AND ACTIVE REMOTE SENSING

A very strong limitation of the passive sensors arises from the fact that the all the signals the radiometer can receive come from the natural emission of the target (soil). Such natural microwave emission is quite weak by itself, thus the sensor antenna has to have a very large effective aperture to achieve a reasonable spatial resolution and noise-to-signal ratio. Or otherwise the satellite has to fly lower, that is, to measure at a closer distance to the target. However, both the large antenna and low orbit can

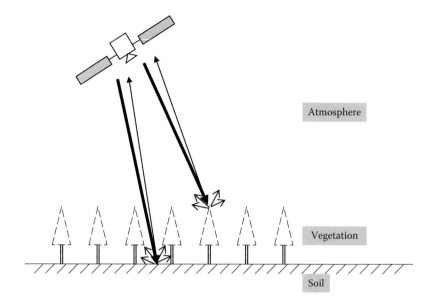

FIGURE 3.3   Active sensor (radar).

dramatically add to the manufacture and operation cost and bring compromises on other aspects such as swath width, revisit time, service life, and so on. The most practical design of a passive sensor (e.g., with an unfolding antenna reflector or synthetic aperture design) and orbit configuration can achieve the ground spatial resolution of several tens of kilometers. To further improve the resolution, active sensors are needed. The active sensor, also known as radar, sends a well-focused microwave beam to the target through a transmitter pretty much like shining a flash light on the target when we need to see it in the dark. As the radar beam hits the target (soil and vegetation), part of the energy is scattered back and measured by the radar (see Figure 3.3). The wetter the soil or vegetation is, the brighter it looks on the radar screen. Because the radar shines a bright beam onto the surface, the returned signals are much stronger than a passive sensor and a significantly higher spatial resolution can be achieve (several kilometers).

The relationship between backscatter signals and soil/vegetation water has also been studied for a long time (Ulaby et al., 1982). The basic theory is that the radar backscatter intensity depends on the soil moisture content (linearly) and other factors such as vegetation water content and viewing angles. At the same time, the viewing angle dependence of backscatter is not caused by soil moisture but vegetation, and this allows the vegetation effect to be removed by analyzing the angular behavior of the backscatter. After soil moisture dynamics are extracted from the backscatter data, the time series can then be normalized to a predefined range (e.g., zero to saturation). One typical example is various versions of the WAter Retrieval Package (WARP) algorithm, for example, WARP4 and WARP5 (Naeimi et al., 2009). As different radars have different sets of viewing angles and sampling patterns (e.g., real aperture vs. synthetic aperture), the retrieval (Dobson and Ulaby, 1986; Wagner et al., 1999) and the algorithm may vary slightly with the specific design of the radar (Naeimi et al., 2009; Entekhabi et al., 2010).

3.3   SATELLITE MISSIONS AND MICROWAVE SENSORS

Even though the research about estimating surface soil moisture from microwave sensors started decades ago (Njoku, 1977; Wang and Schmugge, 1980; Ulaby et al., 1982), there have not been any "dedicated" satellite missions for soil moisture observations until very recently, that is, the launch of the European Space Agency (ESA) Soil Moisture and Ocean Salinity (SMOS) mission on November 2, 2009. Prior to SMOS, nevertheless, many other satellite missions do carry microwave sensors (radiometers and radars). These microwave sensors are deployed primarily for measuring other parameters such as cloud activities, precipitation, water vapor, sea wind, and so on, and the science community has been utilizing these "opportunistic" sensors for soil moisture retrievals. Sensors not specifically designed/built for soil moisture observing purposes may not carry the best channels or optimally configured for soil moisture detections, but very successful efforts have been made by various researchers to derive soil moisture datasets from those sensors.

3.3.1   PASSIVE SENSORS

A number of microwave sensors (generally dedicated to purposes other than soil moisture measurement) have been utilized to estimate soil moisture, from the early Scanning Multichannel Microwave Radiometer (SMMR) sensor on board Nimbus (launched in 1978) to the most recent and fully dedicated Soil Moisture Active and Passive (SMAP) mission (Entekhabi et al., 2010) launched January 31, 2015 (see Table 3.1).

3.3.2   ACTIVE SENSORS

The number of active sensors is much smaller than the passive sensors because the latter usually serve more purposes than the former. Soil moisture retrievals have been made primarily upon only three active sensors, the scatterometers on board the European Remote Sensing satellites (ERS-1 and ERS-2) and the METeorological OPerational satellite (METOP). The ERS-1 and ERS-2 satellites carry a 5.3 GHz scatterometer (SCAT) and METOP carries an improved version of the instrument, the Advanced Scatterometer (ASCAT). The third active sensor flies on board the SMAP mission, a 1.4 GHz synthetic aperture radar.

3.4   SOIL MOISTURE RETRIEVAL PRODUCTS

Prior to SMAP and SMOS, most satellite-based soil moisture estimation was via passive sensors of opportunity, with wavelengths shorter than are desirable for soil moisture sensing. Many studies have therefore been devoted to understanding the radiative transfer processes that link the soil moisture to the brightness temperature measured by space-borne sensors, and these studies have led to a number of soil moisture retrieval algorithms. Therefore, quite a large number of soil moisture

TABLE 3.1

Summary of Satellite Sensors Used for Soil Moisture Estimation

Sensor and Platform	Channel	Resolution (km)	Revisit	Period
Scanning Multichannel Microwave Radiometer (SMMR), Nimbus-7	Multiple, from 6.6 GHz (4.5 cm)	140	Daily	1978–1987
Microwave Imager/Sounder (MIS), WindSat	Multiple, from 6.8 GHz (4.4 cm)	25	Daily	2003–present
TRMM Microwave Imager (TMI), TRMM	Multiple, from 10.7 GHz (2.8 cm)	25 (resampled)	Daily	1997–present <50° latitude
Special Sensor Microwave Imager (SSM/I), DSMP	Multiple, from 19.4 GHz (1.5 cm)	25	Daily	1987–present
Advanced Microwave Scanning Radiometer (AMSR-E), EOS-Aqua	Multiple, from 6.9 GHz (4.3 cm)	25 (resampled)	Daily	2002–2011
Scatterometer (SCAT), ERS	5.3 GHz (5.7 cm)	~50	3–4 days	1992–present
Advanced Scatterometer (ASCAT), MetOp	5.3 GHz (5.7 cm)	25	1–2 days	2006–present
Soil Moisture and Ocean Salinity (SMOS), SMOS	1.4 GHz (21 cm)	~35	2–3 days	2009–present
Advanced Microwave Scanning Radiometer 2 (AMSR-2), GCOM-W	Multiple, from 6.9 GHz (4.3 cm)	25 (resampled)	Daily	2012–present
Soil Moisture Active and Passive (SMAP), SMAP	1.4 GHz (21 cm)	3, 9, 36	2–3 days	2015–present

retrieval products, both operational and experimental, have been created based on different satellite sensors and by many institutions.

Among satellite (as contrasted with airborne) sensors, the AMSR-E is one of the most frequently used for its relatively long service time, modest radio frequency interference in its 10.7 GHz channel, and public availability. The official NASA AMSR-E product (Njoku and Li, 1999), the LPRM (Owe et al., 2001; de Jeu et al., 2008), and the modified LSMEM-based AMSR-E retrievals (Pan et al., 2014) are produced by solving for the vegetation/other parameters separately from multiple polarizations or simultaneously with soil moisture through an iterative scheme. The USDA (Jackson and Hsu, 2001) also offers an AMSR-E retrieval product where a different approach is taken to use predefined vegetation parameters and to treat the soil moisture as the only unknown. The original version of AMSR-E-based LEMEM retrieval product (Gao et al., 2006) also takes a similar approach as the USDA product. See Figure 3.4 for the sample retrieval maps from four different products.

The LPRM model has also been used to retrieve soil moisture from other sensors such as Nimbus 7 SMMR, DMSP SSM/I, TRMM TMI (Liu et al., 2012), and AMSR2 (Parinussa et al., 2015). The USDA algorithm and LSMEM algorithm have

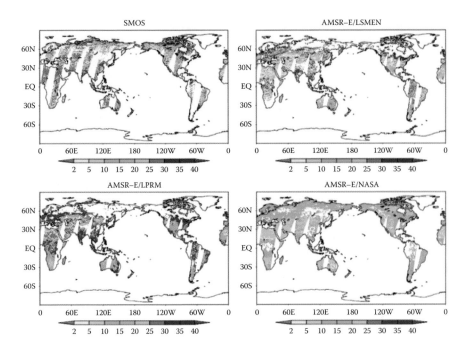

FIGURE 3.4 Soil moisture retrievals on September 6, 2010, from four products: SMOS, AMSR-E/LSMEM, AMSR-E/LPRM, and AMSR-E/NASA.

also been applied to TRMM TMI (Jackson and Hsu, 2001; Gao et al., 2006). For the L-band sensor on board SMOS, the L-MEB model (Wigneron et al., 2007) is used to create the standard SMOS soil moisture product (Kerr et al., 2012). The Japan Aerospace Exploration Agency (JAXA) maintains an AMSR2-based soil moisture retrieval product (Fujii et al., 2009) that uses a similar algorithm for retrieving soil moisture from ASMR-E after some intercalibration between the two sensors (Imaoka et al., 2010).

With SMOS and SMAP products available to the community (see Figure 3.5 for an example of an early SMAP product), the community is beginning to have global soil moisture measurements (albeit of near-surface conditions) that have been targeted to hydrologic users. The designs of these missions are bringing about tremendous advances in the quality of retrieval products. Both SMOS and SMAP carry a radiometer operating at 1.4 GHz, the longest practical wavelength for soil penetration. The SMOS radiometer uses synthetic aperture techniques to achieve large equivalent antenna size and makes measurements at a range of incident angles. The SMAP radiometer has a large (6 m) space-deployed reflector antenna and a fixed angle design that enhances the spatial resolution of its passive product to 36 km. However, a key advance of SMAP is its Synthetic Aperture Radar that provides the ability effectively to disaggregate the passive retrievals to an unprecedentedly high resolution (~3 km). The two missions bring the soil moisture remote sensing to a dramatically higher level and have motivated increased attention to soil moisture

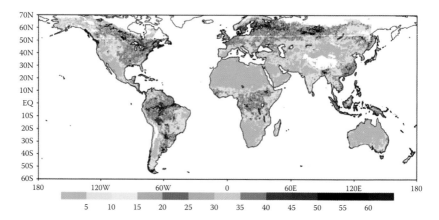

FIGURE 3.5 Global soil moisture map from SMAP's combined radar and radiometer instruments, averaged between May 4 and 11, 2015, during SMAP's commissioning phase. The map has a spatial resolution of 9 km.

product development. Unfortunately, the radar on board the SMAP satellite stopped functioning as of July 7, 2015, though all other operations of SMAP (including the passive sensor) continued normally.

Besides the soil moisture retrieval products from individual sensors, various global long-term remotely sensed soil moisture data sets have been established by blending estimates from different sensors. Examples are the Soil Moisture Essential Climate Variable (ECV) (Liu et al., 2012) produced under the ESA, Climate Change Initiative (CCI), and the Soil Moisture Operational Products System (SMOPS) produced by the NOAA's Satellite and Information Service (NESDIS). The ESA Soil Moisture ECV products (separated into active-only, passive-only, and combined) include and/or will include SMMR, TMI, SSM/I, AMSR-E, WindSat, SMOS, SCAT, ASCAT, and SMAP sensors. Data homogenizations are performed to make it more convenient for long-term global scale analysis. SMOPS currently blends WindSat, ASCAT, and SMOS.

3.5 VALIDATION, UNCERTAINTY ASSESSMENT, AND IMITATIONS OF MICROWAVE SOIL MOISTURE ESTIMATES

For both the passive and active remote sensing of soil moisture, the major noise to the soil moisture signals comes from the vegetation overlaying on top of the soil surface. In the case of passive sensors, the vegetation layer can considerably attenuate the soil emission and the vegetation's own emission also adds noise to the soil signals reaching the satellite. In the case of the active sensors, the vegetation layer generates significant backscatter on hit by the radar beam. Depending on specific conditions, for example, number of channels/polarizations available and vegetation layer thickness, the noise may or may not be effectively filtered out. Both types of soil moisture retrievals will suffer over heavily vegetated areas and snow-covered areas as well.

As many studies confirm the breakdown of the quality of soil moisture retrievals over thicker vegetation (Wagner et al., 2007; Draper et al., 2009; Jackson et al., 2012a), there hasn't been an "absolute" threshold above which the soil moisture retrievals are considered impossible to too unreliable (Pan et al., 2014). Figure 3.6 shows the seasonal variations of vegetation water content (kg/m^2) over the globe and usually the vegetation is considered "thick" when this number is greater than 1–2 kg/m^2.

The validation of soil moisture retrievals are normally performed against *in situ* observations. As discussed earlier, such *in situ* observations are not readily available over most parts of the world. Some areas such as the United States may have more monitoring site available, for example, the SCAN and USCRN (Bell et al., 2013; Diamond et al., 2013). There are also dense sensor networks at local experimental sites, such as the Soil Moisture *In Situ* Sensor Testbed, USDA experimental watersheds at Little Washita, Little River, Reynolds Creek, and Walnut Gulch. These sites provide an excellent source of data for scaling analysis. More recently, the COsmic-ray Soil Moisture Observing System (COSMOS) (Zreda et al., 2012) started to offer a measurement footprint size of tens to hundreds of meters and sensing depth of tens of centimeters (Zreda et al., 2008; Franz et al., 2012). The measurements at COSMOS pivot site reach ~30 m resolution (Franz et al., personal communications). Figure 3.7 shows the ground validation of two AMSR-E retrieval products over SCAN sites the contiguous United States region.

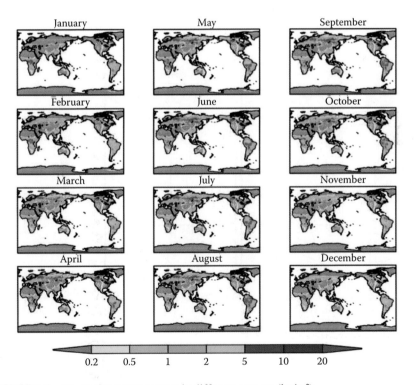

FIGURE 3.6   Vegetation water content in different seasons (kg/m^2).

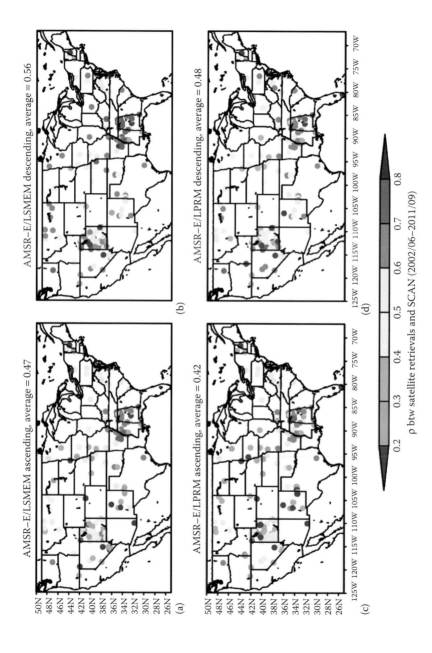

FIGURE 3.7 Pearson correlation between satellite soil moisture retrievals and SCAN *in situ* observations (June 2002 to September 2011): (a) AMSR–E/LSMEM ascending, (b) AMSR–E/LSMEM descending, (c) AMSR-E/LPRM ascending, and (d) AMSR-E/LPRM descending.

The soil moisture retrievals can also be validated against land surface model (LSM) simulations forced with observed meteorological fields. LSM simulations offer a gapless coverage and a reasonable temporal dynamics, even though they often suffer from biases and errors in parameters and input forcing data. Another type of alternative error assessment approach relies on pure statistical manipulations. For example, the triple collocation method (Scipal et al., 2008; Dorigo et al., 2010; Zwieback et al., 2012; Pan et al., 2015) and data assimilation (DA) (Crow et al., 2005). The triple collocation method makes a strong assumption on the error behavior of the soil moisture products to assess, for example, independent errors across products (Scipal et al., 2008), and thus the assessment conclusion can strongly depend on the validity of such assumptions. The DA method (Crow et al., 2005; Pan et al., 2012) tries to measure how much the remote sensing soil moisture can help correcting LSM model errors due to poor-quality precipitation forcing.

The design error level of retrievals is usually 0.02–0.05 (volumetric) (Entekhabi et al., 2010) and *in situ* validations normally support such goals under favorable conditions. As we can see from Figure 3.7, the errors do vary a lot from place to place and the uncertainty level can be quite high over difficult areas.

3.6 APPLICATIONS

Given the relatively large uncertainties in satellite soil moisture retrievals, these datasets are often used in an aggregated way (in time and space) for long-term analysis in climate and hydrologic applications (Liu et al., 2011). For example, drought analysis over crop growing season has been performed using the ECV soil moisture dataset (Yuan et al., 2015).

DA techniques (McLaughlin, 2002; Reichle et al., 2007; Reichle, 2008; Crow et al., 2009) have been a popular choice when we work with uncertain data. Numerous efforts have been made to assimilate satellite soil moisture retrievals into an LSM. Researchers have been assimilating retrievals from all sensors, from the X-band TRMM/TMI, AMSR-E (Pan and Wood, 2009; Pan et al., 2009; De Lannoy et al., 2012; Sahoo et al., 2013; Reichle et al., 2014; Lu et al., 2015) to L-band SMOS (Wanders et al., 2014; Han et al., 2015; Lievens et al., 2015). Some efforts try to use multiple satellites at the same time (Wanders et al., 2015). The goal of these DA studies is usually to improve the model predictions of soil moisture, soil temperature, evapotranspiration, runoff and streamflow, and even the input precipitation forcing (Crow et al., 2011; Wanders et al., 2015).

A lot of different algorithms are used in those DA studies such as simple nudging, variational methods (Reichle et al., 2000, 2001), various types of ensemble Kalman filters (EnKF) (Reichle et al., 2002) and smoothers, and sequential Monte Carlo methods (primarily particle filters) (Kitagawa, 1996; Arulampalam et al., 2002). Among those DA methods, the Kalman filter (Kalman, 1960) and its ensemble variants are the most popular, for example, the EnKF (Evensen, 1994), multiscale EnKF (Zhou et al., 2008), constrained ensemble filter (Pan and Wood, 2006), weakly constrained EnKF (Yilmaz et al., 2012), correlation localized EnKF, scented Kalman filter, and so on.

The Kalman filter works like a tracking system where the LSM derives the trajectory of soil moisture states and the state gets adjusted toward the observation every time the observation is available (Figure 3.8). Essentially, a filter tries to optimally

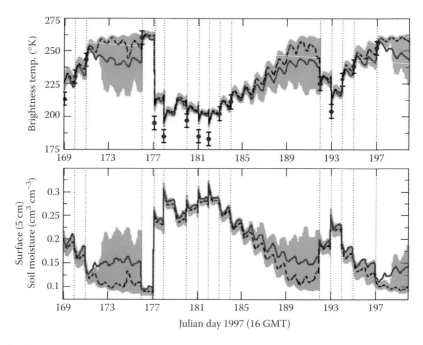

FIGURE 3.8 The workflow of a typical EnKF: The blue dots are the observations, the green shaded areas are the ensemble, the blue dashed lines are the ensemble median, and the red line is the ensemble mean.

merge/blend the information from a dynamic model (e.g., LSM) and the observations together based on error levels in each source of information. Such practice is also called data fusion, and in the case of filtering, such fusion is performed recursively in time to ensure that the state estimates are optimal given all the observations up to the current model time step. When the dynamic model and its errors are simple, that is, a linear system with Gaussian errors, a deterministic filter (e.g., Kalman filter and extended Kalman filter) may be used. However, the terrestrial hydrologic dynamics is inherently nonlinear with non-Gaussian errors, and an ensemble or particle version of the filter is normally used. The use of an ensemble of model replicates helps the filter to quantify the model errors and determine much adjustment should be applied to the model states. Both the ensemble filters and particle filters use Monte Carlo samples to quantify the model uncertainty and the replicates in ensemble filters are not weighted while they are weighted in particle filters.

3.7 SUMMARY

Soil moisture content in the top thin layer of the surface can be estimated from both passive (radiometer) and active (radar) sensors on board satellites. The greatest advantage of satellite sensors is their global coverage, and the technical limitations such as wavelength, antenna size, orbit height, and swath width have resulted in a relatively coarse spatial resolution of the measurements. Passive sensors usually have a footprint size of 30–50 km and active sensors can resolve down to several kilometers. The depth of soil moisture detection is also limited to about 0.01–0.1 m given the wavelength being used. Noise in the soil moisture retrievals comes primarily from the vegetation cover and also from surface water bodies, snow cover, and atmospheric-source emissions such as storm clouds.

A series of passive and active microwave sensors have been utilized by the community since the late 1970s to estimate soil moisture and a large number of retrieval products have been created. These retrieval products have been validated against *in situ* observations, LSM and other model simulations, and through statistical techniques. Validation and assessment studies have found very reasonable performance of the retrievals under favorable conditions, although the uncertainties become large over heavily vegetated as well as snow-covered areas. Long-term global satellite records of soil moisture have been used for in various different studies such as climate and drought analysis. With new satellite missions dedicated to soil moisture observations, we envision a more active and fruitful research into the satellite remote sensing of soil moisture as well as its applications to a wider range of problems in Earth science.

REFERENCES

Albertson, J. D. and M. B. Parlange (1999), Natural integration of scalar fluxes from complex terrain, *Adv. Water Resour.*, 23(3), 239–252, doi:10.1016/S0309-1708(99)00011-1.

Arulampalam, M. S., S. Maskell, N. Gordon, and T. Clapp (2002), A tutorial on particle filters for online nonlinear/non-Gaussian Bayesian tracking, *IEEE Trans. Signal Process.*, 50(2), 174–188, doi:10.1109/78.978374.

Bell, J. E., M. A. Palecki, W. G. Collins, J. H. Lawrimore, R. D. Leeper, M. E. Hall, J. Kochendorfer et al. (2013), Climate reference network soil moisture and temperature observations, *J. Hydrometeorol.*, 14(3), 977–988.

Cahill, A. T., M. B. Parlange, T. J. Jackson, P. O'Neill, and T. J. Schmugge (1999), Evaporation from nonvegetated surfaces: Surface aridity methods and passive microwave remote sensing, *J. Appl. Meteorol.*, 38(9), 1346–1351, doi:10.1175/1520-0450(1999)038<1346:EFNSSA> 2.0.CO;2.

Choudhury, B., T. Schmugge, A. Chang, and R. Newton (1979), Effect of surface-roughness on the microwave emission from soils, *J. Geophys. Res. Oc. Atm.*, 84(9), 5699–5706, doi:10.1029/JC084iC09p05699.

Crow, W. T., M. J. van den Berg, G. J. Huffman, and T. Pellarin (2011), Correcting rainfall using satellite-based surface soil moisture retrievals: The Soil Moisture Analysis Rainfall Tool (SMART), *Water Resour. Res.*, 47, W08521, doi:10.1029/2011WR010576.

Crow, W. T. and E. F. Wood (1999), Multi-scale dynamics of soil moisture variability observed during SGP'97, *Geophys. Res. Lett.*, 26(23), 3485–3488, doi:10.1029/1999GL010880.

Crow, W. T., G. J. Huffman, R. Bindlish, and T. J. Jackson (2009), Improving satellite-based rainfall accumulation estimates using spaceborne surface soil moisture retrievals, *J. Hydrometeorol.*, 10(1), 199–212, doi:10.1175/2008JHM986.1.

Crow, W., R. Bindlish, and T. Jackson (2005), The added value of spaceborne passive microwave soil moisture retrievals for forecasting rainfall-runoff partitioning, *Geophys. Res. Lett.*, 32(18), L18401, doi:10.1029/2005GL023543.

de Jeu, R. A. M., W. Wagner, T. R. H. Holmes, A. J. Dolman, N. C. van de Giesen, and J. Friesen (2008), Global soil moisture patterns observed by space borne microwave radiometers and scatterometers, *Surv. Geophys.*, 29(4–5), 399–420, doi:10.1007/s10712-008-9044-0.

De Lannoy, G. J. M., R. H. Reichle, K. R. Arsenault, P. R. Houser, S. Kumar, N. E. C. Verhoest, and V. R. N. Pauwels (2012), Multiscale assimilation of advanced microwave scanning radiometer-EOS snow water equivalent and moderate resolution imaging spectroradiometer snow cover fraction observations in Northern Colorado, *Water Resour. Res.*, 48, W01522, doi:10.1029/2011WR010588.

Diamond, H. J., T. R. Karl, M. A. Palecki, C. B. Baker, J. E. Bell, R. D. Leeper, D. R. Easterling et al. (2013), U.S. Climate reference network after one decade of operations: Status and assessment, *Bull. Amer. Meteor. Soc.*, 94, 489–498, doi:10.1175/BAMS-D-12-00170.

Dobson, M. C. and F. T. Ulaby (1986), Active microwave soil-moisture research, *IEEE Trans. Geosci. Remote Sens.*, 24(1), 23–36, doi:10.1109/TGRS.1986.289585.

Dobson, M., F. Ulaby, M. Hallikainen, and M. Elrayes (1985), Microwave dielectric behavior of wet soil. 2. Dielectric mixing models, *IEEE Trans. Geosci. Remote Sens.*, 23(1), 35–46, doi:10.1109/TGRS.1985.289498.

Dorigo, W. A., K. Scipal, R. M. Parinussa, Y. Y. Liu, W. Wagner, R. A. M. de Jeu, and V. Naeimi (2010), Error characterisation of global active and passive microwave soil moisture datasets, *Hydrol. Earth Syst. Sci.*, 14(12), 2605–2616, doi:10.5194/hess-14-2605-2010.

Draper, C. S., J. P. Walker, P. J. Steinle, R. A. M. de Jeu, and T. R. H. Holmes (2009), An evaluation of AMSR-E derived soil moisture over Australia, *Remote Sens. Environ.*, 113(4), 703–710, doi:10.1016/j.rse.2008.11.011.

Drusch, M., E. Wood, and T. Jackson (2001), Vegetative and atmospheric corrections for the soil moisture retrieval from passive microwave remote sensing data: Results from the southern great plains hydrology experiment 1997, *J. Hydrometeorol.*, 2(2), 181–192, doi:10.1175/1525-7541(2001)002<0181:VAACFT>2.0.CO;2.

Entekhabi, D., E. G. Njoku, P. E. O'Neill, K. H. Kellogg, W. T. Crow, W. N. Edelstein, J. K. Entin et al. (2010), The soil moisture active passive (SMAP) mission, *Proc. IEEE*, 98(5), 704–716, doi:10.1109/JPROC.2010.2043918.

Evensen, G. (1994), Sequential data assimilation with a nonlinear quasi-geostrophic model using Monte-Carlo methods to forecast error statistics, *J. Geophys. Res. Oceans*, 99(C5), 10143–10162, doi:10.1029/94JC00572.

Franz, T. E., M. Zreda, T. P. A. Ferre, R. Rosolem, C. Zweck, S. Stillman, X. Zeng, and W. J. Shuttleworth (2012), Measurement depth of the cosmic ray soil moisture probe affected by hydrogen from various sources, *Water Resour. Res.*, 48(8), W08515, doi:10.1029/2012WR011871.

Fujii, H., T. Koike, and K. Imaoka (2009), Improvement of the AMSR-E algorithm for soil moisture estimation by introducing a fractional vegetation coverage dataset derived from MODIS data, *J. Remote Sens. Soc. Jpn.*, 29(1), 11.

Gao, H., E. Wood, T. Jackson, M. Drusch, and R. Bindlish (2006), Using TRMM/TMI to retrieve surface soil moisture over the southern United States from 1998 to 2002, *J. Hydrometeorol.*, 7(1), 23–38, doi:10.1175/JHM473.1.

Gao, H., E. Wood, M. Drusch, W. Crow, and T. Jackson (2004), Using a microwave emission model to estimate soil moisture from ESTAR observations during SGP99, *J. Hydrometeorol.*, 5(1), 49–63, doi:10.1175/1525-7541(2004)005<0049:UAMEMT>2.0.CO;2.

Han, X., X. Li, R. Rigon, R. Jin, and S. Endrizzi (2015), Soil moisture estimation by assimilating L-band microwave brightness temperature with geostatistics and observation localization, *Plos One*, 10(1), e0116435, doi:10.1371/journal.pone.0116435.

Imaoka, K., M. Kachi, M. Kasahara, N. Ito, K. Nakagawa, and T. Oki (2010), Instrument performance and calibration of Amsr-E and Amsr2, *Network. World Remote Sens.*, 38, 13–16.

Jackson, T. J. and A. Y. Hsu (2001), Soil moisture and TRMM microwave imager relationships in the Southern great plains 1999 (SGP99) experiment, *IEEE Trans. Geosci. Remote Sens.*, 39(8), 1632–1642, doi:10.1109/36.942541.

Jackson, T. J., D. Le Vine, A. Hsu, A. Oldak, P. Starks, C. Swift, J. Isham, and M. Haken (1999), Soil moisture mapping at regional scales using microwave radiometry: The southern great plains hydrology experiment, *IEEE Trans. Geosci. Remote Sens.*, 37(5), 2136–2151, doi:10.1109/36.789610.

Jackson, T. J. (1993), Measuring surface soil-moisture using passive microwave remote-sensing. 3. *Hydrol. Process.*, 7(2), 139–152, doi:10.1002/hyp.3360070205.

Jackson, T. J., R. Bindlish, M. H. Cosh, T. Zhao, P. J. Starks, D. D. Bosch, M. Seyfried et al. (2012a), Validation of soil moisture and ocean salinity (SMOS) soil moisture over watershed networks in the U.S. *IEEE Trans. Geosci. Remote Sens.*, 50(5), 1530–1543, doi:10.1109/TGRS.2011.2168533.

Jackson, T. J., A. Colliander, J. S. Kimball, R. H. Reichle, W. T. Crow, D. Entekhabi, P. E. O'Neill, and E. G. Njoku (2012b), Soil moisture active passive (SMAP) mission science data calibration and validation plan, Jet Propulsion Laboratory, California Institute of Technology, 1–96. http://smap.jpl.nasa.gov/files/smap2/CalVal_Plan_120706_pub.pdf.

Kalman, R. E. (1960), A new approach to linear filtering and prediction problems, *Trans. ASME J. Basic Eng.*, 82(Series D), 35–45.

Kerr, Y. H., P. Waldteufel, P. Richaume, J. P. Wigneron, P. Ferrazzoli, A. Mahmoodi, A. Al Bitar et al. (2012), The SMOS soil moisture retrieval algorithm, *IEEE Trans. Geosci. Remote Sens.*, 50(5), 1384–1403, doi:10.1109/TGRS.2012.2184548.

Kirdyashev, K. P., A. A. Chukhlantsev, and A. M. Shutko (1979), Microwave radiation of grounds with vegetative cover, *Radiotekh. Elektron.*, 24(2), 256–264.

Kitagawa, G. (1996), Monte Carlo filter and smoother for non-Gaussian nonlinear state space models, *J. Comput. Graph. Statist.*, 5(1), 1–25, doi:10.2307/1390750.

Koster, R. D. and M. J. Suarez (2001), Soil moisture memory in climate models, *J. Hydrometeorol.*, 2(6), 558–570, doi:10.1175/1525–7541(2001)002<0558:SMMICM>2.0.CO;2.

Koster, R. D. and M. J. Suarez (1992), Modeling the land surface boundary in climate models as a composite of independent vegetation stands, *J Geophys Res: Atmos.*, 97(D3), 2697–2715, doi:10.1029/91JD01696.

Lievens, H., S. K. Tomer, A. Al Bitar, G. J. M. De Lannoy, M. Drusch, G. Dumedah, H. -H. Franssen et al. (2015), SMOS soil moisture assimilation for improved hydrologic simulation in the Murray Darling Basin, Australia, *Remote Sens. Environ.*, 168, 146–162, doi:10.1016/j.rse.2015.06.025.

Liu, Y. Y., W. A. Dorigo, R. M. Parinussa, R. A. M. de Jeu, W. Wagner, M. F. McCabe, J. P. Evans, and A. I. J. M. van Dijk (2012), Trend-preserving blending of passive and active microwave soil moisture retrievals, *Remote Sens. Environ.*, 123, 280–297, doi:10.1016/j.rse.2012.03.014.

Liu, Y. Y., R. M. Parinussa, W. A. Dorigo, R. A. M. De Jeu, W. Wagner, A. I. J. M. van Dijk, M. F. McCabe, and J. P. Evans (2011), Developing an improved soil moisture dataset by blending passive and active microwave satellite-based retrievals, *Hydrol. Earth Syst. Sci.*, 15(2), 425–436, doi:10.5194/hess-15-425-2011.

Lu, H., K. Yang, T. Koike, L. Zhao, and J. Qin (2015), An improvement of the radiative transfer model component of a land data assimilation system and its validation on different land characteristics, *Remote Sens.*, 7(5), 6358–6379, doi:10.3390/rs70506358.

Luo, L. and E. F. Wood (2008), Use of Bayesian merging techniques in a multimodel seasonal hydrologic ensemble prediction system for the eastern United States, *J. Hydrometeorol.*, 9(5), 866–884, doi:10.1175/2008JHM980.1.

McLaughlin, D. B. (2002), An integrated approach to hydrologic data assimilation: Interpolation, smoothing, and filtering, *Adv. Water Resour.*, 25(8–12), 1275–1286, doi:10.1016/S0309-1708(02)00055-6.

Mo, K. C. (2008), Model-based drought indices over the United States, *J. Hydrometeorol.*, 9(6), 1212–1230, doi:10.1175/2008JHM1002.1.

Naeimi, V., K. Scipal, Z. Bartalis, S. Hasenauer, and W. Wagner (2009), An improved soil moisture retrieval algorithm for ers and metop scatterometer observations, *IEEE Trans. Geosci. Remote Sens.*, 47(7), 1999–2013, doi:10.1109/TGRS.2009.2011617.

Njoku, E. (1977), Theory for passive microwave remote-sensing of near-surface soil-moisture, *J. Geophys. Res.*, 82(20), 3108–3118, doi:10.1029/JB082i020p03108.

Njoku, E. and S. Chan (2006), Vegetation and surface roughness effects on AMSR-E land observations, *Remote Sens. Environ.*, 100(2), 190–199, doi:10.1016/j.rse.2005.10.017.

Njoku, E., T. Jackson, V. Lakshmi, T. Chan, and S. Nghiem (2003), Soil moisture retrieval from AMSR-E, *IEEE Trans. Geosci. Remote Sens.*, 41(2), 215–229, doi:10.1109/TGRS.2002.808243.

Njoku, E. and L. Li (1999), Retrieval of land surface parameters using passive microwave measurements at 6–18 GHz, *IEEE Trans. Geosci. Remote Sens.*, 37(1), 79–93, doi:10.1109/36.739125.

Owe, M., R. de Jeu, and J. Walker (2001), A methodology for surface soil moisture and vegetation optical depth retrieval using the microwave polarization difference index, *IEEE Trans. Geosci. Remote Sens.*, 39(8), 1643–1654, doi:10.1109/36.942542.

Owe, M., R. de Jeu, and T. Holmes (2008), Multisensor historical climatology of satellite-derived global land surface moisture, *J. Geophys. Res. Earth Surf.*, 113(F1), F01002, doi:10.1029/2007JF000769.

Pan, M., A. K. Sahoo, and E. F. Wood (2014), Improving global soil moisture retrievals from a physically based radiative transfer model, *Remote Sens. Environ.*, 140, 130–140, doi:10.1016/j.rse.2013.08.020.

Pan, M., C. K. Fisher, N. W. Chaney, W. Zhan, W. T. Crow, F. Aires, D. Entekhabi, and E. F. Wood (2015), Triple collocation: Beyond three estimates and separation of structural/non-structural errors, *Remote Sens. Environ.*, 171, 299–310, doi:http://dx.doi.org/10.1016/j.rse.2015.10.028.

Pan, M., A. K. Sahoo, and E. F. Wood (2014), Improving soil moisture retrievals from a physically-based radiative transfer model, *Remote Sens. Environ.*, 140, 130–140, doi:10.1016/j.rse.2013.08.020.

Pan, M., A. K. Sahoo, E. F. Wood, A. Al Bitar, D. Leroux, and Y. H. Kerr (2012), An initial assessment of smos derived soil moisture over the continental United States, *IEEE J. Sel. Topics Appl. Earth Observ.*, 5(5), 1448–1457, doi:10.1109/JSTARS.2012.2194477.

Pan, M. and E. F. Wood (2006), Data assimilation for estimating the terrestrial water budget using a constrained ensemble Kalman filter, *J. Hydrometeorol.*, 7(3), 534–547, doi:10.1175/JHM495.1.

Pan, M. and E. F. Wood (2009), A multiscale ensemble filtering system for hydrologic data assimilation. Part II: Application to land surface modeling with satellite rainfall forcing, *J. Hydrometeorol.*, 10(6), 1493–1506, doi:10.1175/2009JHM1155.1.

Pan, M., E. F. Wood, D. B. McLaughlin, D. Entekhabi, and L. Luo (2009), A multiscale ensemble filtering system for hydrologic data assimilation. Part I: Implementation and synthetic experiment, *J. Hydrometeorol.*, 10(3), 794–806, doi:10.1175/2009JHM1088.1.

Parinussa, R. M., T. R. H. Holmes, N. Wanders, W. A. Dorigo, and R. A. M. de Jeu (2015), A preliminary study toward consistent soil moisture from AMSR2, *J. Hydrometeorol.*, 16(2), 932–947, doi:10.1175/JHM-D-13-0200.1.

Reichle, R. H., D. McLaughlin, and D. Entekhabi (2000), Variational data assimilation of soil moisture and temperature from remote sensing observations. In *Calibration and Reliability in Groundwater Modelling: Coping with Uncertainty*, Stauffer, F., Kinzelbach, W., Kovar, K., Hoehn, E. (eds.), pp. 353–359. International Association of Hydrological Sciences, Wallingford, UK.

Reichle, R. H., D. B. McLaughlin, and D. Entekhabi (2001), Variational data assimilation of microwave radiobrightness observations for land surface hydrology applications, *IEEE Trans. Geosci. Remote Sens.*, 39(8), 1708–1718, doi:10.1109/36.942549.

Reichle, R. H., D. B. McLaughlin, and D. Entekhabi (2002), Hydrologic data assimilation with the ensemble Kalman filter, *Mon. Weather Rev.*, 130(1), 103–114, doi:10.1175/1520-0493(2002)130<0103:HDAWTE>2.0.CO;2.

Reichle, R. H. (2008), Data assimilation methods in the earth sciences, *Adv. Water Resour.*, 31(11), 1411–1418, doi:10.1016/j.advwatres.2008.01.001.

Reichle, R. H., G. J. M. De Lannoy, B. A. Forman, C. S. Draper, and Q. Liu (2014), Connecting satellite observations with water cycle variables through land data assimilation: Examples using the NASA GEOS-5 LDAS, *Surv. Geophys.*, 35(3), 577–606, doi:10.1007/s10712-013-9220-8.

Reichle, R. H., R. D. Koster, P. Liu, S. P. P. Mahanama, E. G. Njoku, and M. Owe (2007), Comparison and assimilation of global soil moisture retrievals from the advanced microwave scanning radiometer for the earth observing system (AMSR-E) and the scanning multichannel microwave radiometer (SMMR), *J Geophys Res. Atmos.*, 112(D9), D09108, doi:10.1029/2006JD008033.

Robock, A., K. Y. Vinnikov, G. Srinivasan, J. K. Entin, S. E. Hollinger, N. A. Speranskaya, S. X. Liu, and A. Namkhai (2000), The Global Soil Moisture Data Bank, *Bull. Am. Meteorol. Soc.*, 81(6), 1281–1299, doi:10.1175/1520-0477(2000)081<1281:TGSMDB>2.3.CO;2.

Sahoo, A. K., G. J. M. De Lannoy, R. H. Reichle, and P. R. Houser (2013), Assimilation and downscaling of satellite observed soil moisture over the little river experimental watershed in Georgia, USA, *Adv. Water Resour.*, 52, 19–33, doi:10.1016/j.advwatres.2012.08.007.

Schaake, J. C., Q. Y. Duan, V. Koren, K. E. Mitchell, P. R. Houser, E. F. Wood, A. Robock et al. (2004), An intercomparison of soil moisture fields in the North American land data assimilation system (NLDAS), *J Geophys Res. Atmos.*, 109(D1), D01S90, doi:10.1029/2002JD003309.

Scipal, K., T. Holmes, R. de Jeu, V. Naeimi, and W. Wagner (2008), A possible solution for the problem of estimating the error structure of global soil moisture data sets, *Geophys. Res. Lett.*, 35(24), L24403, doi:10.1029/2008GL035599.

Seyfried, M. S., L. E. Grant, E. Du, and K. Humes (2005), Dielectric loss and calibration of the hydra probe soil water sensor, *Vadose Zone J.*, 4(4), 1070–1079, doi:10.2136/vzj2004.0148.

Sheffield, J. and E. F. Wood (2007), Characteristics of global and regional drought, 1950–2000: Analysis of soil moisture data from off-line simulation of the terrestrial hydrologic cycle, *J Geophys Res. Atmos.*, 112(D17), D17115, doi:10.1029/2006JD008288.

Ulaby, F. T., R. K. Moore, and A. K. Fung (1982), Radar remote sensing and surface scattering and emission theory. In *Microwave Remote Sensing: Active and Passive*, Vol. II, Addison-Wesley, Reading, MA.

Wagner, W., G. Lemoine, and H. Rott (1999), A method for estimating soil moisture from ERS scatterometer and soil data, *Remote Sens. Environ.*, 70(2), 191–207, doi:10.1016/S0034-4257(99)00036-X.

Wagner, W., V. Naeimi, K. Scipal, R. de Jeu, and J. Martinez-Fernandez (2007), Soil moisture from operational meteorological satellites, *Hydrogeol. J.*, 15(1), 121–131, doi:10.1007/s10040-006-0104-6.

Wanders, N., D. Karssenberg, A. de Roo, S. M. de Jong, and M. F. P. Bierkens (2014), The suitability of remotely sensed soil moisture for improving operational flood forecasting, *Hydrol. Earth Syst. Sci.*, 18(6), 2343–2357, doi:10.5194/hess-18-2343-2014.

Wanders, N., M. Pan, and E. F. Wood (2015), Correction of real-time satellite precipitation with multi-sensor satellite observations of land surface variables, *Remote Sens. Environ.*, 160, 206–221, doi:10.1016/j.rse.2015.01.016.

Wang, J. and B. Choudhury (1981), Remote-sensing of soil-moisture content over bare field at 1.4 GHz frequency, *J. Geophys. Res. Oc. Atm.*, 86(NC6), 5277–5282, doi:10.1029/JC086iC06p05277.

Wang, J. and T. Schmugge (1980), An empirical-model for the complex dielectric permittivity of soils as a function of water-content, *IEEE Trans. Geosci. Remote Sens.*, 18(4), 288–295, doi:10.1109/TGRS.1980.350304.

Wigneron, J. -P., Y. Kerr, P. Waldteufel, K. Saleh, M. -J. Escorihuela, P. Richaume, P. Ferrazzoli et al. (2007), L-band microwave emission of the biosphere (L-MEB) model: Description and calibration against experimental data sets over crop fields, *Remote Sens. Environ.*, 107(4), 639–655, doi:10.1016/j.rse.2006.10.014.

Yilmaz, M. T., W. T. Crow, M. C. Anderson, and C. Hain (2012), An objective methodology for merging satellite- and model-based soil moisture products, *Water Resour. Res.*, 48, W11502, doi:10.1029/2011WR011682.

Yuan, X., Z. Ma, M. Pan, and C. Shi (2015), Microwave remote sensing of short-term droughts during crop growing seasons, *Geophys. Res. Lett.*, 42(11), 4394–4401, doi:10.1002/2015GL064125.

Zhou, Y., D. McLaughlin, D. Entekhabi, and G. C. Ng (2008), An ensemble multiscale filter for large nonlinear data assimilation problems, *Mon. Weather Rev.*, 136(2), 678–698, doi:10.1175/2007MWR2064.1.

Zreda, M., W. J. Shuttleworth, X. Zeng, C. Zweck, D. Desilets, T. E. Franz, and R. Rosolem (2012), COSMOS: The COsmic-ray soil moisture observing system, *Hydrol. Earth Syst. Sci.*, 16(11), 4079–4099, doi:10.5194/hess-16-4079-2012.

Zreda, M., D. Desilets, T. P. A. Ferré, and R. L. Scott (2008), Measuring soil moisture content non-invasively at intermediate spatial scale using cosmic-ray neutrons, *Geophys. Res. Lett.*, 35(21), L21402, doi:10.1029/2008GL035655.

Zwieback, S., K. Scipal, W. Dorigo, and W. Wagner (2012), Structural and statistical properties of the collocation technique for error characterization, *Nonlinear Proc. Geoph.*, 19(1), 69–80, doi:10.5194/npg-19-69-2012.

4 Satellite Remote Sensing of Lakes and Wetlands

Huilin Gao, Shuai Zhang, Michael Durand,
and Hyongki Lee

CONTENTS

4.1 INTRODUCTION

Stored in lakes, wetlands, and rivers, surface freshwater is the most critical resource that sustains environments and the lives of the organisms within them. In addition, it plays a large role in the water cycle—both by contributing to the storage component (i.e., lake and wetland storage) and by serving as a key flux term (i.e., river discharge) that links the land to the ocean. For many scientific disciplines, surface freshwater is of great significance. In water resources management, reservoir impounding has been used for hydropower generation, flood control, irrigation, and recreational purposes. Lakes and rivers host most of the inland fisheries, and wetlands are home to a wide variety of wildlife. Floods and droughts, the two most life-threatening and financially expensive natural disasters, are directly associated with extreme conditions within these water bodies.

Geographically, inland water bodies are distributed very unevenly. The highest concentration of natural lakes occurs in the de-glaciated areas between 50°N and 70°N, including Alaska, Canada, Scandinavia, and northern Russia (Lehner and Doll, 2004). For reservoirs (i.e., man-made lakes), the largest combined storage volumes are concentrated in Canada, Russia, the United States, Brazil, and China (Lehner et al., 2011). The world's largest wetlands are located in Eurasia, South America, and North America—oftentimes near some of the largest rivers, such as the Amazon, the Congo, and the Mississippi (Keddy et al., 2009). Over the years, a number of comprehensive databases have been constructed to characterize the size and location of these water bodies (Lehner and Doll, 2004; Lehner et al., 2011; Yamazaki et al., 2014). However, knowledge about lake and wetland storage change, as well as about river discharges, is extremely limited—especially over regions where observations are few or data are not shared (Zhang et al., 2013). This situation has been exacerbated with the number of existing gauge observations continuously decreasing over the last few decades.

Since the 1980s, satellite remote sensing has opened up tremendous opportunities to earth scientists—with its contribution to hydrologic sciences initiated and accelerated during the last two decades. Although satellite remote sensing cannot provide storage estimations as precise as traditional *in situ* measurements, it does complement *in situ* data in several ways. First, it collects data at a large scale (oftentimes globally). This is especially advantageous over remote areas where natural lakes are not typically observed, as well as over regions where reservoir data sharing is uncommon. Second, it consistently takes measurements regularly at near real time. Data collected through traditional approaches—if shared—are typically lagged several months behind. Third, considering the amount of data that can be collected (especially over remote regions), it is very cost efficient. Many of the satellite missions, such as TOPEX/Poseidon, have exceeded their planned life spans by many years.

In this chapter, we provide an inclusive review of past, current, and future remote sensing techniques for observing lakes and wetlands. We also discuss exciting opportunities that will be brought on by future satellite missions.

4.2   LAKES AND RESERVOIRS

Containing about 87% of worldwide fresh surface water, large lakes and reservoirs cover over 1.9 million km² of the global land surface area (Lehner and Doll, 2004). Space-borne lake elevation observations are typically acquired from satellite altimetry missions, while lake surface area estimations are primarily obtained from optical sensors. Lake storage variations are inferred either from elevation and area together, or from elevation (or area) alone with prior knowledge of lake bathymetry (i.e., elevation–area relationship). This section reviews the sensors and approaches for monitoring elevations, areas, and storage variation of lakes and reservoirs. In addition, benefits and limitations of past and current data sources and approaches are discussed.

4.2.1   Surface Elevation

Satellite altimetry is used for estimating surface water elevation by measuring the time taken by a radar (or laser) pulse to travel from the satellite antenna to the surface and back to the satellite receiver. Measurement errors are attributed to instrument errors, as well as to geophysical errors (e.g., atmospheric attenuation). Radar altimetry is the most commonly used technique for observing large lakes and reservoirs, while laser altimetry is more often used for tracking the changes of smaller lakes.

Earlier applications of satellite radar altimetry were mainly for observing sea-surface heights, such as the observations made by the Geosat satellite owned and operated by the US Navy (1985–1989). Since the launches of the European Remote Sensing Satellite (ERS-1) in 1991 and TOPEX/Poseidon in 1992, satellite radar altimeters have been used for monitoring inland large water bodies. By the time this chapter was written, water levels for nearly 300 lakes and reservoirs (as shown in Figure 4.1) had been documented by the US Department of Agriculture's (USDA) Global Reservoir and Lake Elevation Database (http://www.pecad.fas.usda.gov/cropexplorer/global_reservoir/), and by the French Space Agency Centre National d'Etudes Spatiales' (CNES) HYDROWEB lake database (http://www.legos.obs-mip.fr/soa/hydrologie/hydroweb/). Specifically, lake and reservoir elevation values from satellite radar altimeters have the following characteristics.

1. Radar altimeters, typically at Ku and (more recently) Ka bands, usually have footprint sizes ranging from 2 to 10 km. To minimize the errors associated with each single measurement, the average of the observations over one lake from one track is used to represent the lake elevation. Although some small lakes can be monitored accurately, Gao et al. (2012) suggested a cross-section length of 10 km as a threshold for observable lakes. The satellite repeat period varies from 10 to 35 days.
2. Radar altimeters are nonimaging sensors—meaning the swath of a track is the same as the spatial resolution of the footprint. Consequently, only a small portion of the large number of global large lakes (about 15%) can be monitored using radar altimeters (Biancamaria et al., 2010).
3. The accuracy of lake elevation estimates depends on multiple factors. For very large lakes with no interference of the radar echo from land (e.g., the Great Lakes), one remotely sensed elevation value is a result of the average

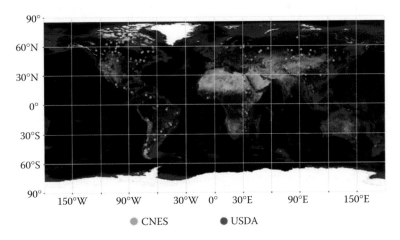

FIGURE 4.1 Locations of the large lakes and reservoirs where radar altimetry-based elevations are provided by the CNES and USDA databases. A total of 274 lakes and reservoirs are reported on between the two databases (166 by CNES, 193 by USDA, and 85 overlap). (From Gao, H., *Wiley Interdiscip. Rev.*, 2, 147–157, 2015.)

of many data points from one track. In such cases, the satellite measurement error can be as small as a few centimeters. However, for smaller lakes, larger lakes with interference of the radar echo from land, or lakes with low surface roughness conditions (i.e., a calm surface), the errors can be as large as a few meters. The elevation accuracy also varies by instrument. For instance, in a study conducted over Lake Argyle (Jarihani et al., 2013), it was found that the Root-Mean-Square Errors from Jason-2, Envisat, Jason-1, and TOPEX/Poseidon were 0.28, 0.42, 1.07, and 1.5 m, respectively.

Radar altimetry has been the most commonly used approach for monitoring lakes and reservoirs, such as studies of the Great Lakes (Morris and Gill, 1994), Amazonian lakes (Alsdorf et al., 2001), lakes in central Asia (Crétaux et al., 2009), Tibetan lakes (Lee et al., 2011b), and continental and global inland waters (Ponchaut and Cazenave, 1998; Berry et al., 2005). Altimetry data have also been applied in water resources studies (Coe and Birkett, 2004), hydrologic models (Gao et al., 2011), ecosystem response studies (Sarch and Birkett, 2000), and studies about ocean–atmosphere interactions (Birkett et al., 1999; Mercier et al., 2002).

While there have been dozens of satellite radar altimetry systems used for measuring surface water elevations over the past few decades, the Geoscience Laser Altimeter System (GLAS) on board the Ice Cloud and Land Elevation Satellite (ICESat) is the only space-borne laser device that has been used for this purpose. During its service from 2003 to 2010, the ICESat/GLAS instrument made significant contributions in observing trends of small lakes over mountainous and/or cold regions. For instance, Smith et al. (2009) presented an inventory of 124 subglacial lakes (among which 15 are large lakes with a volume larger than 1 km^3) in Antarctica—all detected by ICESat/GLAS. Over the Tibetan Plateau, ICESat/GLAS data were used for several studies examining lake level elevation changes and the causes of those changes (Zhang et al., 2011; Phan et al., 2012). Another study investigated water level fluctuations from 56 of the 100 largest lakes in China (Wang et al., 2013).

4.2.2 SURFACE AREA

Across the electromagnetic spectrum, water has distinct radiative properties at many different wavelengths. For instance, the emissivity of water—as observed by passive microwave radiometers—is much smaller than that from other land cover types, and the backscatter from water is lower than that from other land features. However, microwave frequencies are seldom used for quantifying the size of lakes/reservoirs. Because of the low spatial resolution of passive microwave radiometers, most lakes only cover one or a few pixels partially (with the exception of very large lakes, such as the Great Lakes). Active microwave sensors, such as Synthetic Aperture Radar (SAR) instruments, are known both for their high resolution and their sensitivity to surface roughness (Alsdorf et al., 2007). Because of their unique capabilities for penetrating thick canopies, L-band SAR sensors have been used more than other instruments for measuring forest covered wetland areas (see Section 4.3.2). For accurately quantifying lake areas, however, the overall best choice is the use of instruments that work in the visible/infrared bands—whose high spatial resolution and easy accessibility outweigh their constraints due to cloudy/rainy conditions.

Passive visible/infrared sensors that have been used for estimating water area include Meteosat, the Advanced Very High Resolution Radiometer (AVHRR), the Moderate resolution Imaging Spectroradiometer (MODIS), Landsat, and many others. Using Lake Chad (Africa) as an example, Landsat images from the 1970s to recent have been used for identifying its dramatic loss (Gao et al., 2011). Meteosat and AVHRR were utilized for some of the early studies involving the lake (Birkett, 2000), and MODIS images have been employed for mapping its coverage operationally (http://floodobservatory.colorado.edu/Customized/LakeChad.html).

As summarized by Gao (2015), area estimation algorithms largely fall into three categories: threshold-based approaches, image-classification-based approaches, and multiple-step hybrid approaches. Threshold approaches usually employ a criterion based on the optical properties of a single band, or from a spectral water index. Because of its simplicity, it has been applied in many studies for delineating water. In Birkett (2000), the midpoint of the AVHRR channel 2 reflectance (between the water and land mean values) was selected for thresholding the image and identifying the water area within the entire Lake Chad basin. Taking advantage of the multichannel information obtained, different spectral indices—such as the Normalized Difference Vegetation Index (NDVI) and the Normalized Difference Water Indices (NDWIs)—were used. A comprehensive comparison among different forms of NDWIs was made by Ji et al. (2009). It was found that the NDWI defined as (green − SWIR)/ (green + SWIR) performed the best. The drawback of the threshold method is that the selected criterion is often case specific, which inhibits it from generating consistent results over multiple lakes under various land cover and/or climatic conditions.

The image classification approaches typically adopt either a supervised or an unsupervised algorithm to extract the water area from the surrounding land surface. The key difference between the supervised and unsupervised algorithms is that a prior selection of training areas is needed for the former but not for the latter. Using statistical parameters from the training area, the supervised algorithm typically produces higher accuracy results. It has been often used for identifying high spatial resolution data, such as Landsat imageries. In Gao et al. (2011), Lake Chad's surface area was delineated from multiple Landsat images to validate model simulations. Results from such algorithms, however, may vary depending on which areas/classes are chosen to "supervise" the classification process. By minimizing human-induced errors, unsupervised algorithms are able to extract water surface area over multiple lakes and reservoirs (Cheema and Bastiaanssen, 2010; Lu et al., 2011). The accuracy of results from the unsupervised algorithm can be compromised, however, if the iteration for minimizing the intracluster sum of squares becomes trapped in local optima (Maulik and Saha, 2010; Zhong et al., 2013).

The multiple-step hybrid approaches intend to improve upon the single-step approaches (e.g., thresholding and supervised/unsupervised classifications) via iterative classification techniques. In Wang et al. (2014), the decadal lake dynamics over the Yangtze River Basin was investigated by classifying MODIS images using a combination of an NIR-based single-band threshold, an NDWI index threshold, and a set of supervised classification algorithms. In Zhang et al. (2014), a post-classification

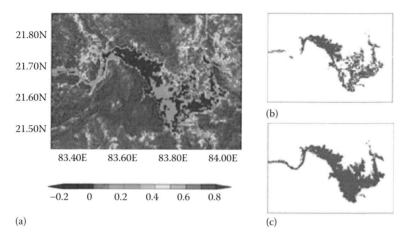

FIGURE 4.2 The comparison between classification before and after image enhancement is shown. (a) is the MODIS NDVI image for the Hirakud reservoir in India on the 273–289 days of 2005, (b) is from the K-means classification, and (c) is the result of K-means after enhancement.

image enhancement procedure was applied after a MODIS unsupervised classification. By leveraging the water coverage percentile, the enhancement allowed for the correction of misclassifications due to the image quality and pixel purity problems. Figure 4.2 shows an example of how the results can be improved when the original MODIS NDVI image is impaired.

4.2.3 STORAGE VARIATIONS

Water storage of reservoirs and lakes can be inferred from remotely sensed surface area and elevation data. Equation 4.1 from Gao et al. (2012) shows the storage function, where V_c, A_c, and h_c represent storage, area, and water elevation at capacity; and V_o, A_o, and h_o are the observed storage, area, and water elevation. For large reservoirs, information at capacity can be acquired from databases such as the Global Reservoir and Dam Database (Lehner et al., 2011) and the International Commission on Large Dam's World Register of Dams (http://www.icold-cigb.net). For natural lakes, such data are generally unavailable. In such cases, the storage variation (ΔV) from time 1 to time 2 can be inferred from the change of elevation (ΔH) and the water surface area (A) (averaged during the period) according to Equation 4.2

$$V_o = V_c - (h_c - h_o)(h_c + A_o)/2 \tag{4.1}$$

$$\Delta V = \Delta h \cdot (A_1 + A_2)/2 \tag{4.2}$$

If the lake bathymetry (e.g., elevation–area relationship, h–V curve, or A–V curve) is known, the storage or storage variation equation can be simplified as a function of either elevation or area—depending on the data availability. Because of the limitations with satellite data availability and precision, early studies relied more on the *in situ* data and/or bathymetry. A study by Gupta and Banerji (1985) estimated the storage of an Indian reservoir using Landsat water surface area and the *A–V* curve. In Zhang et al. (2006), water levels from TOPEX/Poseidon were combined with *in situ* water storage to derive an *h–V* curve, which allows Lake Dongting storage variations to be estimated from radar altimetry measurements. More recent studies have utilized both satellite imagery and altimetry for storage variations. Crétaux et al. (2011) combined the surface area from a variety of sensors with radar altimetry data to create the first global lakes/reservoir storage dataset. Gao et al. (2012) developed a multistep classification algorithm using MODIS NDVI to estimate globally distributed reservoir areas consistently, and then leveraged the MODIS area and radar elevation to generate a 19-year record representing 15% of the capacity of all reservoirs. Zhang et al. (2014) further improved the area classification algorithm by Gao et al. (2012), and used ICESat elevations to capture 21 reservoirs within South Asia (this work created the first regional scale remotely sensed data product). Using ICESat and Landsat, Song et al. (2013) reconstructed the time series of water volumes for 30 Tibetan lakes from the 1970s to 2011. Duan and Bastiaanssen (2013) estimated storage variations in three lakes/reservoirs from four satellite altimetry sources (both radar and laser) and from Landsat imagery data. Furthermore, a study by Crétaux et al. (2015) showed that errors of water storage variations for the Chardarya and the Toktogul reservoirs using only satellite data (altimetry and MODIS/Landsat images) are within 2%–4% (compared to annual variations).

To derive the elevation–area relationship from remote sensing data, a linear relationship has been used the most often (as shown in Figure 4.3). The linear elevation–area relationships are based on the assumption that the reservoir catchment cross-sectional curve follows a second-order polynomial. If the catchment slope is constant, then the area–elevation relationship will be a second-order polynomial. Figure 4.4 shows how the area–elevation relationship assumptions are made under these two different scenarios.

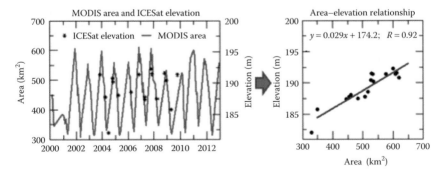

FIGURE 4.3 An example of deriving the elevation–area relationship using remotely sensed data. (From Zhang, S. et al., *Water Res. Res.*, 50, 8927–8943, 2014. With permission.)

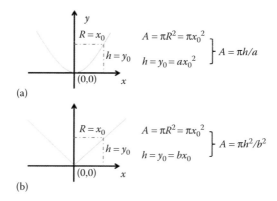

FIGURE 4.4 Area-elevation relationship assumptions under two scenarios: (a) The area-elevation relationship is linear, with the reservoir cross section as a second order polynomial and (b) the area-elevation relationship is a second order polynomial, with the reservoir cross section as a linear function.

4.3 WETLANDS

Although the remote sensing of wetlands essentially focuses on measuring the same set of variables as those used for lakes and reservoirs—elevation, area, and storage—it is more challenging. Regulated by many factors, such as soil texture, topography, weather and climate, vegetation, and human disturbance, wetlands vary considerably at both regional and local scales. From a remote sensing technical perspective, data from wetlands are difficult to extract because wetlands are often vegetated and are known for their ill-defined boundaries.

4.3.1 SURFACE ELEVATION

Wetland elevations have primarily been estimated using data from satellite radar altimeters and interferometric SAR (InSAR) techniques. Water levels obtained from TOPEX and Poseidon radar altimeters were used for quantifying the inundated Negro River basin and the Mekong River basin (Frappart et al., 2005, 2006). In a study by Lee et al. (2009), researchers developed an innovative technique to estimate water level variations using TOPEX radar altimetry (in the Louisiana wetlands). The study utilized the stacking of retracked TOPEX data over wetlands, and demonstrated that the water levels observed by each sample (of 10-Hz data with along-track sampling of ~660 m) exhibited variations. Wetland dynamics were therefore detected. However, because satellite altimeter is a profiling instrument, water elevations can be obtained for a water body only if a satellite passes over it. Furthermore, it is challenging to obtain water elevation changes over a small water body due to altimeters' coarse along-track sampling (~330 m for 20-Hz data, ~165 m for 40-Hz data).

InSAR, typically with L-band sensors such as Japanese Earth Resources Satellite (JERS-1), phased array type L-band synthetic aperture radar (PALSAR), and PALSAR-2, is a unique technique that has allowed us to map two-dimensional water level changes beneath flooded vegetation with high spatial resolution (of ~40 m) from standard strip-mode data collected from different SAR acquisitions

(Alsdorf et al., 2000; Wdowinski et al., 2008; Kim et al., 2009; Jung et al., 2010; Lee et al., 2015). One of the method's limitations, however, is that it requires a stable tree trunk (or other body) that can act as a corner-reflector, which results in the so-called double-bounce backscattering. In other words, interferometric phase cannot be obtained over open water bodies (and often times not over herbaceous wetlands as well). Another limitation is its susceptibility to interferometric phase decorrelations (Zebker and Villasenor, 1992).

Recently, there have been attempts to estimate wetland water level changes using the SAR backscattering coefficient (σ_0)—based on the fact that water level variation is one of the important factors contributing to the radar backscatter power (Kim et al., 2014; Lee et al., 2015; Yuan et al., 2015). Kim et al. (2014) have investigated how L-band PALSAR σ_0 varies with the water level over the Everglades wetlands in south Florida. They estimated a reference water level from σ_0, which was then used to convert InSAR-derived relative water level changes to absolute water level changes. Lee et al. (2015) and Yuan et al. (2015) considered the effect of vegetation density using the MODIS Vegetation Continuous Field product in estimating water level changes over the Congo wetlands. Their technique was based on the relationship between PALSAR σ_0 and Envisat altimetry-derived water level changes. However, these methods can be applied only if a dense network of *in situ* gauges or altimetry tracks are available in the wetlands.

Through synthetic aperture processing, SAR is able to provide finer spatial resolution imagery than conventional beam-scanning radar. A differential interferometric SAR (differential InSAR) is a sophisticated processing approach that "co-registers" a pair of SAR images to form an interferogram. This interferogram can be used to estimate water elevation changes beneath the forest that occurred between the SAR acquisition times. Both differential InSAR and radar altimetry techniques have their advantages and disadvantages. On one hand, water elevation estimations from differential InSAR are advantageous (over those obtained from radar altimetry) due to the superior spatial resolution; on the other hand, SAR instruments are not commonly operated continuously, which makes it even harder to acquire images suitable for interferometry. Consequently, it is difficult to monitor wetland elevation variations on a regular basis using the differential InSAR technique (at least not with current capabilities).

4.3.2 INUNDATED AREA

The estimation of wetland area can be achieved through remote sensing observations at visible/infrared—passive or active—microwave frequencies. Each frequency has its benefits and constraints. The algorithm selection and development is tailored both to the wetland characteristic of interest (e.g., spatial distribution or temporal variability) and the data availability.

At visible and infrared frequencies, decision trees based on multiple thresholds have been used the most often. Islam et al. (2010) employed several vegetation and water indices to develop flood inundation maps from MODIS data. A similar approach—with different indices for thresholding—has been used for monitoring floods globally by NASA (http://oas.gsfc.nasa.gov/floodmap/index.html). Because of the irregular shapes and ambiguous boundaries, image classifications (as explained in Section 4.2.2) are most ideal for delineating wetlands from the surrounding terrain (especially floodplains). Although data at visible/infrared frequencies are exceedingly abundant, these wavelengths are unable to penetrate through clouds. In addition, they cannot reach under the canopy, where some of the largest wetlands lie (e.g., the Amazon floodplains).

At passive microwave frequencies, the surface emissivity of water is much lower than that of soil, which allows for satellite radiometers to monitor wetlands. Because the spatial resolution of passive sensors is very low (e.g., 10–40 km depending on frequencies and instruments), they have been typically used for representing overall water coverage within a grid cell. For instance, a 37-GHz brightness temperature polarization ratio has been used for studying the Amazon River floodplain (Giddings and Choudhury, 1989; Sippe et al., 1998; Arnesen et al., 2013).

For active microwave measurements, SAR has been utilized to map wetlands with high accuracy. For instance, the Global Rain Forest Mapping (Rosenqvist et al., 2000) mission derived mosaics of inundated areas throughout tropical regions using L-band SAR data from the JERS-1. Smith and Alsdorf (1998) delineated inundated water from land surface over the Ob River from SAR images collected by the European Remote Sensing Satellites, ERS-1, and ERS-2. Although free of the limitations resulting from vegetation cover and cloud/precipitation conditions, SAR imageries are constrained by acquisition times and difficulties with wind-induced waves on the water surface.

To overcome the limitations from each, some of the most successful algorithms are developed for leveraging multifrequency, multisatellite data (Prigent et al., 2001; Papa et al., 2006; Schroeder et al., 2010). In a study by Aires et al. (2013), a long-term, high-resolution wetland dataset was generated with observations from a suite of satellite sensors, including passive and active instruments operating in both the microwave and visible/infrared bands. The low spatial resolution results from multiwavelength retrievals were downscaled by SAR data, making the most of the benefits from the different spatial and temporal sensing coverage. Figure 4.5 shows an example of the improved accuracy achieved by this technique.

4.3.3 STORAGE CHANGE

To estimate the wetland storage change, the different approaches can be divided and summarized into two groups. The first group of methods solely relies on elevation and area measurements. In Frappart et al. (2005), spatiotemporal variations of water volume in inundated areas located on the Negro River basin were quantified. JERS-1 SAR images were utilized to identify the type of surface (open water, inundated areas, or forest), and water elevations were obtained from TOPEX/Poseidon and from *in situ* gauges. Linear interpolation was used to generate water level maps for low and high water stages in the flooded zones, which allowed for the calculation of storage variation from radar altimetry elevations. Similarly, Frappart et al. (2006) computed monthly surface storage changes in the Mekong River basin over 6 years, by combining flood maps delineated from SPOT NDVI data with water elevations from ERS-2/ENVISAT data.

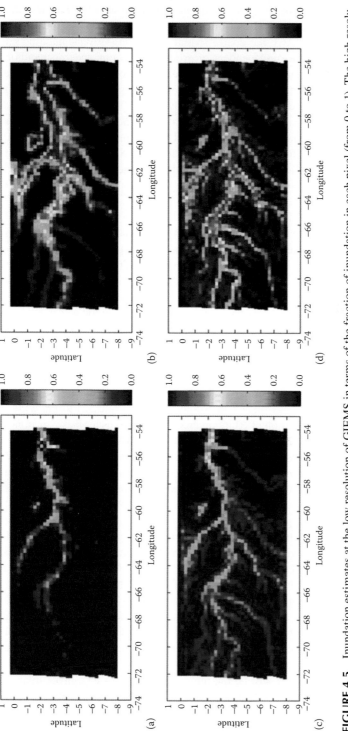

FIGURE 4.5 Inundation estimates at the low resolution of GIEMS in terms of the fraction of inundation in each pixel (from 0 to 1). The high resolution of the SAR estimates is upscaled using a simple compositing of every SAR pixel into the GIEMS boxes. (a) Minimum of GIEMS, (b) maximum of GIEMS, (c) minimum of the high-resolution SAR, and (d) maximum of the SAR. The minimum and maximum for GIEMS have been estimated independently for each box (so GIEMS maps are not for a particular month). (From Aires, F. et al., *J. Hydrometeorol.*, 14, 594–607, 2013. With permission.)

The second group involves methods that infer wetland storage from the water budget equation combined with Gravity Recovery and Climate Experiment (GRACE) data. Although the GRACE has coarse spatial resolution (~several hundred kilometers), its observations are independent of ground conditions and can monitor total water storage change consistently at large scales. Therefore, GRACE offers an excellent opportunity to examine floodplain variations (Famiglietti and Rodell, 2013; Lee et al., 2011a). A good example is the study by Alsdorf et al. (2010) that produced Amazon floodplain storage estimations. With prior knowledge of the basin's precipitation (P), evapotranspiration (ET), and GRACE terrestrial storage change values (ΔS), the net inflows and outflows (Q) were inferred as $\Delta S - (P - ET)$. The floodplain storage was then identified by further subtracting channel storage from Q.

4.4   FUTURE VISION

Although a number of studies (as described in Sections 4.2 and 4.3) have been conducted successfully, our knowledge about the dynamics of lakes, reservoirs, and wetlands is still very limited. As noted above, current altimeters do not provide global coverage of water bodies. Future satellite missions are expected to provide new knowledge related to surface water—both by continuing the use of existing observation systems, and by incorporating new technologies and new measurements. In particular, the Surface Water & Ocean Topography (SWOT), Jason-3, Sentinel-3, ICESat-2, and GRACE Follow-On (GRACE-FO) missions are expected to play a leading role in surface hydrology studies.

SWOT, which is to be launched in 2020, will contribute to a fundamental understanding of the global water cycle. The SWOT satellite will use the innovative Ka-band Radar Interferometer instrument, which features 120-km-wide swath images at a spatial resolution of 10–60 m cross-track, and as high as 2 m along-track (Biancamaria et al., 2015). Its wide swath will make SWOT the first of its kind for covering all global lakes/reservoirs, wetlands, and rivers. Accurate and precise storage change products are envisioned for lakes and reservoirs with areas as small as (250 m²). Considering its spatial resolution, Biancamaria et al. (2010) estimated that about 65% of global lake storage can be detected by SWOT. Lee et al. (2010) estimated that SWOT will generally provide less than 5% relative storage change errors for lakes larger than 1 km². SWOT's repeat period is designed to be 21 days; and most midlatitude locations will be observed 3 times during each 21-day cycle. For rivers wider than 100 m (and possibly as small as 50 m), SWOT will also provide measurements of river height, slope, and width—from which water storage can be measured and river discharge can be estimated (Paiva et al., 2015; Pavelsky et al., 2014). Sentinel-3 is primarily an ocean-focused mission by the European Space Agency, which is composed of three versatile satellites—Sentinel-3A, Sentinel-3B, and Sentinel-3C. The planned launch date for Sentinel-3A is late 2015, with Sentinel-3B following 18 months later. Sentinel-3C is then scheduled for launch sometime before 2020. The SAR Altimeter will be the main instrument for measuring inland lakes and rivers. The repeat cycle of Sentinel-3 is 27 days. The second generation of ICESat (ICESat-2) is scheduled for launch in 2017. Combining the observations from the previous and future ICESat satellites will allow us to better understand the responses of the cryosphere to climate change. GRACE-FO is a successor to GRACE, which is expected to be launched in 2017. Combined usage among GRACE, GRACE-FO, and other satellite sensors will help us understand decadal surface water storage variations at a large scale.

REFERENCES

Aires, F., Papa, F., and Prigent, C. (2013), A long-term, high-resolution wetland dataset over the Amazon basin, downscaled from a multiwavelength retrieval using SAR data. *Journal of Hydrometeorology,* 14, 594–607.

Alsdorf, D. E., Birkett, C. M., Dunne, T., Melack, J., and Hess, L. (2001), Water level changes in a large Amazon lake measured with spaceborne radar interferometry and altimetry. *Geophysical Research Letters*, 28, 2671–2674.

Alsdorf, D., Han, S. C., Bates, P., and Melack, J. (2010), Seasonal water storage on the Amazon floodplain measured from satellites. *Remote Sensing of Environment*, 114, 2448–2456.

Alsdorf, D. E., Melack, J. M., Dunne, T., Mertes, L. A. K., Hess, L. L., and Smith, L. C. (2000), Interferometric radar measurements of water level changes on the Amazon floodplain. *Nature*, 404, 174–177.

Alsdorf, D. E., Rodríguez, E., and Lettenmaier, D. P. (2007), Measuring surface water from space. *Reviews of Geophysics*, 45, RG2002.

Arnesen, A. S., Silva, T. S., Hess, L. L., Novo, E. M., Rudorff, C. M., Chapman, B. D., and McDonald, K. C. (2013), Monitoring flood extent in the lower Amazon River flood-plain using ALOS/PALSAR ScanSAR images. *Remote Sensing of Environment*, 130, 51–61.

Berry, P., Garlick, J., Freeman, J., and Mathers, E. (2005), Global inland water monitoring from multi-mission altimetry. *Geophysical Research Letters*, 32, L16401.

Biancamaria, S., Andreadis, K. M., Durand, M., Clark, E. A., Rodriguez, E., Mognard, N. M., Alsdorf, D. E., Lettenmaier, D. P., and Oudin, Y. (2010), Preliminary characterization of SWOT hydrology error budget and global capabilities. *IEEE Journal of Selected Topics in Applied Earth Observations and Remote Sensing*, 3, 6–19.

Biancamaria, S., Lettenmaier, D. P., and Pavelsky, T. M. (2015), The SWOT mission and its capabilities for land hydrology. *Surveys in Geophysics*, 37, 307–337, doi:10.1007/s10712-015-9346-y.

Birkett, C. M. (2000), Synergistic remote sensing of Lake Chad, variability of basin inunda-tion. *Remote Sensing of Environment*, 72, 218–236.

Birkett, C. M., Murtugudde, R., and Allan, T. (1999), Indian Ocean climate event brings floods to East Africa's lakes and the Sudd Marsh. *Geophysical Research Letters*, 26, 1031–1034.

Cheema, M. and Bastiaanssen, W. (2010), Land use and land cover classification in the irri-gated Indus Basin using growth phenology information from satellite data to support water management analysis. *Agricultural water management*, 97, 1541–1552.

Coe, M. and Birkett, C. (2004), Calculation of river discharge and prediction of lake height from satellite radar altimetry, example for the Lake Chad basin. *Water Resources Research*, 40, W10205.

Crétaux, J. F., Biancamaria, S., Arsen, A., Bergé-Nguyen, M., and Becker, M. (2015), Global surveys of reservoirs and lakes from satellites and regional application to the Syrdarya river basin. *Environmental Research Letters*, 10, 015002.

Crétaux, J. F., Calmant, S., Romanovski, V., Shabunin, A., Lyard, F., Bergé-Nguyen, M., Cazenave, A., Hernandez, F., and Perosanz, F. (2009), An absolute calibration site for radar altimeters in the continental domain, Lake Issykkul in Central Asia. *Journal of Geodesy*, 83, 723–735.

Crétaux, J. F., Jelinski, W., Calmant, S., Kouraev, A., Vuglinski, V., Bergé-Nguyen, M., Gennero, M. C., Nino, F., Abarca Del Rio, R., and Cazenave, A. (2011), SOLS: A lake database to monitor in the near real time water level and storage variations from remote sensing data. *Advances in Space Research*, 47, 1497–1507.

Duan, Z. and Bastiaanssen, W. G. M. (2013), Estimating water volume variations in lakes and reservoirs from four operational satellite altimetry databases and satellite imagery data. *Remote Sensing of Environment*, 134, 403–416.

Famiglietti, J. S. and Rodell, M. (2013), Water in the Balance. *Science,* 340, 1300–1301.

Frappart, F., Do Minh, K., L'Hermitte, J., Cazenave, A., Ramillien, G., Le Toan, T., and Mognard-Campbell, N. (2006), Water volume change in the lower Mekong from satellite altimetry and imagery data. *Geophysical Journal International*, 167, 570–584.

Frappart, F., Seyler, F., Martinez, J. M., León, J. G., and Cazenave, A. (2005), Floodplain water storage in the Negro River basin estimated from microwave remote sensing of inundation area and water levels. *Remote Sensing of Environment*, 99, 387–399.

Gao, H. (2015), Satellite remote sensing of large lakes and reservoirs, from elevation and area to storage. *Wiley Interdisciplinary Reviews: Water*, 2, 147–157.

Gao, H., Birkett, C., and Lettenmaier, D. P. (2012), Global monitoring of large reservoir storage from satellite remote sensing. *Water Resources Research*, 48, WR012063.

Gao, H., Bohn, T., Podest, E., McDonald, K., and Lettenmaier, D. P. (2011), On the causes of the shrinking of Lake Chad. *Environmental Research Letters*, 6, 034021.

Giddings, L. and Choudhury B. J. (1989), Observation of hydrological feature with Nimbus-7 37 GHz data applied to South America. *International Journal of Remote Sensing*, 10, 1673–1686.

Gupta, R. and Banerji, S. (1985), Monitoring of reservoir volume using LANDSAT data. *Journal of Hydrology*, 77, 159–170.

Islam, A., Bala, S., and Haque, M. (2010), Flood inundation map of Bangladesh using MODIS time-series images. *Journal of Flood Risk Management*, 3, 210–222.

Jarihani, A. A., Callow, J. N., Johansen, K., and Gouweleeuw, B. (2013), Evaluation of multiple satellite altimetry data for studying inland water bodies and river floods. *Journal of Hydrology*, 505, 78–90.

Ji, L., Zhang, L., and Wylie, B. (2009), Analysis of dynamic thresholds for the normalized difference water index. *Photogrammetric Engineering and Remote Sensing*, 75, 11, 1307–1317.

Jung, H., Hamski, J., Durand, M., Alsdorf, D., Hossain, F., Lee, H., Hossain, A. K. M. A., Hasan, K., Khan A. S., and Hoque, A. K. M. Z. (2010), Characterization of complex fluvial systems via remote sensing of spatial and temporal water level variations. *Earth Surface Processes and Landforms*, 35, 294–304.

Keddy, P. A., Fraser, L. H., Solomeshch, A. I., Junk, W. J., Campbell, D. R., Arroyo, M. T., and Alho, C. J. (2009), Wet and wonderful, the world's largest wetlands are conservation priorities. *Bioscience*, 59, 39–51.

Kim, J. W., Lu, Z., Jones, J. W., Shum, C., Lee, H., and Jia, Y. (2014), Monitoring Everglades freshwater marsh water level using L-band synthetic aperture radar backscatter. *Remote Sensing of Environment*, 150, 66–81.

Kim, J. W., Lu, Z., Lee, H., Shum, C., Swarzenski, C. M., Doyle, T. W., and Baek S. H. (2009), Integrated analysis of PALSAR/Radarsat-1 InSAR and ENVISAT altimeter data for mapping absolute water level changes in Louisiana wetlands. *Remote Sensing of Environment*, 113, 2356–2365.

Lee, H., Beighley, R. E., Alsdorf, D., Jung, H. C., Shum, C., Duan, J., Guo, J., Yamazaki, D., and Andreadis, K. (2011a), Characterization of terrestrial water dynamics in the Congo Basin using GRACE and satellite radar altimetry. *Remote Sensing of Environment*, 115, 3530–3538.

Lee, H., Durand, M., Jung, H. C., Alsdorf, D., Shum, C. K., and Sheng, Y. (2010), Characterization of surface water storage changes in Arctic lakes using simulated SWOT measurements. *International Journal of Remote Sensing*, 31, 3931–3953.

Lee, H., Shum, C., Tseng, K. H., Guo, J. Y., and Kuo, C. Y. (2011b), Present-day lake level variation from Envisat altimetry over the Northeastern Qinghai-Tibetan plateau, links with precipitation and temperature. *Terrestrial, Atmospheric and Oceanic Sciences*, 22, 169–175.

Lee, H., Shum, C., Yi, Y., Ibaraki, M., Kim, J. W., Braun, A., Kuo, C. Y., and Lu, Z. (2009), Louisiana wetland water level monitoring using retracked TOPEX/Poseidon altimetry. *Marine Geodesy*, 32, 284–302.

Lee, H., Yuan, T., Jung, H. C., and Beighley, R. E. (2015), Mapping wetland water depths over the central Congo Basin using PALSAR ScanSAR, Envisat altimetry, and MODIS VCF data. *Remote Sensing of Environment*, 159, 70–79.

Lehner, B. and Doll, P. (2004), Development and validation of a global database of lakes, reservoirs and wetlands. *Journal of Hydrology*, 296, 1–22.

Lehner, B., Liermann, C. R., Revenga, C., Vörösmarty, C., Fekete, B., Crouzet, P., Döll, P., Endejan, M., Frenken, K., and Magome, J. (2011), High-resolution mapping of the world's reservoirs and dams for sustainable river-flow management. *Frontiers in Ecology and the Environment*, 9, 494–502.

Lu, S., Wu, B., Yan, N., and Wang, H. (2011), Water body mapping method with HJ-1A/B satellite imagery. *International Journal of Applied Earth Observation and Geoinformation*, 13, 428–434.

Maulik, U. and Saha, I. (2010), Automatic fuzzy clustering using modified differential evolution for image classification. *IEEE Transactions on Geoscience and Remote Sensing*, 48, 3503–3510.

Mercier, F., Cazenave, A., and Maheu, C. (2002), Interannual lake level fluctuations (1993–1999) in Africa from Topex/Poseidon, connections with ocean–atmosphere interactions over the Indian Ocean. *Global and Planetary Change*, 32, 141–163.

Morris, C. S. and Gill, S. K. (1994), Evaluation of the TOPEX/POSEIDON altimeter system over the Great Lakes. *Journal of Geophysical Research: Oceans (1978–2012)*, 99, 24527–24539.

Paiva, R. C., Durand, M. T., and Hossain, F. (2015), Spatiotemporal interpolation of discharge across a river network by using synthetic SWOT satellite data. *Water Resources Research*, 51, 430–449.

Papa, F., Prigent, C., Durand, F., and Rossow, W. B. (2006), Wetland dynamics using a suite of satellite observations, A case study of application and evaluation for the Indian Subcontinent. *Geophysical Research Letters*, 33, L08401.

Pavelsky, T. M., Durand, M. T., Andreadis, K. M., Beighley, R. E., Paiva, R. C., Allen, G. H., and Miller, Z. F. (2014), Assessing the potential global extent of SWOT river discharge observations. *Journal of Hydrology*, 519, 1516–1525.

Phan, V. H., Lindenbergh, R., and Menenti M. (2012), ICESat derived elevation changes of Tibetan lakes between 2003 and 2009. *International Journal of Applied Earth Observation and Geoinformation*, 17, 12–22.

Ponchaut, F. and. Cazenave, A. (1998), Continental lake level variations from Topex/Poseidon (1993–1996). *Comptes Rendus de l'Académie des Sciences-Series IIA-Earth and Planetary Science*, 326, 13–20.

Prigent, C., Matthews, E., Aires, F., and Rossow, W. B. (2001), Remote sensing of global wetland dynamics with multiple satellite data sets. *Geophysical Research Letters,* 28, 4631–4634.

Rosenqvist, A., Shimada, M., Chapman, B., Freeman, A., De Grandi, G., Saatchi, S., and Rauste, Y. (2000), The global rain forest mapping project-a review. *International Journal of Remote Sensing*, 21, 1375–1387.

Sarch, M. T. and Birkett, C. (2000), Fishing and farming at Lake Chad, responses to lake-level fluctuations. *The Geographical Journal*, 166, 156–172.

Schroeder, R., Rawlins, M. A., McDonald, K. C., Podest, E., Zimmermann, R., and Kueppers, M. (2010), Satellite microwave remote sensing of North Eurasian inundation dynamics, development of coarse-resolution products and comparison with high-resolution synthetic aperture radar data. *Environmental Research Letters*, 5(1), 015003, doi:10.1088/1748-9326/5/1/015003.

Sippe, S., Hamilton, S., Melack, J., and Novo, E. (1998), Passive microwave observations of inundation area and the area/stage relation in the Amazon River floodplain. *International Journal of Remote Sensing*, 19, 3055–3074.

Smith, B. E., Fricker, H. A., Joughin, I. R., and Tulaczyk, S. (2009), An inventory of active subglacial lakes in Antarctica detected by ICESat (2003–2008). *Journal of Glaciology*, 55, 573–595.

Smith, L. C. and Alsdorf, D. E. (1998), Control on sediment and organic carbon delivery to the Arctic Ocean revealed with spaceborne synthetic aperture radar, Ob' River, Siberia. *Geology*, 26, 395–398.

Song, C., Huang, B., and Ke, L. (2013), Modeling and analysis of lake water storage changes on the Tibetan Plateau using multi-mission satellite data. *Remote Sensing of Environment*, 135, 25–35.

Wang, J., Sheng, Y., and Tong, T. S. D. (2014), Monitoring decadal lake dynamics across the Yangtze Basin downstream of three Gorges Dam. *Remote Sensing of Environment*, 152, 251–269.

Wang, X., Gong, P., Zhao, Y., Xu, Y., Cheng, X., Niu, Z., Luo, Z., Huang, H., Sun, F., and Li, X. (2013), Water-level changes in China's large lakes determined from ICESat/ GLAS data. *Remote Sensing of Environment*, 132, 131–144.

Wdowinski, S., Kim, S. W., Amelung, F., Dixon, T. H., Miralles-Wilhelm, F., and Sonenshein, R. (2008), Space-based detection of wetlands' surface water level changes from L-band SAR interferometry. *Remote Sensing of Environment*, 112, 681–696.

Yamazaki, D., O'Loughlin, F., Trigg, M. A., Miller, Z. F., Pavelsky, T. M., and Bates, P. D. (2014), Development of the global width database for large rivers. *Water Resources Research*, 50, 3467–3480.

Yuan, T., Lee. H., and Jung, H. C. (2015), Toward estimating wetland water level changes based on hydrologic sensitivity analysis of PALSAR backscattering coefficients over different vegetation fields. *Remote Sensing*, 7, 3153–3183.

Zebker, H. and Villasenor, J. (1992), Decorrelation in interferometric radar echoes. *IEEE Transactions on Geoscience and Remote Sensing*, 30, 950–959.

Zhang, G., Xie, H., Kang, S., Yi, D., and Ackley, S. F. (2011), Monitoring lake level changes on the Tibetan Plateau using ICESat altimetry data (2003–2009). *Remote Sensing of Environment*, 115, 1733–1742.

Zhang, S., Gao, H., and Naz, B. S. (2014), Monitoring reservoir storage in South Asia from multisatellite remote sensing. *Water Resources Research*, 50, 8927–8943.

Zhang, J., Xu, K., Yang, Y., Qi, L., Hayashi, S., and Watanabe, M. (2006), Measuring water storage fluctuations in Lake Dongting, China, by Topex/Poseidon satellite altimetry. *Environmental monitoring and assessment*, 115, 23–37.

Zhang, Y., Hong, Y. Wang, X., Gourley, J. J., Gao, J., Vergara, H., and Yong, B. (2013), Assimilation of passive microwave streamflow signals for improving flood forecasting, A first study in Cubango river basin, Africa. *IEEE Journal of Selected Topics in Applied Earth Observations and Remote Sensing*, 6, 2375–2390.

Zhong, Y., Zhang, S., and Zhang, L. (2013), Automatic fuzzy clustering based on adaptive multi-objective differential evolution for remote sensing imagery. *IEEE Journal of Selected Topics in Applied Earth Observations and Remote Sensing*, 6, 2290–2301.

5 Drought and Flood Monitoring for a Large Karst Plateau in Southwest China Using Extended GRACE Data

Di Long, Yanjun Shen, Alexander Sun,
Yang Hong, Laurent Longuevergne,
Yuting Yang, Bin Li, and Lu Chen

CONTENTS

Gravity Recovery and Climate Experiment (GRACE) data have a relatively short time span (since its launch in 2002), making most analyses of GRACE data subjected to the recent decade. A longer time series of total water storage (TWS) changes, for example, 30 years, should be more helpful in examining TWS variations on a climatic timescale. This section elaborates an approach to quantify how climate change impacts TWS in the long term, and how climate extremes (e.g., droughts and floods) impact TWS in the short term. Droughts and floods alternately occur over a large karst plateau (Yun–Gui Plateau) in Southwest China. Here we show that both the frequency and severity of droughts and floods over the plateau are intensified during the recent decade from three-decade total water storage anomalies (TWSA) generated by GRACE satellite data and artificial neural network (ANN) models. The developed ANN models performed well in hindcasting TWSA for the plateau and its three subregions (i.e., the upper Mekong River, Pearl River, and Wujiang River basins), showing coefficients of determination (R^2) of 0.91, 0.83, 0.76, and 0.57, respectively. The intensified climate extremes are indicative of large changes in the hydrologic cycle and brought great challenges in water resources management there. The TWSA of the plateau remained fairly stable during the 1980s, and featured an increasing trend at a rate of 5.9 $\pm$ 0.5 mm/a in the 1990s interspersed extreme flooding in 1991 and during the second half of the 1990s. Since 2000, the TWSA fluctuated drastically, featuring severe spring droughts from 2003 to 2006, the most extreme spring drought on record in 2010, and severe flooding in 2008. The TWSA of the upper Mekong has decreased by ~100 mm (~15 km^3) compared with that at the end of the 1990s. In addition to hindcasting TWSA, the developed approach could be effective in generating future TWSA and potentially bridging the gap between the current GRACE satellites and the GRACE Follow-On Mission expected to launch in 2017.

5.1 INTRODUCTION

The Yun–Gui (an abbreviation of Yunnan and Guizhou Provinces) Plateau in Southwest China characterized by large karst landforms (~159,000 km^2) is plagued by massive drought and flood hazards, leading to tremendous economic, societal, and ecological losses over the past decades. Controlled primarily by the South Asian monsoon, and additionally impacted by the East Asian monsoon, plateau monsoon, and westerlies (Li et al., 2012), the climate is humid and shows large seasonal variability, with the major portion of annual precipitation in summer and relatively lower precipitation in winter and spring. According to the Bulletin of Water Resources of Yunnan Province (Department of Water Resources of Yunnan Province, 2011), the most extreme spring drought in Southwest China in 2010 resulted in drying up of 744 streams, 564 small reservoirs, and 7599 ponds in Yunnan alone. There were ~27 million people being affected, ~9.65 million experienced shortages of drinking water, and \$2.5-billion worth of crops failed in Yunnan according to the Ministry of Civil Affairs (Qiu, 2010). On the other hand, the plateau is a flood-prone region due to the combined effect of subtropical/tropical climate and karst formation. Karst surface features (e.g., sinkholes and thin soil layer) allow rainfall to enter the subsurface and percolate down rapidly, resulting in a high runoff ratio and interconnected surface and subsurface systems (Guo et al., 2013). Mountain rainfall could bring mud, sand, silt, and waste deposits to the mouth of sinkholes, reducing cross-sectional flow and the discharge capacity of conduits, and consequently increasing the potential for flooding (Guo and Jiang, 2010). It is therefore especially important to monitor droughts and floods to mitigate associated losses over this region.

The plateau situates major branches of the upper Yangtze and Pearl Rivers in Southwest China, and several transboundary rivers, for example, the Mekong River flowing through China, Burma, Laos, Thailand, Cambodia, and Vietnam. There have been fierce debates over whether the existing and planned hydropower plants and construction of large dams on the mainstream of the Mekong influence the low-flow hydrology and reduce the availability of water in Southeast Asian countries in the lower reaches (Grumbine and Xu, 2011). The quantification of water storage changes of these transboundary river basins could be helpful in isolating climate change impact on the water resources from the one induced by human activities, and for negotiations among stakeholders. Data sharing across political boundaries is critical for such purposes, but national hydrologic records are often withheld due to political, socioeconomic, and national security reasons (Famiglietti and Rodell, 2013).

GRACE satellites have been used to monitor droughts and floods (Chen et al., 2010; Leblanc et al., 2009; Reager and Famiglietti, 2009; Xavier et al., 2010; Yirdaw et al., 2008) by observing temporal variations in Earth's gravity field (Tapley et al., 2004), which are controlled primarily by the TWS change (TWSC). TWSC for a certain time interval (e.g., 1 month) can be derived from GRACE TWSA. For the karst plateau, it is a great challenge to disaggregate changes in SWS, SMS, and GWS from TWSC as was done in other regions because of the complex interconnections between surface and subsurface water. Given the large spatial extent, special geology, and inaccessible hydrologic records for the transboundary rivers, GRACE TWSA could be valuable in drought and flood monitoring and prediction for the plateau.

GRACE monthly TWSA data are only available since 2002 and there is 2- to 6-month latency before the data are released (Famiglietti and Rodell, 2013; Tapley et al., 2004). Monitoring of droughts and floods was mostly done in the recent decade (Famiglietti et al., 2011; Voss et al., 2013). However, TWSA data before GRACE launch are critical to developing an understanding of long-term trends in TWSC, and to predicting future TWSC. In addition, there are likely no reliable TWSA data between the period of the decaying orbit of current GRACE satellites and the planned launch of the GRACE Follow-On Mission in 2017. Developing methods of generating TWSA data beyond the GRACE period is hence an urgent need.

Becker et al. (2011) reconstructed GRACE TWSA data over the Amazon basin back to 1980 by examining the correlation between GRACE TWSA and water level observations from interannual to multidecadal timescales using Empirical Orthogonal Function decomposition. Pan et al. (2012) integrated multiple sources of water cycle estimates, including in situ observations, remote sensing retrievals, land surface model (LSM) simulations, and global reanalyses, into one consistent set of water-cycle data records over 32 major river basins globally from 1984 to 2006 using data assimilation techniques. In this data set, total water storage beyond the GRACE period was provided by the Variable Infiltration Capacity (VIC) model (Liang et al., 1994) that precludes simulation of GWS change and therefore might not be capable of simulating the interactions between surface water and groundwater systems over karst regions. de Linage et al. (2013) developed simple statistical models to predict TWSA over Amazon by examining variations in sea surface temperature (SST) anomalies from the equatorial central Pacific (represented by Niño 4) and tropical North Atlantic (represented by the Tropical North Atlantic Index, TNAI), with R^2 between TWSA and a combination of Niño 4 and TNAI of 0.43 for the entire Amazon basin. Note that complete water level data are not available for the Yun–Gui Plateau due to limited data sharing. LSMs (e.g., VIC) alone are not able to recover TWSC for the large karst region where the surface and subsurface water are interconnected. Predictability of TWSC could be further improved as compared with R^2 (0.43) between TWSC and Niño or TNAI.

The ANN is a nonparametric modeling technique that is especially suited for modeling input–output relationships when lacking well-calibrated, physics-based models. ANNs have been extensively used in streamflow forecasting (House-Peters and Chang, 2011; Hsu et al., 1995) and recently in predicting groundwater level changes (Sun, 2013). A major strength of ANN models lies in its universal approximation property—an ANN model can be trained to approximate the causal relationship of any nonlinear dynamic system without a priori assumptions about the underlying physical system (Haykin, 2009).

The objectives of this study are to (1) hindcast TWSA for the Yun–Gui Plateau and its subregions using GRACE satellite data and ANN models in combination with other in situ and modeling data over the past three decades, (2) examine long-term trends in TWSA for the study regions, and (3) monitor droughts and floods across the large karst plateau based on the three-decade TWSA time series, including the most extreme drought in 2010 and severe flooding in 2008. The hindcasted long-term TWSA data could be valuable for understanding climate change impacts on the hydrologic cycle and providing implications of water resources management over this region.

5.2 STUDY REGION

5.2.1 Yun–Gui Plateau

The Yun–Gui Plateau, the fourth largest plateau in China, encompasses eastern Yunnan Province, entire Guizhou Province, some portions of northwestern Guangxi Zhuang Autonomous Region, and frontiers of Sichuan, Hubei, and Hunan Provinces. It is in the confluence of north–south and northeast–southwest oriented mountain ranges of China, with elevation generally decreasing from the northwest to the southeast and ranging from ~2000 m down to ~1000 m. Here, the study region was chosen to coincide with provincial boundaries, that is, both Yunnan and Guizhou Provinces totaling 559,000 km², for simplicity of discussion (Figure 5.1a).

Summer (June–August) precipitation of the Yun–Gui Plateau is ~580 mm (derived from monitored daily precipitation at 55 meteorological stations, see Section 5.3.4.1), accounting for ~52% of mean annual precipitation of 1125 mm across this region. Fall (September–November), spring (March–May), and winter (December–February) precipitation is ~250 mm, ~230 mm, and ~60 mm, accounting for ~22%, ~20%, and ~5% of mean annual precipitation, respectively. This region is characterized primarily

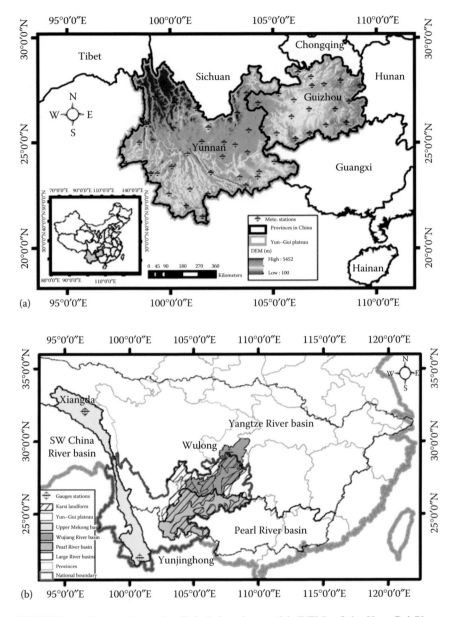

FIGURE 5.1 (a) Location and a digital elevation model (DEM) of the Yun–Gui Plateau and distribution of 55 meteorological stations across this region and (b) location of the upper Mekong River basin, the Pearl River basin within the Yun–Gui Plateau, and the Wujiang River basin, with showing gauging stations of the upper Mekong River (Xiangda and Yunjinghong stations) and the Wujiang River basin (Wulong station). Hatched areas denote karst landform.

by typical karst geological formation (159,000 km², accounting for ~28% of the study region, Figure 5.1b), with an extensive distribution of limestone in the form of caves, stalactites, stalagmites, stone-bar, underground streams, and peak-clusters. Karst formations are often cavernous and therefore have high rates of permeability. Rainfall can quickly move through crevices into ground and underground, which results in surface water with few to no rivers or lakes and close coupling of surface water and underground water. Surface soil is often parched between rains and during spring and winter when there is no much precipitation.

5.2.2 INDIVIDUAL BASINS

Both the Wujiang River, the largest tributary of the right bank of the upper Yangtze River, and the Xijiang River, the longest river (~2214 km) of the Pearl River, originate from the Yun–Gui Plateau (Figure 5.1b). Trends in TWSA of the Wujiang River basin (area: 83,000 km² with reference to the Wulong gauging station in Chongqing Municipality and the runoff ratio is greater than 0.5) characterized primarily by a karst geological formation, and the Pearl River basin within the Yun–Gui Plateau (area: 124,000 km²) will be examined (Figure 5.1b). In addition, there are several transboundary rivers (i.e., termed China's Southwest Rivers) originating from the Qinghai–Tibet Plateau, flowing through Yunnan and then into Southeast Asian countries. For instance, the Nujiang River originates from the southern foot of the Tanggula Mountains on the Qinghai–Tibet Plateau, flows through Tibet and Yunnan in China and Burma (the Salween River), and finally flows into the Andaman Sea in the Indian Ocean. The Lancang river (China's section of the Mekong River or the upper Mekong River) originates from melting snow in Zadoi County on the Qinghai–Tibet Plateau, flows through Qinghai, Tibet, and Yunnan, extends into the countries of Laos, Burma, Thailand, Cambodia, and Vietnam, and finally flows into the South China Sea. These transboundary river basins are steep and narrow, for example, the area of the upper Mekong River basin (151,000 km²) accounts for ~19% of the total area of the Mekong River basin of ~810,000 km² (Figure 5.1b).

5.3 METHODS AND MATERIALS

5.3.1 EVALUATION OF GRACE TOTAL WATER STORAGE CHANGE USING A WATER BALANCE EQUATION

GRACE-derived TWSC was compared with water budget estimates of TWSC to evaluate reliability of GRACE TWSC estimates for the study region:

$$dS/dt = P + R_{in} - R_{out} - ET \tag{5.1}$$

where:
    dS/dt is TWSC at the monthly scale (mm/month)
    P is monthly precipitation (mm/month)
    R_{in} and R_{out} are inflow and outflow for a specific region of interest

For the Yun–Gui Plateau, the inflow mainly comes from snowmelt from Qinghai Province and flows into the study region north of Yunnan (Figure 5.1b). The Yun–Gui Plateau is one of the source regions of the upper Yangtze River and the Pearl River (the Xijiang River). The outflows leave this region from the north, east, and south to the upper Yangtze River and the Pearl River basins. The characteristics of the geological and topographic conditions, and inaccessible data in most parts of the Southwest China Rivers, make estimation of inflow and outflow of this region extremely difficult. In this study, we assumed that streamflow over the entire plateau could be represented by streamflow measurements of the individual basins of distinct hydrologic and geological characteristics, that is, streamflow measurements at the Wulong station of the Wujiang River basin were used to represent the outflow for the karst region (Figure 5.1b). Inflow at the Xiangda station and outflow at the Yunjinghong station of the upper Mekong River were jointly used to approximate the net flow flux ($R_{out} - R_{in}$) for nonkarst areas. The net flow term in Equation 5.1 for the plateau is therefore the weighted sum of the net flow for the karst region and the net flow of the nonkarst region in the unit of mm. ET is monthly evapotranspiration (mm/month). There is limited irrigation in this region due to a humid climate, geological (e.g., it is difficult to access groundwater in the karst region) and economic reasons (e.g., infrastructure in many mountainous areas is backward). Therefore, the impact of anthropogenic activities on TWSC is assumed to be negligible in Equation 5.1. GRACE TWSC is computed as the backward difference of TWSA (mm) whose reference is the mean gravity field for a calculation period (e.g., January 2003–September 2012 in this study):

$$dS/dt = \frac{\text{TWSA}(t) - \text{TWSA}(t-1)}{t} \qquad (5.2)$$

5.3.2 Artificial Neural Network

Time series hindcasting (prediction) problems seek to learn a functional mapping between a set of predictors **x** and the target variable y (Sun, 2013):

$$y = f(\mathbf{x}) + \varepsilon \qquad (5.3)$$

where:
  f is the mapping
  ε is the process noise

We developed multilayer perceptron (MLP) ANN models to learn the functional mapping and hindcast (predict) monthly GRACE TWSA 1 month ahead from February 1979–December 2002. There are one input layer, one hidden layer, and one output layer (Figure 5.2). Inputs (predictors) of the MLP models consist of SMS, monthly precipitation, and monthly mean temperature. A multitude of studies showed that GRACE TWSA is highly correlated with SMS anomalies from LSMs (Long et al., 2013; Sun et al., 2010). Here, we examined the correlation between SMS of the total depth from four LSMs, that is, Noah (2 m), Mosaic (3.5 m), VIC (1.9 m), and Community Land Model (CLM) (3.4 m) in Global Land Data Assimilation System-1 (GLDAS-1), and GRACE TWSA for the Yun–Gui Plateau. It was found that the SMS anomaly from Noah shows the highest correlation with GRACE TWSA (r = ~0.9) for the period

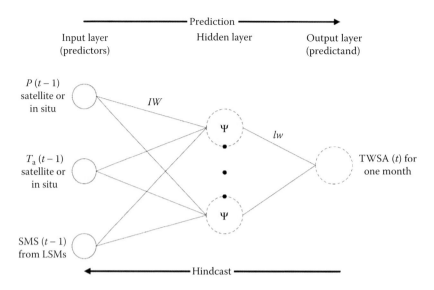

FIGURE 5.2 Structure of MLP for hindcasting (prediction) 1 month ahead developed in this study. The input layer (predictors) consists of two or three neurons, including in situ or satellite precipitation, SMS from GLDAS-1 Noah, and/or in situ or satellite air temperature. There are three neurons in the hidden layer, and the output layer (predictant) consists of TWSA. $t-1$ and t means the last month and the current month, respectively. IW and lw represent connection weights; and ψ represents the sigmoid transfer function.

2003–2012. The use of precipitation as a predictor was intended to reflect the rapid response of SWS and GWS to precipitation across the large karst plateau. Monthly mean temperature may be able to indirectly reflect evaporation from water bodies and soil layers in the humid region and therefore be linked to SWS and GWS (Meng et al., 2014). Therefore, it was also used as a predictor. Three combinations of the predictors were tested in this study. The first combination (Prediction 1) comprises Noah SMS and in situ precipitation, the second combination (Prediction 2) includes Noah SMS and in situ monthly mean temperature, and the third combination (Prediction 3) consists of Noah SMS, in situ precipitation, and monthly mean temperature.

The hidden layer comprises K hidden neurons and each neuron is a weighted sum of predictors (Bishop, 2006):

$$a_k = \sum_{i=1}^{M} w_{ki}^{(1)} x_i + w_{k0}^{(1)}, k = 1, \ldots, K, \tag{5.4}$$

where:

a_k is a hidden neuron

$\{w_{ki}^{(1)}\}_{i=1}^{M}$ are unknown weights associated with each input neuron

$w_{k0}^{(1)}$ is an unknown bias term used for correcting the estimation bias

Superscripts in Equation 5.4 represent the layer number. When developing MLP networks, a rule-of-thumb is that the number of hidden neurons should be about half the number of predictors and should never be more than twice as large (Berry and Linoff, 2004). Therefore, we set the number of hidden neurons to 3 (i.e., $K = 3$) through a trial-and-error process, in which MLPs having different numbers of hidden neurons were tested. Equation 5.4 is subsequently passed to a transfer function to compute outputs from hidden neurons:

$$z_k = \psi(a_k), k = 1,\ldots,K,$$ (5.5)

where:

z_k is the output

ψ is the logistic sigmoid transfer function ranging from 0 to 1

Connections from the hidden layer to the output layer are built through a linear transfer function:

$$\text{TWSA} = \sum_{k=1}^{K} w'^{(2)}_{jk} z_k + w'^{(2)}_{j0}$$ (5.6)

The output neuron (predictand) is monthly TWSA. Subscript j is the number of output neurons (=1 in this study). $\{w'^{(2)}_{jk}\}_{k=1}^{K}$ and $w'^{(1)}_{j0}$ are the unknown weights and the bias term of the output layer, respectively. During the training phase, the unknowns in Equations 5.4 and 5.6 are solved through backpropagation, which is a process of propagating fitting errors backward through the network to obtain the optimal weights in each layer. GRACE TWSA data from January 2003 to November 2010 totaling 96 months (~80% of the samples) were used for training (60%) and validating (20%) the MLP, and that from December 2010 to September 2012 totaling 22 months (~20% of the samples) were used for testing the performance of the trained MLP. The resulting MLP was then applied to hindcasting TWSA for the period February 1979–December 2002 for the plateau and three subregions.

5.3.3   FLOOD POTENTIAL AMOUNT

Flood potential amount (FPA) was calculated as the following (Reager and Famiglietti, 2009):

$$\text{FPA}(t) = P(t) - (\text{TWSA}_{\max} - \text{TWSA}(t-1))/dt$$ (5.7)

where $\text{TWSA}_{\max}$ is the regionally observed historic maxima of TWSA. Deviating from (Reager and Famiglietti, 2009), $\text{TWSA}_{\max}$ in this study is considered to be of statistical meaning, instead of a single value derived from a short time period (~10 years for traditional analysis). The maximum values of TWSA for each year from 1979 to 2012 were used to fit parameters of the probability density function (y) for the extreme distribution as following:

$$y = f(\text{TWSA}_{\max} | \mu, \sigma) = \sigma^{-1} \exp\left(\frac{\text{TWSA}_{\max} - \mu}{\sigma}\right) \exp\left(-\exp\left(\frac{\text{TWSA}_{\max} - \mu}{\sigma}\right)\right)$$ (5.8)

where:

μ is the location

σ is the scale parameters

TWSA_{max} for a certain return period (e.g., 10a, 25a, 50a, and 100a) was subsequently derived from the extreme value inverse cumulative distribution function. FPA for a certain month and location can be calculated by using the TWSA_{max} for a certain return period and monthly precipitation derived from the nearest meteorological station of a grid cell based on Equation 5.7. If a grid cell has multiple stations (i.e., ≥ 1), averaged monthly P across stations was used.

5.3.4 DATA AND PROCESSING

5.3.4.1 Hydrometeorological Data

Precipitation and mean temperature observations were obtained from 55 meteorological stations (Figure 5.1a) in Yunnan and Guizhou Provinces at a daily timescale from the 1950s to September 2012. Monthly precipitation of the entire study region and individual river basins was obtained by spatial averaging and aggregated from the daily precipitation records. Comparison of the spatially averaged in situ precipitation with the satellite precipitation product (TRMM 3B43) (Huffman et al., 2007) indicated high consistency with a Root Mean Square Difference (RMSD) of 8 mm/month and nil bias for the GRACE period. This gives confidence in reliability of the in situ precipitation over the plateau (Figure 5.3). Streamflow measurements at a few gauging stations, including Wulong in the Wujiang River basin (data span from 1952 to 2011 at the daily scale) and Xiangda and Yunjinghong in the upper Mekong River basin (data span from 1956 to 2000 at the monthly scale), were obtained for use in estimating net flow of the study regions (see Section 5.3.1 and Figure 5.3). It is noted that monthly outflow measurements of the upper Mekong River basin from 2001 to the present were inaccessible due to the sensitivity of water resources of the transboundary rivers. Bulletins of Water Resources of Yunnan Province from 2000 to 2010 provide annual streamflow anomalies (1956–1979) of the upper Mekong

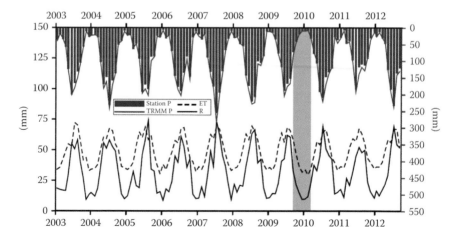

FIGURE 5.3 In situ precipitation from 55 stations and the satellite precipitation product (TRMM 3B43), observed streamflow inferred from individual basins, and the AVHRR ET product for the Yun–Gui Plateau from January 2003 to September 2012.

River basin. Therefore, monthly outflow of the upper Mekong River basin from 2001 to 2012 was estimated from a simple precipitation–runoff relationship constructed based on data from 1956 to 2000 and corrected by a scaling factor of the observed annual streamflow over the estimated annual streamflow.

ET in Equation 5.1 was obtained from two remotely sensed products based on (1) the National Oceanic and Atmospheric Administration-Advanced Very High Resolution Radiometer satellite (NOAA-AVHRR, ET_{AVHRR} spanning from 1983 to 2006) (Zhang et al., 2010) and (2) Moderate Resolution Imaging Spectroradiometer sensor (MODIS, ET_{MODIS} spanning from 2000 to 2012) (Mu et al., 2011). In addition, ET outputs from four LSMs (i.e., Noah, Mosaic, VIC, and CLM) in Global Land Data Assimilation System-1 (GLDAS-1) (Rodell et al., 2004) were used to provide estimates of ET over this region. All these ET products were compared with annual ET derived from a water budget at the annual scale ($ET \approx P-R$) in the Wujiang River basin where there are the most complete streamflow measurements in this study. There is an assumption of negligible TWSC at the annual timescale in the water budget estimates of ET. The Root Mean Square (RMS) of annual TWSC derived from GRACE satellites is ~60 mm accounting for only 5% of mean annual precipitation in the plateau, which indicates the validity of the assumption of negligible annual TWSC in the humid region. The ET output of the least discrepancy with the water budget estimates of annual ET will be used as an input of Equation 5.1 to evaluate GRACE-derived TWSC. Uncertainties in P and $R_{out} - R_{in}$ are not known and were prespecified as 10% and 20%, respectively. Uncertainties in ET were estimated by the standard deviation of all ET products being used in this study for each month. Uncertainties in P, $R_{out} - R_{in}$, and ET were then added in quadrature to obtain the total uncertainty estimates of TWSC from the water budget. The global Palmer Drought Severity Index (PDSI) data set (Dai et al., 2004) was obtained for defining the duration of the extreme drought in this study (PDSI < -3).

5.3.4.2 GRACE Data

Both spherical harmonic solutions and postprocessed gridded GRACE products were used in this study to derive TWSA for the plateau and three subregions. GRACE spherical harmonic solutions were provided by the Center for Space Research at the University of Texas at Austin (CSR) and Groupe de Recherche de Géodésie Spatiale (GRGS) analysis center. CSR and GRGS represent two end members in terms of GRACE processing. CSR is one of the least constrained solutions and GRGS is one of the most constrained solutions. The latest release of CSR data (Release number 5, RL05) was used in the analysis. Spherical harmonics for CSR RL05 were truncated at the maximum degree and order of 60, destriped (Swenson and Wahr, 2006), and filtered using a 300 km Gaussian filter to suppress GRACE measurement noise of high-degree and order spherical harmonics. Spherical harmonics for GRGS RL02 were truncated at the maximum degree and order of 50. Truncation is a low-pass filter and the lower degree and order applied to GRGS constitutes a regularized solution and no further filtering was required.

Bias and leakage effects of filtered GRACE TWSA were corrected using the additive approach that calculates bias (signal loss due to low pass filtering applied to

GRACE data in a study region) and leakage (signal gain from the surrounding due to low pass filtering) using synthetic data from LSMs (Longuevergne et al., 2010). The resulting GRACE TWSA estimate is therefore the filtered GRACE TWSA adding the bias and subtracting the leakage. There is another approach for bias and leakage corrections termed as the multiplicative approach (Longuevergne et al., 2010), which computes a multiplicative factor based on a filtered and unfiltered basin function under the assumption of uniform distribution of TWSC in a study region. Given hydrologic heterogeneity (e.g., karst and nonkarst) in the plateau that may not meet the assumption of uniform distribution of TWSC in the multiplicative approach, we used the additive approach for bias and leakage corrections for the filtered GRACE TWSA. Bias and leakage of GRACE signals were computed using SMS from the Noah model in GLDAS-1 (0.25° × 0.25°) as a priori knowledge of the global SMS variation. Uncertainties in GRACE TWSA estimates include: (1) uncertainties in GRACE L1 measurements and (2) bias and leakage corrections due to uncertainties in SMS changes from GLDAS-1 Noah. Uncertainties in GRACE L1 measurements were estimated by residuals over the Pacific and Atlantic Oceans at the same latitudes as the plateau and its subregions. Uncertainties in bias and leakage corrections were estimated from the standard deviation of SMS from four LSMs (Noah, Mosaic, VIC, and CLM) in GLDAS-1.

Gridded TWSA products (1° × 1°) derived from spherical harmonics from CSR, Jet Propulsion Laboratory (JPL), and German Research Center for Geoscience (GFZ) centers with signals being restored by gain factors computed by CLM4 output (Landerer and Swenson, 2012; Swenson and Wahr, 2006) were obtained for comparison with TWSA time series processed by the additive approach. Use of spherical harmonics from different processing centers (i.e., CSR, GRGS, JPL, and GFZ) and approaches (i.e., the additive and gain factor) provides valuable information on uncertainty in GRACE TWSA.

It should be emphasized that for relatively small areas (e.g., <150,000 km²), the fraction of filtered GRACE signal in realistic TWSC decreases significantly. Bias and leakage corrections for small areas will therefore make the resulting TWSA estimates biased toward output of LSMs that may, however, not be accurate due to the missing hydrologic components (e.g., groundwater and lateral flow) and uncertainties associated with forcing and parameters of the models.

5.4 RESULTS AND DISCUSSION

5.4.1 EVALUATION OF GRACE TOTAL WATER STORAGE CHANGE AND ITS SPATIOTEMPORAL VARIABILITY

Comparison of TWSA time series derived from different data processing centers and approaches of the Yun–Gui Plateau indicates a high consistency (Figure 5.4) with a mean correlation coefficient of ~0.95 between two different TWSA time series among the five TWSA time series derived. In general, the TWSA derived from the GRGS center and three gridded TWSA products (CSR, JPL, and GFZ) are within uncertainties (shading areas in Figure 5.4) in the TWSA derived from CSR RL05. The consistency provides confidence in use of GRACE TWSA for monitoring

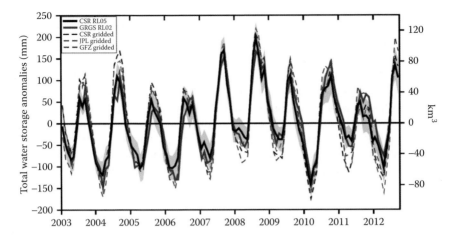

FIGURE 5.4 TWSA time series of the Yun–Gui Plateau for the period January 2003 to September 2012 from spherical harmonics solutions of CSR RL05 and GRGS RL02 with signal restored by the additive approach (Longuevergne et al., 2010) and three gridded products from spherical harmonics solutions of CSR, JPL, and GFZ centers (Landerer and Swenson, 2012). Shading areas show uncertainties in TWSA from CSR spherical harmonics.

droughts and floods over the plateau and training the ANN model for hindcasting TWSA beyond the GRACE period. Relatively large differences among these TWSA time series occur in high or low amplitudes in summer or spring, with the gridded TWSA products showing higher amplitudes. In addition, slight differences in these TWSA time series may also arise from the masks of the study region defined by different approaches. The TWSA time series derived by the additive approach used a basin function on a spatial resolution of $0.5° \times 0.5°$, which is higher than the one of $1° \times 1°$ for the gridded products and may better represent the TWSA for the plateau and subregions. Differences may therefore result partly from approximating the boundary of the study region by different GRACE data sets. The TWSA derived from CSR RL05 using the additive approach for bias and leakage corrections will be used for the following analysis and discussion.

Annual ET_{AVHRR} in the Wujiang River basin exhibited the lowest RMSD of ~50 mm/a relative to the water budget estimates of annual ET in all ET products being examined. Annual ET estimates from LSMs were significantly overestimated relative to the water budget estimates of ET, suggesting relatively higher RMSD ranging from ~190 mm/a (CLM) to ~400 mm/a (VIC). Similarly, annual ET_{MODIS} showed an RMSD as large as ~390 mm/a compared with the ET derived from the water budget. These results indicate that the LSMs may not be able to reasonably partition available energy into latent and sensible heat fluxes over the karst region where underground drainage systems have been well developed and the surface and underground flows are closely coupled. LSMs lack the mechanisms to describe and simulate these processes and lateral flow, resulting in unrealistic streamflow and ET estimates. In addition, ET_{MODIS} was found to overestimate ET over this region, which warrants further study of the forcing data and its algorithm. Therefore, in the water budget calculation for the entire Yun–Gui Plateau, monthly ET_{AVHRR} (2003–2006) was used as a surrogate of observed ET in Equation 5.1 to drive TWSC. Given the high correlation ($r > 0.94$) between monthly ET_{AVHRR} and ET_{MODIS}, monthly ET in Equation 5.1 from 2006 to 2012 was generated by using ET_{MODIS} and the ratio of ET_{AVHRR} to ET_{MODIS} for the overlapping period (2000–2006) (Figure 5.3).

Evaluation of GRACE-derived TWSC against ground-observed TWSC from a water budget with in situ precipitation, inflow and outflow, and ET_{AVHRR} was subsequently performed. Results demonstrate the validity of GRACE satellites to track the amplitude and timing of TWSC from the water budget estimate over the study region (Figure 5.5). GRACE-derived TWSC estimates are generally consistent ($r = 0.82$) with the water budget estimates of TWSC for the period January 2003–September 2012, with showing an RMSD of ~25 mm/month and a bias of ~ −6 mm/month. The slightly higher water budget estimates of TWSC in cold seasons (e.g., October to December) appear to result from the underestimation of outflow in the water budget calculation. The underground drainage system has been well developed over the large karst plateau; however, there is no accounting for underground flow in the water budget estimates of TWSC. The RMS error in the water budget estimates of TWSC is ~20 mm/month, which is lower than the one of GRACE-derived TWSC of ~45 mm/month (i.e., the red shading areas are generally within the blue shading areas in Figure 5.5). This means that in general, uncertainty in GRACE-derived TWSC is relatively higher than the one in water budget estimates of TWSC (Long et al., 2014), though substantial uncertainties in estimates of ET and outflow could also occur.

Both TWSC estimates from the water budget and GRACE are generally positive (total water storage surplus) from April to September, and negative (total water storage deficit) in other months (Figure 5.5). This characteristic is different from

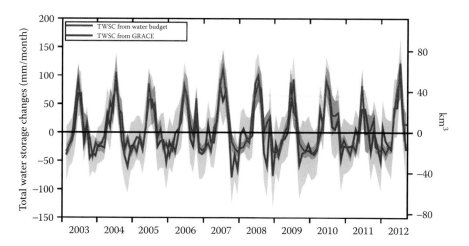

FIGURE 5.5 Comparison of GRACE-derived TWSC (CSR) and water budget estimates of TWSC of the Yun–Gui Plateau from January 2003 to September 2012. Red shading areas show uncertainties in water budget estimates of TWSC and blue shading areas show uncertainties in GRACE-derived TWSC.

those systems where total water storage deficits occur in the warm season and total water storage surpluses occur in the cold season, for example, in Texas (Long et al., 2013), and is likely driven by the combined effect of the monsoon climate and geological conditions. Because the precipitation of the Yun–Gui Plateau is concentrated from April to September, total water storage increases markedly in response to the increasing precipitation. After the rainy season, total water storage starts to decline due mostly to decreasing precipitation but sustaining surface and underground flows.

Spatial patterns of GRACE TWSA exhibit a dominant seasonality and prominent gradients of decreasing TWSA from the south to the north and from the west to the east, which are likely to be associated with the impact of monsoons and topography. We averaged TWSA at the 0.5° × 0.5° over each longitudinal zone (Figure 5.6) and latitudinal zone (Figure 5.7) of the Yun–Gui Plateau from 2003 to 2012. The TWSA time series over the latitudinal and longitudinal zones exhibit a marked seasonal cycle, for example, total water storage starts to increase in response to increasing precipitation in April/May and decrease after the end of rainy season in September through spring next year. This is a strong indication that droughts frequently occur in spring over this region. Because of the seasonality in TWSC and the difficulty in accessing surface and groundwater in mountainous and/or karst areas, spring (March–May) droughts have a large impact on agricultural, industrial, and municipal water use. Spatially, amplitudes of TWSA dampen gradually from the south to the north and from the west to the east of the Yun–Gui Plateau. However, gradients of TWSA across the latitudinal zones (Figure 5.6) are larger than those across the

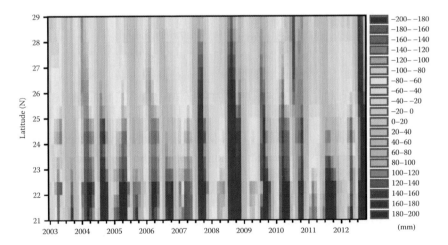

FIGURE 5.6 GRACE total water storage anomalies (CSR) averaged over longitudes for each latitudinal zone of the Yun–Gui Plateau from January 2003 to September 2012. The TWSA were truncated at the maximum degree and order of 60, destriped (Swenson and Wahr, 2006) and filtered using a 300 km Gaussian filter to suppress GRACE measurement noise of high-degree and order spherical harmonics (Longuevergne et al., 2010).

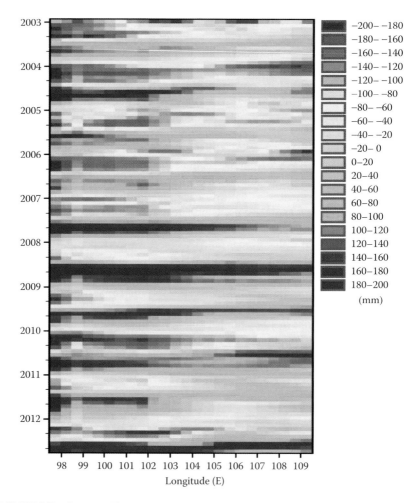

FIGURE 5.7   Same as Figure 5.6 but GRACETWSA averaged over latitudes for each longitudinal zone.

longitudinal zones (Figure 5.7). For instance, from 2004 to 2006, the TWSA of the western and eastern parts of the study region appears to be higher than those in the central longitudes (Figure 5.7).

5.4.2   EVALUATION OF TOTAL WATER STORAGE ANOMALIES PREDICTED FROM ANN MODELS

The developed ANN model with predictors of SMS from GLDAS-1 Noah and in situ precipitation can generate TWSA of the plateau reasonably well (referring to prediction 1 in Figure 5.8 and Section 5.3.2), showing a coefficient of determination (R^2) of 0.91, a bias of about −21 mm, and an RMSD of ~28 mm between the predicted and the GRACE TWSA. The RMSD is lower than the RMS error of GRACE TWSA

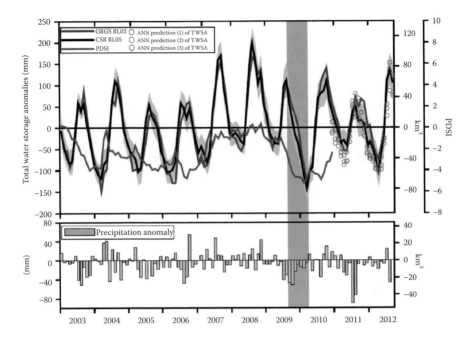

FIGURE 5.8 GRACE TWSA from CSR RL05 and GRGS RL02, the Palmer Drought Severity Index (PDSI), and precipitation anomaly (with reference to the period January 1958 to September 2012) of the Yun–Gui Plateau for the period January 2003 to September 2012 (referring to Section 5.3.4). ANN-predicted TWSA with different combinations of predictors from December 2010 to September 2012 is shown in open circle. Red circles denote the prediction with predictors of SMS from Noah in GLDAS-1 and monitored precipitation, pink circles denotes the prediction with predictors of SMS and monitored monthly mean temperature, and green circles denote the prediction with predictors of SMS, precipitation, and temperature. The shading areas show uncertainties in CSR TWSA, and the shading rectangle shows the period of the extreme drought in Southwest China from September 2009 to March 2010.

of ~33 mm, and generally within the range of uncertainties in GRACE TWSA for the testing period (shading areas in Figure 5.8). Prediction 1 ($R^2 = 0.91$; RMSD = 28) outperformed prediction 2 ($R^2 = 0.78$; RMSD = 35 mm) and prediction 3 for the Yun–Gui Plateau ($R^2 = 0.86$; RMSD = 30 mm, Figure 5.8).

The developed ANN model also performed relatively well in subregions, showing R^2 for the upper Mekong, Pearl River, and Wujiang River basins of 0.83, 0.76, and 0.57, and RMSD of 21, 44, and 58 mm, respectively (Figure 5.9). The best performance of the developed ANN model by taking the plateau as a whole could be due mostly to the lower uncertainties in GRACE TWSA than relatively small subregions, though hydrogeological heterogeneity does exist across the entire plateau. By the contrary, the worst performance of the developed ANN model for the Wujiang River basin could be ascribed to the fact that uncertainties in GRACE TWSA tend to be higher than the other two river basins due to the relatively small area of the Wujiang River basin (~83,000 km²) and the relatively large footprint of GRACE signal (~200,000 km²). Bias and leakage corrections for GRACE TWSA for the Wujiang River basin tend to make the resulting TWSA estimates biased

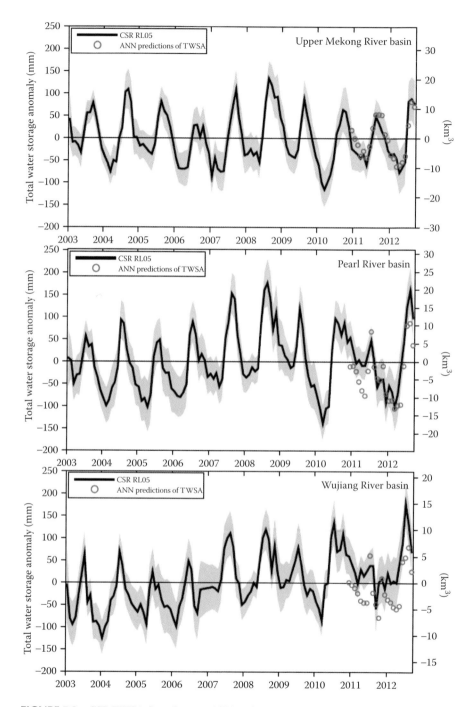

FIGURE 5.9 CSR TWSA from January 2003 to September 2012, and ANN-predicted TWSA from December 2010 to September 2012 for the upper Mekong River basin, the Pearl River basin, and the Wujiang River basin. Gray shading areas indicate uncertainties in CSR TWSA.

toward the Noah output. However, the correlation between SMS from GLDAS-1 Noah and GRACE TWSA in the Wujiang River basin during the GRACE record period is ~0.7, which is far lower than the one for the entire Yun–Gui Plateau of ~0.9. The Noah model (and also many other LSMs) is not able to depict surface and subsurface hydrologic systems well in karst regions. The lack of mechanisms of simulating lateral flow could be an important source of error in Noah output in this region.

5.4.3 Long-Term Trends in Total Water Storage Anomalies of the Yun–Gui Plateau

Frequencies of droughts and floods across the plateau are intensified for the recent decade as evidenced by changes in trends of three-decade TWSA time series from GRACE data and the proposed ANN model. Different low-pass filters intended to remove seasonal variability and high frequency noise of the TWSA gave similar results (Table 5.1), indicating that there are generally five episodes of variations in

TABLE 5.1
Trends and Associated Standard Errors in TWSA for Five Episodes of the Past Three Decades

Study Region	Period	Moving Average		Butterworth		HP Filter	
		Trend (mm/a)	Error (mm/a)	Trend (mm/a)	Error (mm/a)	Trend (mm/a)	Error (mm/a)
Yun–Gui	9/1979–8/1990	−1.8	0.2	−1.9	0.2	−1.3	0.1
Plateau	8/1990–9/2000	5.9	0.5	5.8	0.5	6.4	0.3
	9/2000–8/2004	−31.2	0.8	−32.1	0.7	−25.9	0.3
	8/2004–8/2008	22.2	1.8	25.1	1.5	16.9	0.6
	8/2008–8/2011	−15.6	3.2	−18.3	2.1	−10.5	0.4
Upper Mekong	9/1982–9/2000	1.8	0.1	1.8	0.1	1.7	0.0
River basin	9/2000–8/2006	−19.4	0.7	−19.9	0.7	−17.6	0.4
	8/2006–8/2011	−1.9	1.6	−3.1	1.4	−1.9	0.5
Pearl River	8/1979–9/1983	4.2	0.2	4.9	0.6	3.5	0.1
basin	9/1983–7/1989	−5.4	0.3	−5.8	0.3	−5.4	0.1
	7/1989–7/2000	4.7	0.3	4.5	0.3	4.6	0.1
	7/2000–8/2005	−18.1	0.4	−19.2	0.3	−15.6	0.4
	8/2005–8/2008	31.0	0.3	33.2	0.9	19.5	0.4
	8/2008–7/2010	−45.1	4.2	−34.4	0.9	−15.4	0.3
Wujiang River	8/1979–8/1983	3.9	0.5	4.0	0.5	3.1	0.1
basin	8/1983–9/1990	−2.2	0.3	−2.5	0.3	−2.3	0.1
	7/1990–7/2000	2.7	0.2	2.7	0.2	2.8	0.1
	7/2000–7/2004	−24.3	0.9	−24.2	0.4	−20.0	0.3
	7/2004–7/2012	10.4	0.6	10.9	0.6	11.3	0.3

Note: The long-term TWSA time series were filtered by three low pass filters (see Figure 5.11).

the TWSA (Figure 5.10). During the 1980s, the TWSA remained generally stable, with severe summer droughts occurring in 1982 and 1989. Since the 1990s, the TWSA increased at a rate of 5.9 ± 0.5 mm/a (the moving average filter). An abrupt decrease in the TWSA at a rate of −31.2 ± 0.8 mm/a was subsequently followed after 2000, reaching a low value in spring 2004. The fourth episode features a substantial increase at a rate of 22.2 ± 1.8 mm/a by summer 2008 featuring a severe flooding in Yunnan. Afterward, the TWSA dropped at a rate of −15.6 ± 3.2 mm/a, and reached the most extreme spring drought in 2010 on record. After the 2010 drought, the TWSA appeared to recover to some degree, but severe droughts hit this region in summer 2011 and spring and summer 2013. Apparently, variations in the TWSA became drastic after 2000. Variations in TWSA of the three subregions are also

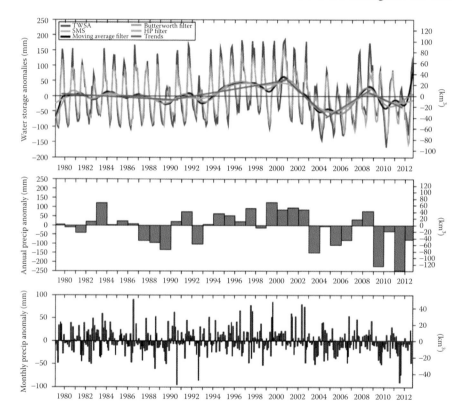

FIGURE 5.10 TWSA (the ANN-generated TWSA from February 1979 to December 2002 combined with GRACETWSA from January 2003 to September 2012) and soil moisture (Noah in GLDAS-1) anomalies for the Yun–Gui Plateau from February 1979 to September 2012, with showing linear trends for five episodes of the study period based on the TWSA time series filtered by the moving average filter. TWSA filtered by the Butterworth filter (order 4 and normalized cutoff frequency 0.05) and Hodrick–Prescott filter (HP, a parameter of 14,400 for monthly time series) are also shown. It is noted that the TWSA from February 1979 to December 2002 hindcasted by the ANN model was based on the mean gravity field from January 2003 to September 2012. The final TWSA time series from February 1979 to September 2012 was derived by subtracting the mean of the TWSA for the three decades.

intensified compared with those in the 1980s and 1990s (Table 5.1 and Figure 5.11). Particularly, TWS in the upper Mekong has decreased by ~100 mm (~15 km^3) relative to the end of the 1990s. The reduced TWS was mostly subjected to climate, which may have resulted in decreases in outflow and changes in the flow regime.

Amplitudes of soil moisture variations are generally lower than TWSA variations, especially during peaks of TWSA in each year (Figure 5.10). The timing of TWSA and SMS variations appears to be synchronized. During rainy season, SWS and GWS of the plateau could respond quickly to substantially increased precipitation, thereby

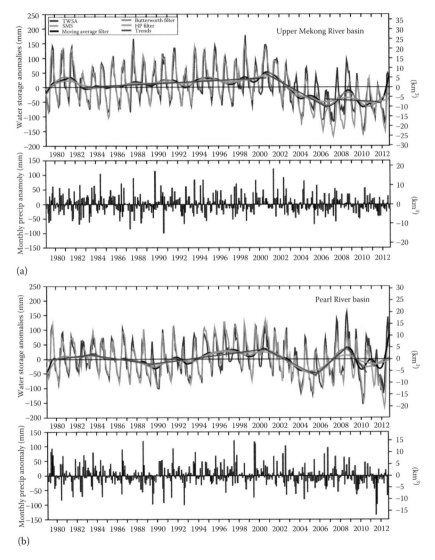

FIGURE 5.11   Time series of total water storage and soil moisture (Noah in GLDAS-1) anomalies for (a) the upper Mekong River basin and (b) the Pearl River basin.   (*Continued*)

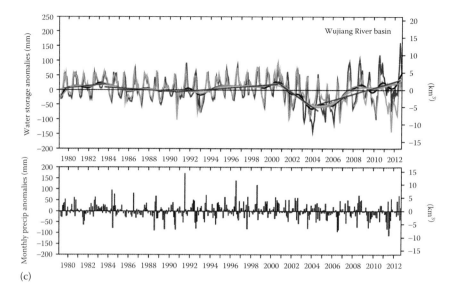

FIGURE 5.11 (Continued) Time series of total water storage and soil moisture (Noah in GLDAS-1) anomalies for (c) the Wujiang River basin from February 1979 to September 2012, with showing linear trends for five episodes of the study period based on the time series of TWSA filtered by the moving average filter. TWSA filtered by the Butterworth filter and Hodrick–Prescott (HP) filter are also shown.

contributing to an appreciable portion of TWSA signals, for example, in years 2001, 2002, 2007, and 2008. This demonstrates that the sole use of SMS from LSMs to depict TWSC over this region could deviate greatly from reality. The developed ANN model could, however, provide a straightforward and effective way to obtain a complete picture of TWSC in the long run. This could also be useful in making prediction of TWSA forward when GRACE data are not available or the 2- to 6-month latency before the release of GRACE data is longer than what we need in practice.

5.4.4 Monitoring of the Most Extreme Drought in 2010 across the Yun–Gui Plateau

Severe droughts and floods over the recent decade were also recorded. The duration of the 2010 drought was defined from September 2009 (the PDSI < -3)–Mar 2010 (the lowest TWSA is -143 mm, Figure 5.8). TWS depletion during the 2010 drought was estimated to be about -255 ± 51 mm ($\sim143 \pm 29$ km^3), that is, the difference in TWSA between March 2010 and August 2009. The PDSI declined to -5.2 in February 2010, the lowest value since 1937. Furthermore, the PDSI responded to low TWSA values well; however, the correspondence between the PDSI and GRACE TWSA was generally low ($r = \sim0.1$). This could be ascribed to the thin soil layer of the karst area. The PDSI depicts primarily variations in SMS of the upper meter (Dai et al., 2004). However, changes in SWS in the plateau could also play a critical role in

TWSC especially during the rainy season. Therefore, there are prominent differences in timing between the PDSI and TWSA as reflected by a low correlation coefficient.

Large precipitation deficits are responsible primarily for the 2010 drought. Mean annual precipitation of the plateau for the period 1958–2010 is ~1130 mm. Precipitation in 2009 was only ~900 mm, the second lowest value on record. The lowest record (~877 mm) occurred in 2011 due mostly to the large precipitation deficit in summer 2011 resulting in an extreme summer drought in central Guizhou. All months (except April) in 2009 and January–March in 2010 show precipitation deficits (Figure 5.8), which may be caused by negative anomalies in the 500 hPa geopotential height in the area of south 22°N and 70–110°E (Wang and Li, 2010) or negative anomalies of the northern hemisphere annular mode (NAM), less likely by El Niño (Jiang and Li, 2010). Figure 5.12 shows evolution and extent of TWSA during the 2010 drought. The largest TWS depletion was centered in western and southern Yunnan and spread out to Guizhou. Precipitation anomaly was ~13 mm/month in April 2010 and TWS depletion gradually recovered since then (Figure 5.11).

5.4.5 Monitoring of Severe Flooding in 2008 for the Yun–Gui Plateau

The initial TWS status is one of the critical factors determining the formation of flooding (Reager and Famiglietti, 2009). Higher TWSA could likely cause large potential for flooding during rainy season due to the excess of water that cannot be stored further. In addition, rolling terrain has relatively short time of concentration of watersheds, contributing to larger potential for flooding for higher TWSA. The TWSA for the plateau peaks in August 2008 for the GRACE period (Figures 5.8 and 5.13). This was due in part to relatively low TWS depletion during winter (January–February) 2008 when Southwest China experienced extremely and persistently low temperature and snowing and freezing weather over the past five decades, marking the highest TWSA (only ~ −40 mm in May 2008) and the PDSI of ~0 in spring (Figure 5.8). Similar situations without severe spring droughts also occurred in years 1996, 1997, 1998, and 2000 when TWSA was ~ −50 mm (Figure 5.10).

FPA calculated by monitored precipitation and regional TWSA maxima from the developed approaches are consistent with the reality of flood disasters during summer 2008. FPA with the regional $TWSA_{max}$ having a 10-year return period from June to November 2008 was calculated (referring to Section 5.3.3). Results indicate that in June, most parts of the plateau have relatively lower FPA than other months, with relatively low values of about −100 mm/month in eastern Yunnan and central-north Guizhou (Figure 5.14). FPA increases throughout the study region in July, especially over central Guizhou and southern Yunnan. The southern parts of Yunnan along the national boundary and central Guizhou mark high FPA values ranging from ~200 mm/month to ~300 mm/month in August. These high FPA values could be ascribed to the combined effect of high TWSA and precipitation. For instance, the highest monthly precipitation of ~505 mm in August occurred at the meteorological station near the national boundary in southern Yunnan. This corresponds to the highest FPA for the grid cell situating the meteorological station due to the relatively

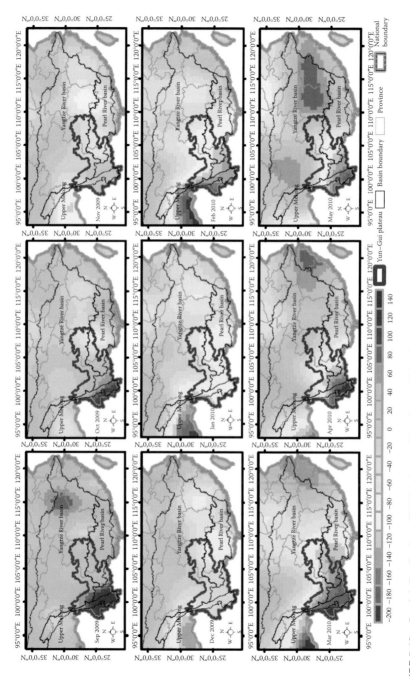

FIGURE 5.12 Spatial distribution of TWSA (mm) of Southwest China from September 2009 to May 2010 at a spatial resolution of 0.5° × 0.5° derived from CSR RL05 spherical harmonics (SH) solutions. The SH solutions were truncated at the maximum degree and order of 60, destriped (Swenson and Wahr, 2006) and filtered using a 300 km Gaussian filter.

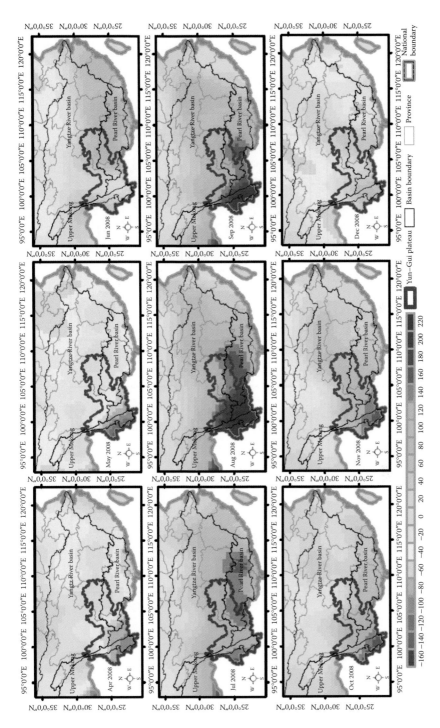

FIGURE 5.13 Same as Figure 5.12 but from April to December 2008.

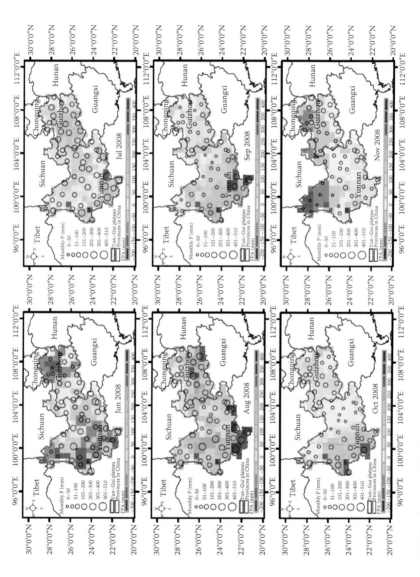

FIGURE 5.14 Spatial distribution of flood potential amount (mm/month) with a 10-year return period and precipitation (mm/month) of the Yun–Gui Plateau from June to November 2008.

high TWSA developed in the previous months and the highest monthly precipitation in August. Relatively low FPA values were found in east-central Yunnan. Spatial patterns of FPA for both September and October are similar, that is, most parts of the Yun–Gui Plateau show FPA lower than 50 mm/month. Exceedingly high FPA values are scattered along the national boundary in southern and western Yunnan. In December, FPA across the entire region tends to be further dampened, especially in the northern region where FPA values range from −150 to 0 mm/month.

According to the 2008 Bulletin of Water Recourses of Yunnan Province (Department of Water Resources of Yunnan Province, 2009), rainstorms and relatively high annual precipitation occurred mostly over western and southern Yunnan along the national boundary. The direct economic loss due to flood disasters is highest (~5.5 billion Yuan or ~$0.92 billion) since 2001, with 129 counties, 1216 villages and towns, a population of 8.8 billion people, and 450,000 ha of agricultural crops being affected. Cities totaling 23 were inundated to varying degrees. There was a death toll of 129 due to floods and associated mountain hazards. The numbers of rain events with intensity greater than 50 mm/24 h and between 25 and 49.9 mm/24 h are 395 and 1793, respectively, which rank first during the recent decade. In addition, there were 52 times for all six major rivers in Yunnan that had stages higher than the warning levels.

In combination with real time precipitation and/or precipitation forecast (e.g., Climate Forecast System, CFSv2) and the initial condition of SMS from a LSM, GRACE-derived FPA has the potential to detect flooding-prone areas and would be valuable in improving flood prediction. In addition, GRACE-derived FPA could also be useful in assessment of disaster losses. With improving the latency of GRACE data release and spatial resolution, GRACE satellites would be more useful in predicting floods in the future (Famiglietti and Rodell, 2013) (Figure 5.14).

5.5 CONCLUSIONS

Quantification of climate extreme impact on water resources over karst regions is a big challenge.

In this study, we show that both the frequency and severity of droughts and floods over the Yun–Gui Plateau in Southwest China are intensified during the recent decade of three-decade TWSA generated using GRACE satellite data and ANN models. The intensified climate extremes are indicative of large changes in the hydrologic cycle and brought great challenges in water resources management there. The TWSA of the plateau remained fairly stable during the 1980s, and featured an increasing trend at a rate of 5.9 ± 0.5 mm/a in the 1990s interspersed extreme flooding in 1991 and during the second half of the 1990s. Since 2000, the TWSA fluctuated drastically, featuring severe spring droughts from 2003 to 2006, the most extreme spring drought on record in 2010, and severe flooding in 2008. The TWSA of the upper Mekong has decreased by ~100 mm (~15 km^3) compared with the end of the 1990s.

Given the intensification of climate extremes and special geological conditions, enhancing and developing water conservancy projects especially over rural areas and isolated villages would be an urgent need to improve water supply during droughts and reduce losses associated with flooding disasters. Further, improving water use efficiency as economy and population grow and developing monitoring and early warning systems for climate extremes, including droughts and floods, could be extremely important. The developed ANN approach could serve as a powerful tool to reconstruct long-term TWSA beyond the GRACE period and potentially bridge the gap between the current GRACE satellites and the GRACE Follow-On Mission expected to launch in 2017. The extended TWSA data could be more valuable in drought and flood monitoring and prediction over the Yun–Gui Plateau and similar regions of the globe.

REFERENCES

Becker, M., Meyssignac, B., Xavier, L., Cazenave, A., Alkama, R., and Decharme, B. (2011). Past terrestrial water storage (1980–2008) in the Amazon Basin reconstructed from GRACE and in situ river gauging data. *Hydrology and Earth System Sciences,* 15(2), 533–546. doi: 10.5194/hess-15-533-2011.

Berry, M. J. and Linoff, G. S. (2004). *Data Mining Techniques: For Marketing, Sales, and Customer Relationship Management.* Wiley Computer Publishing, Hoboken, NJ.

Bishop, C. M. (2006). *Pattern Recognition and Machine Learning.* Springer, New York.

Chen, J. L., Wilson, C. R., and Tapley, B. D. (2010). The 2009 exceptional Amazon flood and interannual terrestrial water storage change observed by GRACE. *Water Resources Research,* 46(12). doi: 10.1029/2010WR009383.

Dai, A., Trenberth, K. E., and Qian, T. T. (2004). A global dataset of palmer drought severity index for 1870–2002: Relationship with soil moisture and effects of surface warming. *Journal of Hydrometeorology,* 5(6), 1117–1130.

de Linage, C., Famiglietti, J. S., and Randerson, J. T. (2013). Forecasting terrestrial water storage changes in the Amazon Basin using Atlantic and Pacific sea surface temperatures. *Hydrology and Earth System Sciences Discussions,* 10(10), 12453–12483.

Department of Water Resources of Yunnan Province (2009). *2008 Bulletin of Water Resources in Yunnan Province*, Kunming, Yunnan, China.

Department of Water Resources of Yunnan Province (2011). *2010 Bulletin of Water Resources in Yunnan Province*, Kunming, Yunnan, China.

Famiglietti, J. S., Lo, M., Ho, S. L., Bethune, J., Anderson, K. J., Syed, T. H., Swenson, S. C., de Linage, C. R., and Rodell, M. (2011). Satellites measure recent rates of groundwater depletion in California's Central Valley. *Geophysical Research Letters,* 38(3), L03403. doi: 10.1029/2010gl046442.

Famiglietti, J. S. and Rodell, M. (2013). Water in the balance. *Science,* 340(6138), 1300–1301. doi: 10.1126/science.1236460.

Grumbine, R. E. and Xu, J. (2011). Mekong hydropower development. *Science,* 332(6026), 178–179.

Guo, F. and Jiang, G. (2010). Problems of flood and drought in a typical peak cluster depression karst area (SW China). In B. Andreo, F. Carrasco, J. J. Durán, and J. W. LaMoreaux (eds.), *Advances in Research in Karst Media* (Vol. Advances in Research in Karst Media, pp. 107–113). Environmental Earth Sciences. Springer, Berlin, Germany.

Guo, F., Jiang, G. H., Yuan, D. X., and Polk, J. S. (2013). Evolution of major environmental geological problems in karst areas of Southwestern China. *Environmental Earth Sciences,* 69(7), 2427–2435.

Haykin, S. S. (2009). *Neural Networks and Learning Machines.* Prentice Hall, New York.

House-Peters, L. A. and Chang, H. (2011). Urban water demand modeling: Review of concepts, methods, and organizing principles. *Water Resources Research,* 47, 15.

Hsu, K. L., Gupta, H. V., and Sorooshian, S. (1995). Artificial neural-network modeling of the rainfall-runoff process. *Water Resources Research,* 31(10), 2517–2530.

Huffman, G. J., Bolvin, D. T., Nelkin, E. J., Wolff, D. B., Adler, R. F., Gu, G., Bowman, K. P., Hong, Y., and Stocker, E. F. (2007). The TRMM multisatellite precipitation analysis (TMPA): Quasi-global, multiyear, combined-sensor precipitation estimates at fine scales. *Journal of Hydrometeorology,* 8(1), 38–55. doi: 10.1175/jhm560.1.

Jiang, X. W. and Li, Y. Q. (2010). The spatio-temporal variation of winter climate anomalies in Southwestern China and the possible influencing factors. *Acta Geographica Sinica,* 65(11), 1325–1335.

Landerer, F. W. and Swenson, S. C. (2012). Accuracy of scaled GRACE terrestrial water storage estimates. *Water Resources Research,* 48(4), W04531. doi: 10.1029/2011wr011453.

Leblanc, M. J., Tregoning, P., Ramillien, G., Tweed, S. O., and Fakes, A. (2009). Basin-scale, integrated observations of the early 21st century multiyear drought in southeast Australia. *Water Resources Research,* 45(4), W04408. doi: 10.1029/2008wr007333.

Li, Z., He, Y., Wang, P., Theakstoned, W. H., An, W., Wang, X., Lu, A., Zhang, W., and Cao, W. (2012). Changes of daily climate extremes in southwestern China during 1961–2008. *Global and Planetary Change,* 80–81, 255–272.

Liang, X., Lettenmaier, D. P., Wood, E. F., and Burges, S. J. (1994). A simple hydrologically based model of land-surface water and energy fluxes for general-circulation models. *Journal of Geophysical Research-Atmospheres,* 99(D7), 14415–14428. doi: 10.1029/94jd00483.

Long, D., Longuevergne, L., and Scanlon, B. R. (2014). Uncertainty in evapotranspiration from land surface modeling, remote sensing, and GRACE satellites. *Water Resources Research,* 50(2), 1131–1151. doi: 10.1002/2013wr014581.

Long, D., Scanlon, B. R., Longuevergne, L., Sun, A. Y., Fernando, D. N., and Save, H. (2013). GRACE satellite monitoring of large depletion in water storage in response to the 2011 drought in Texas. *Geophysical Research Letters,* 40(13), 3395–3401. doi: 10.1002/grl.50655.

Longuevergne, L., Scanlon, B. R., and Wilson, C. R. (2010). GRACE Hydrological estimates for small basins: Evaluating processing approaches on the High Plains Aquifer, USA. *Water Resources Research,* 46(11), W11517. doi: 10.1029/2009wr008564.

Meng, L., Long, D., Quiring, S. M., and Shen, Y. (2014). Statistical analysis of the relationship between spring soil moisture and summer precipitation in East China. *International Journal of Climatology,* 34, 1511–1523. doi: 10.1002/joc.3780.

Mu, Q., Zhao, M., and Running, S. W. (2011). Improvements to a MODIS global terrestrial evapotranspiration algorithm. *Remote Sensing of Environment,* 115(8), 1781–1800. doi: 10.1016/j.rse.2011.02.019.

Pan, M., Sahoo, A. K., Troy, T. J., Vinukollu, R. K., Sheffield, J., and Wood, E. F. (2012). Multisource estimation of long-term terrestrial water budget for major global river basins. *Journal of Climate,* 25(9), 3191–3206.

Qiu, J. (2010). China drought highlights future climate threats. *Nature Climate Change,* 465(7295), 142–143.

Reager, J. T. and Famiglietti, J. S. (2009). Global terrestrial water storage capacity and flood potential using GRACE. *Geophysical Research Letters,* 36(23). doi: 10.1029/2009GL040826.

Rodell, M., Houser, P. R., Jambor, U., Gottschalck, J., Mitchell, K., Meng, C. J., Arsenault, K. et al. (2004). The global land data assimilation system. *Bulletin of the American Meteorological Society,* 85(3), 381–394. doi: 10.1175/bams-85-3-381.

Sun, A. Y. (2013). Predicting groundwater level changes using GRACE data. *Water Resources Research,* 49(9), 5900–5912. doi: 10.1002/wrcr.20421.

Sun, A. Y., Green, R., Rodell, M., and Swenson, S. (2010). Inferring aquifer storage parameters using satellite and in situ measurements: Estimation under uncertainty. *Geophysical Research Letters,* 37(10), L10401. doi: 10.1029/2010gl043231.

Swenson, S. and Wahr, J. (2006). Post-processing removal of correlated errors in GRACE data. *Geophysical Research Letters,* 33, L08402. doi: 10.1029/2005gl025285.

Tapley, B. D., Bettadpur, S., Ries, J. C., Thompson, P. F., and Watkins, M. M. (2004). GRACE measurements of mass variability in the earth system. *Science,* 305, 503–506. doi: 10.1126/science.1099192.

Voss, K. A., Famiglietti, J. S., Lo, M., Linage, C., Rodell, M., and Swenson, S. C. (2013). Groundwater depletion in the Middle East from GRACE with implications for transboundary water management in the Tigris-Euphrates-Western Iran region. *Water Resources Research,* 49(2), 904–914. doi: 10.1002/wrcr.20078.

Wang, B. and Li, Y. Q. (2010). Relationship analysis between the South branch trough and severe drought of southwest China during autumn and winter 2009/2010. *Plateau and Mountain Meteorology Research,* 130(4), 26–35.

Xavier, L., Becker, M., Cazenave, A., Longuevergne, L., Llovel, W., and Filho, O. C. R. (2010). Interannual variability in water storage over 2003–2008 in the Amazon Basin from GRACE space gravimetry, in situ river level and precipitation data. *Remote Sensing of Environment,* 114(8), 1629–1637. doi: 10.1016/j.rse.2010.02.005.

Yirdaw, S. Z., Snelgrove, K. R., and Agboma, C. O. (2008). GRACE satellite observations of terrestrial moisture changes for drought characterization in the Canadian Prairie. *Journal of Hydrology,* 356(1–2), 84–92. doi: 10.1016/j.jhydrol.2008.04.004.

Zhang, K., Kimball, J. S., Nemani, R. R., and Running, S. W. (2010). A continuous satellite-derived global record of land surface evapotranspiration from 1983 to 2006. *Water Resources Research,* 46(9), W09522. doi:10.1029/2009WR008800.

Section II

Modeling, Data Assimilation, and Analysis

6 Statistical and Hydrologic Evaluation of TRMM-Based Multisatellite Precipitation Analysis over the Wangchu Basin of Bhutan

Are the Latest Satellite Precipitation Products 3B42V7 Ready for Use in Ungauged Basins?

Xianwu Xue, Yang Hong, Ashutosh S. Limaye, Jonathan Gourley, George. J. Huffman, Sadiq Ibrahim Khan, Chhimi Dorji, and Sheng Chen

CONTENTS

6.1   INTRODUCTION

Precipitation is among the most important forcing data for hydrologic models. It has been arguably nearly impossible for hydrologists to simulate the water cycles over regions with no or sparse precipitation gauge networks, especially over complex terrain or remote areas. Until recently, the satellite precipitation products (such as TMPA [Huffman et al., 2007], CMORPH [Joyce et al., 2004], PERSIANN [Sorooshian et al., 2000], and PERSIANN-CCS [Hong et al., 2004]) are starting to provide alternatives for estimating rainfall data and also pose new challenges for hydrologists in understanding and applying the remotely sensed information.

Tropical Rainfall Measurement Mission (TRMM) Multisatellite Precipitation Analysis (TMPA), developed by the National Aeronautics and Space Administration (NASA) Goddard Space Flight Center (GSFC), provides a calibration-based sequential scheme for combining precipitation estimates from multiple satellites, as well as monthly gauge analyses where feasible, at fine spatial and temporal scales ($0.25° \times 0.25°$ and 3 h) over 50°N–50°S (Huffman et al., 2007). TMPA is computed for two products: near-real-time version (TMPA 3B42RT, hereafter referred to as 3B42RT) and post-real-time research version (TMPA 3B42 V6, hereafter referred to as 3B42V6). 3B42V6 has been widely used in hydrologic applications (Su et al., 2008; Stisen and Sandholt, 2010; Bitew and Gebremichael, 2011; Bitew et al., 2011; Khan et al., 2011a, 2011b; Li et al., 2012; Xue et al., 2013); however, its computation ended June 30, 2011, and 3B42V6 was replaced by the new version (TMPA 3B42 V7, hereafter referred to as 3B42V7), which has been reprocessed and available from 1998 to present. Previously, 3B42V6 has been validated by several studies (Islam and Uyeda, 2007; Chokngamwong and Chiu, 2008; Su et al., 2008; Mishra et al., 2010; Stisen and Sandholt, 2010; Yong et al., 2010, 2012; Bitew and Gebremichael, 2011; Bitew et al., 2011; Jiang et al., 2012; Li et al., 2012; Jamandre and Narisma, 2013; Xue et al., 2013), while the newly available 3B42V7 has not been extensively validated and applied, especially in mountainous South Asian regions.

Therefore, the objectives of this study are designed (1) to evaluate the widely used globally available, high-resolution TMPA satellite precipitation products over the mountainous medium-sized Wangchu basin (3550 km²) in Bhutan, and more importantly (2) to assess improvements of the latest upgrade version (3B42V7) relative to its predecessor in terms of statistical performance and hydrologic utility using the

Coupled Routing and Excess Storage (CREST) hydrologic model. Additionally, this study aims to shed light on the suitability of recalibrating a hydrologic model with the remotely sensed rainfall information.

6.2   STUDY AREA, DATA, AND METHODOLOGY

6.2.1   Study Area

The Wangchu basin, with a total drainage area of approximately 3550 km² is located within 89°6′–89°46′E and 27°6′–27°51′N in the west of Bhutan (Figure 6.1). Wangchu basin is the most populous part of the country with about 3/5 of the population living in 1/5 of the basin area. The basin is equipped with one streamflow gauge at the outlet Chhukha Dam Hydrologic station and five rain gauge stations. The soil types are dominated by Sandy Clay Loam (75.1%) and Loam (24.9%) based the Harmonized World Soil Database (HWSD V1.1) (FAO/IIASA/ISRIC/ISSCAS/JRC, 2009). The various vegetation types of this basin are composed of evergreen needleleaf forest (48.1%), woodland (17.8%), open shrubland (9.7%), wooded grassland (8.2%), grassland (7.6%), and other land-use types (<10%) (Hansen et al., 2000).

The northern periphery of the Wangchu basin in the Himalayas has elevations over 6000 m, and maintains an annual snowpack. Lower portions of the basin are drastically different and are subject to a summer monsoon from May to October (Bookhagen and Burbank, 2010). On average, the annual month with the greatest precipitation is July or August with 160.82–545.79 mm/month based on the five rain

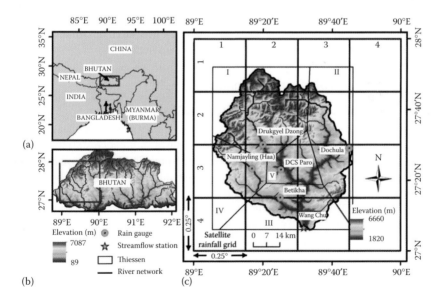

FIGURE 6.1   Wangchu basin map. (a) Location of Bhutan and the surrounding countries. (b) Location of Wangchu basin in Bhutan and its elevation and network of Bhutan. (c) Map of Wangchu basin, rain gauges, streamflow station, topography, Thiessen polygon of the rainfall gauge, and the 0.25° × 0.25° grids of the satellite rainfall estimates.

TABLE 6.1

Monthly Observed Precipitation and Runoff Averaged from 2001 to 2010

	Rain Gauge (mm/month)					Streamflow Station
Month	Betikha	Dochula	Drukgyel Dzong	Namjayling Haa	DSC_Paro	Chhukha $m^3 s^{-1}$
January	14.22	19.05	0.00	12.12	8.32	26.48
February	49.27	17.95	8.50	23.93	17.99	23.45
March	176.94	16.83	26.23	32.77	15.43	24.62
April	346.32	52.95	29.16	53.97	34.08	37.66
May	367.88	104.90	60.13	68.81	57.45	55.38
June	389.67	279.20	123.12	124.37	80.92	111.10
July	545.79	359.17	185.39	182.80	199.11	221.80
August	382.62	367.57	190.92	160.82	103.04	250.65
September	325.61	217.08	108.34	120.09	76.65	180.38
October	182.48	115.70	70.90	76.89	63.19	109.47
November	9.70	11.10	3.40	3.81	2.89	52.00
December	3.62	8.50	0.50	1.65	0.86	34.37

gauge station data showing in (Table 6.1), and the largest resulting streamflow occurs in June or August with 250.65 $m^3 s^{-1}$. It is possible that snowmelt contributes to a portion of this peak streamflow, but the majority is driven by the summer monsoon rains. In this study, neither the precipitation products nor the model explicitly deal with frozen precipitation. These are subjects requiring additional investigation, especially in light of the forthcoming Global Precipitation Measurement mission, which aims to quantitatively estimate frozen precipitation amounts.

6.2.2 IN SITU AND SATELLITE PRECIPITATION DATASETS

6.2.2.1 Gauged Precipitation and Discharge Data

Daily observed precipitation data are obtained from the Hydro-Met Services Department of Bhutan from 2001 to 2010 for the five rain gauge stations located within the Wangchu basin. In winter, the observed precipitation account for snowfall and hail precipitation in the form of water equivalent and computed by melting the ice/snow with hot water in the standard vessel and deducting the same from the total volume. Thiessen polygon method is used to interpolate the rain gauge data to the spatial distributed grid data fitting the model grid resolution (30 arcsec) (Figure 6.1). We also obtained the daily discharge data at the basin outlet for the same time period.

6.2.2.2 TMPA 3B42 Research Products

TMPA precipitation products are available in two versions: near-real-time version (3B42RT) and post-real-time research version (3B42) adjusted by monthly rain gauge data. The 3B42 products have two successive versions: Version 6 and the latest Version 7 (3B42V6 and 3B42V7). In this study, we evaluated and compared the two high-resolution (3 h and $0.25° \times 0.25°$) satellite precipitation products: 3B42V6 and 3B42V7.

The TMPA algorithm (Huffman et al., 2007) calibrates and combines microwave (MW) precipitation estimates, and then creates the infrared (IR) precipitation estimates using the calibrated MW. After this, it combines the MW and IR estimates to create the TMPA precipitation estimates. MW data used in Version 6 are from the TRMM Microwave Imager (TMI), Special Sensor Microwave Imager (SSM/I) F13, F14, and F15 on Defense Meteorological Satellite Program satellites, and the Advanced Microwave Scanning Radiometer–Earth Observing System on Aqua, and the Advanced Microwave Sounding Unit-B N15, N16, and N17 on the NOAA satellite; IR data collected by geosynchronous earth orbit (GEO) satellites, GEO-IR. The 3B42V6 also use other data sources: TRMM Combined Instrument employed from TMI and PR, monthly rain gauge data from GPCP ($1° \times 1°$) and the Climate Assessment and Monitoring System $0.5° \times 0.5°$ developed by CPC. Based on the lessons learned in 3B42V6, 3B42V7 includes consistently reprocessed versions for the data sources used in 3B42V6 and introduces additional datasets, including the Special Sensor Microwave Imager/Sounder (SSMIS) F16–17 and Microwave Humidity Sounder (MHS) (N18 and N19) and Meteorological Operational satellite programme (MetOp) and the $0.07°$ Grisat-B1 infrared data. The 3B42V7 is also designed to accept future satellites to improve the data quality (Huffman et al., 2011). All of these data can be freely downloaded from the website: http://trmm.gsfc.nasa.gov/ and http://mirador.gsfc.nasa.gov.

6.2.2.3 Evapotranspiration

The potential evapotranspiration (PET) data used in this study are from the global daily PET database provided by the Famine Early Warning Systems Network (hereafter referred to as FEWSPET) global data portal (see http://earlywarning.usgs.gov/fews/global/web/readme.php?symbol=pt). FEWSPET is calculated from climate parameter extracted from global data assimilation system analysis fields and has $1°$ ground resolution and covers the entire globe from 2001 to the present.

6.2.3 CREST MODEL

The CREST model (Khan et al., 2011a, b; Wang et al., 2011) is a grid-based distributed hydrologic model developed by the University of Oklahoma (http://hydro.ou.edu) and NASA-SERVIR Project Team (www.servir.net). It computes the runoff generation components (e.g., surface runoff and infiltration) using the variable infiltration capacity (VIC) curve, a concept originally contained in the Xinanjiang model (Zhao, 1992; Zhao et al., 1980) and later represented in the VIC model (Liang et al., 1994, 1996). Multilinear reservoirs are used to simulate cell-to-cell routing of surface and subsurface runoff separately. The CREST model couples the runoff generation component and cell-to-cell routing scheme described above, to reproduce the interaction between surface and subsurface water flow processes. Besides the hydrologic and basic data (DEM, flow direction, flow accumulation, slope, etc.), the CREST model employs gridded precipitation and PET data as its forcing data. CREST Version 1.6 model has been applied at both global (Wu et al., 2012) and regional scales (Khan et al., 2011a, 2011b) (more applications can be found at website: http://eos.ou.edu and http://www.servir.net).

The CREST model used in this study is the upgraded version CREST V2.0. The main features of CREST V2.0 (Xue et al., 2013) are as follows: (1) enhancement of the computation capability using matrix manipulation techniques to make the model more efficient than the previous version (Wang et al., 2011); (2) model implementation with options of either spatially uniform, semidistributed, or distributed parameter values; (3) automatic extraction of *a priori* model parameter estimates from high-resolution land surface and soil texture data; (4) a modular design framework to accommodate research, development, and system enhancements (see Figure 6.2); and (5) inclusion of the optimization scheme SCE-UA (Duan et al., 1992, 1993) to enable automatic calibration of the CREST model parameters. Table 6.2 shows 11 parameters and their description, ranges, and default values.

6.2.4 EVALUATION STATISTICS

To quantitatively analyze the performance of 3B42V6 and 3B42V7 precipitation products against rain gauge observations and the effect to the streamflow, three widely used validation statistical indices were selected in this study. The relative Bias (%) was used to measure the agreement between the averaged value of simulated data (in this study, we call both TMPA products and simulated streamflow as "simulated data," "SIM" was used in the formulae) and observed data (such as rain gauge and observed streamflow in this study, "OBS" was used in the formulae). The root-mean-square error (RMSE) was selected to evaluate the average error magnitude between simulated and observed data. We also use correlation coefficient (CC) to assess the agreement between simulated and observed data.

$$\text{Bias} = \left[\frac{\sum_{i=1}^{n} \text{SIM}_i - \sum_{i=1}^{n} \text{OBS}_i}{\sum_{i=1}^{n} \text{OBS}_i} \right] \times 100 \tag{6.1}$$

$$\text{RMSE} = \sqrt{\frac{\sum_{i=1}^{n} \left(\text{OBS}_i - \text{SIM}_i \right)^2}{n}} \tag{6.2}$$

$$\text{CC} = \frac{\sum_{i=1}^{n} \left(\text{OBS}_i - \overline{\text{OBS}} \right) \left(\text{SIM}_i - \overline{\text{SIM}} \right)}{\sqrt{\sum_{i=1}^{n} \left(\text{OBS}_i - \overline{\text{OBS}} \right)^2 \sum_{i=1}^{n} \left(\text{SIM}_i - \overline{\text{SIM}} \right)^2}} \tag{6.3}$$

where:

n is the total number of pairs of simulated and observed data

i is the ith values of the simulated and observed data

$\overline{\text{SIM}}$ and $\overline{\text{OBS}}$ are the mean values of simulated and observed data, respectively

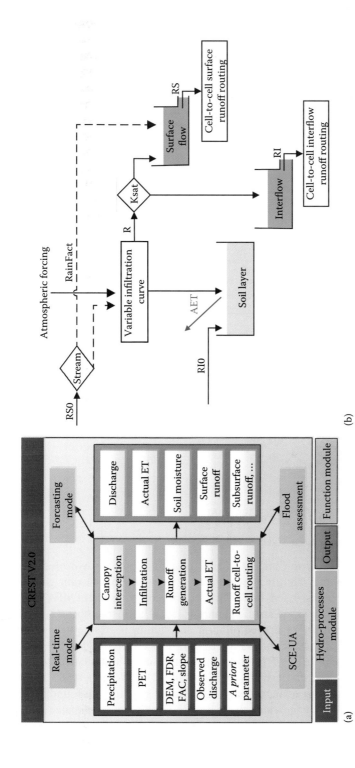

FIGURE 6.2 (a) The framework of the CREST model version 2.0 and (b) vertical profile of hydrologic processes in a grid cell.

TABLE 6.2
Parameters to Be Calibrated in CREST V2.0, Their Meanings, Ranges, and Default Values

Parameter	Description	Numeric Range	Default Value
Ksat	The soil saturate hydraulic conductivity (mm/d)	0–2827.2	500
RainFact	The multiplier on the precipitation field	0.5–1.2	1.0
WM	The mean water capacity	80–200	120
B	The exponent of the variable infiltration curve	0.05–1.5	0.25
IM	Impervious area ratio	0–0.2	0.05
KE	The ratio of the PET to actual evapotranspiration	0.1–1.5	1.0
coeM	Overland runoff velocity coefficient	1–150	90
coeR	Multiplier used to convert overland flow speed to channel flow speed	1–3	2
coeS	Multiplier used to convert overland flow speed to interflow flow speed	0.001–1	0.3
KS	Overland reservoir discharge parameter	0–1	0.6
KI	Interflow reservoir discharge parameter	0–1	0.25

Nash–Sutcliffe coefficient of efficiency (NSCE) is also used to assess the performance of model simulation and observation.

$$\text{NSCE} = 1 - \frac{\sum_{i=1}^{n}\left(\text{OBS}_i - \text{SIM}_i\right)^2}{\sum_{i=1}^{n}\left(\text{OBS}_i - \overline{\text{OBS}}\right)^2} \tag{6.4}$$

6.3 RESULTS AND DISCUSSION

6.3.1 COMPARISON OF PRECIPITATION INPUTS

To better understand the impact of precipitation inputs on hydrologic models, the accuracy of the satellite precipitation against the in situ rain gauge observations should be assessed first. This section compares the TMPA and gauge observations over the time span of January 1, 2001, to December 31, 2010, considering the basin-average precipitation and within a grid-cell containing the dense rain gauge observations (Figure 6.1), respectively. Figure 6.3 shows that both 3B42V6 and 3B42V7 basin-average time series systematically underestimate though at different levels, with biases of −41.15% and −8.38% and CCs of 0.36 and 0.40 at daily scale, respectively. Similar statistics are found at 0.25 grid cell-scale. Figure 6.4 indicates that 3B43V6 largely underestimates with a bias of −40.25% and low CC of 0.37, while 3B42V7 has practically no bias (0.04%) and a relatively higher CC value 0.41.

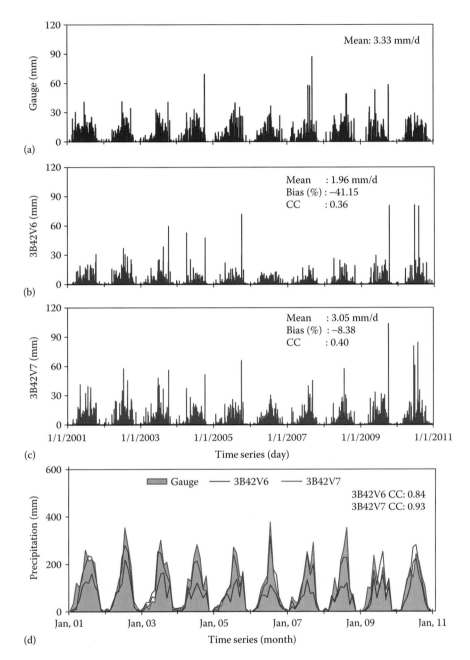

FIGURE 6.3 Basin-averaged precipitation time series of gauge, 3B42V6 and 3B42V7 for the period January 2001–December 2010: (a) gauge daily time series; (b) 3B42V6 daily time series; (c) 3B42V7 daily time series; and (d) monthly time series of gauge, 3B42V6 and 3B42V7.

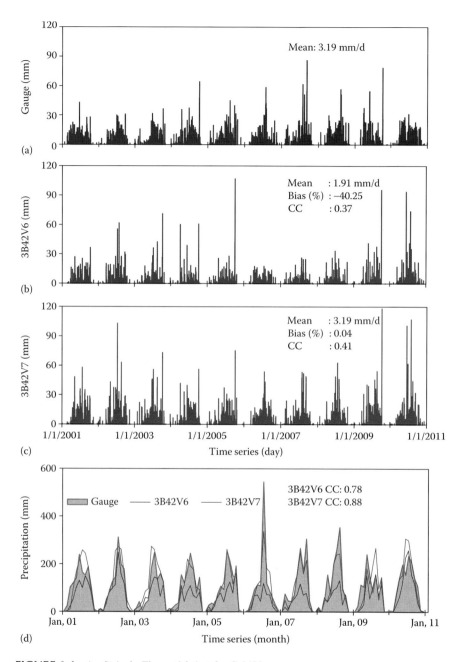

FIGURE 6.4 (a–d) As in Figure 6.3, but for Grid32.

Figures 6.3d and 6.4d present the intercomparison of monthly precipitation estimates to gain further information about the precision and variations at longer timescales. The monthly data for both basin-based and grid-cell based analyses were accumulated from daily data over the same time span from January 2001 to December

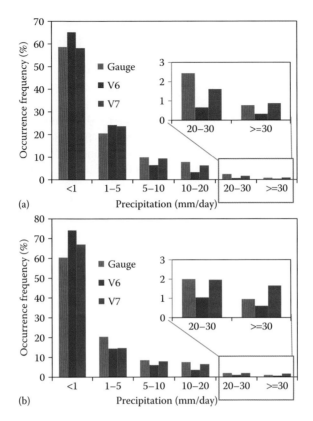

FIGURE 6.5 Occurrence frequencies of rain gauge, 3B42V6 and 3B42V7 data: (a) for the basin-based averaged data and (b) for the Grid32 averaged data.

2010. At monthly timescale, both the basin-based and grid-cell based data show that 3B42V7 has better agreement with the monthly rain gauge data. Both Figures 6.3 and 6.4 indicate that the latest V7 algorithm significantly corrects the underestimation bias in its predecessor version V6.

Figure 6.5a and b shows the frequency distribution of daily precipitation for different precipitation intensities (PI) for the basin-averaged and the grid cell-based precipitation time series, respectively. Figure 6.5a shows that for the basin-averaged data, both 3B42V6 and 3B42V7 overestimate at the low PI range (< 5 mm/day), but they underestimate at the medium and high PI ranges. However, 3B42V7 is in better agreement with the rain gauge observations than 3B42V6 for the basin-averaged comparison across all PIs. Similarly, better agreement has been found in Figure 6.5b for the new Version-7 products at the grid cell scale, except for values greater than 30 mm/day where there is overestimation.

6.3.2 STREAMFLOW SIMULATION SCENARIOS

Although different precipitation products vary in accuracy and spatiotemporal resolutions, they might have similar hydrologic prediction (i.e., streamflow simulation) skill after re-calibrating the model using the respective precipitation products (Stisen and Sandholt, 2010; Jiang et al., 2012). In the previous section, we compared the 3B42V6 and 3B42V7 precipitation products against the rain gauge observations; the next step is to evaluate how these two TMPA products affect streamflow simulations. Their hydrologic evaluation is performed under two scenarios:

    I. *In situ gauge benchmarking*: Calibrate the CREST model with five years of rain gauge data (January 2001–December 2005). Then, replace the rain gauge forcing with precipitation from 3B42V6 and 3B42V7 for an independent validation period from January 2006 through December 2010 using the rain gauge-calibrated model parameters.

    II. *Product-specific calibration*: Recalibrate the CREST model using the 3B42V6 and 3B42V7 precipitation data, respectively, over the same calibration period and then use the product-specific parameter sets to simulate streamflow over the same validation period as Scenario I.

Scenario I, gauge benchmarking, is widely used by the hydrologic community especially over gauged basins, while Scenario II is arguably deemed as an alternative for application to ungauged basins where only rainfall from remote sensing platforms are available for use.

6.3.2.1 Scenario I: CREST Benchmarked by In Situ Gauge Data

    1. Rain gauge calibration and validation: The CREST model parameters are calibrated using rain gauge inputs for the period from January 2001 to December 2005 using the automatic calibration method (SCE-UA) by maximizing the NSCE value between the simulated and observed daily streamflow. The calibrated model is subsequently validated for the period from January 2006 to December 2010. Figure 6.6 compares the simulated streamflow forced by rain gauge data with the observed streamflow in terms of time series plots and exceedance probability plots at daily and monthly scales. Figure 6.6a and b shows that general agreement exists between the observed and simulated streamflow, although the simulated flows underestimated the high peaks and in relatively low flow seasons. The exceedance probabilities in Figure 6.6c and d also show underestimation at low and high streamflow observations, while simulation match relatively well in the intermediate ranges. As summarized in Tables 6.3 and 6.4, the statistical indices show that there is very good agreement between the simulated and observed hydrographs in the calibration period for both daily and monthly timescale, and reasonable simulations occurred in the validation period as well. Based on the criteria of the statistical indices in Moriasi et al. (2007), the model calibration and validation results indicate that the CREST model is well benchmarked by the in situ data at the daily and monthly timescale, so it can be used to evaluate the utility of the satellite precipitation products for hydrologic prediction (i.e., streamflow) in this basin.

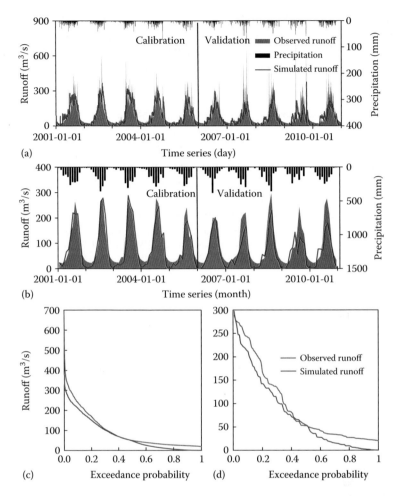

FIGURE 6.6 Comparison of observed and simulated streamflow using gauge data as input: (a) daily calibration (January 1, 2001–December 31, 2005) and validation (January 1, 2006–December 31, 2010), (b) monthly time series, (c) exceedance probabilities using daily data (January 1, 2001–December 31, 2010), and (d) exceedance probabilities using monthly data.

2. *Impacts of satellite precipitation forcing:* The gauge-benchmarked model is then forced by the TMPA 3B42V6 and 3B42V7 products from 2001 to 2010 using the model parameters calibrated by rain gauge data during the period from 2001 to 2005. Figures 6.7 and 6.8 compare the daily and monthly time series of the simulated and observed hydrographs for both the calibration and validation periods. While 3B42V6 largely missed the high peak flows at both daily and monthly time series, 3B42V7 adequately captured a majority of the peak flows, especially at the smoothed monthly scale. The daily and monthly statistical comparisons in Tables 6.3 and 6.4

TABLE 6.3
Comparison of Daily Observed and Simulated Streamflow for Model Simulation under Two Scenarios

Precipitation Products	Scenario I				Scenario II			
	NSCE	Bias (%)	CC	RMSE	NSCE	Bias (%)	CC	RMSE
			Calibration Period					
Gauge	0.76	−9.73	0.89	45.38	–	–	–	–
3B42V6	0.23	−52.94	0.80	81.99	0.63	−1.70	0.80	56.55
3B42V7	0.66	−26.98	0.86	54.65	0.78	−4.81	0.88	43.94
			Validation Period					
Gauge	0.59	−29.59	0.83	57.85	–	–	–	–
3B42V6	0.17	−57.78	0.78	82.98	0.65	−8.67	0.81	54.00
3B42V7	0.63	−25.15	0.83	55.26	0.72	−3.02	0.86	47.72

TABLE 6.4
As in Table 6.3, but for Monthly Data

Precipitation Products	Scenario I				Scenario II			
	NSCE	Bias (%)	CC	RMSE	NSCE	Bias (%)	CC	RMSE
			Calibration Period					
Gauge	0.91	−9.75	0.96	25.18	–	–	–	–
3B42V6	0.25	−53.01	0.88	72.08	0.75	−1.66	0.87	41.41
3B42V7	0.77	−27.06	0.94	39.76	0.91	−4.83	0.95	25.41
			Validation Period					
Gauge	0.70	−29.59	0.88	43.63	–	–	–	–
3B42V6	0.19	−57.81	0.89	71.35	0.79	−8.65	0.89	36.29
3B42V7	0.80	−25.25	0.94	35.58	0.89	−3.08	0.95	26.53

show that the daily and monthly simulations forced by rain gauge data had better skill (NSCE = 0.76/0.91, BIAS = −9.73%/−9.75%, CC = 0.89/0.96) than those based on 3B42V6 and 3B42V7 in calibration period, which is expected. Interestingly, the 3B42V7-forced model simulations had very similar to and slightly better performance compared to the rain gauge-forced simulations in the validation period. A likely explanation is one of the rain gauge stations (i.e., the Dochula) had missing data from September 2006 to December 2010, which apparently degrades the hydrologic skill of this product. Overall, simulations forced by 3B42V7 were a significant improvement over 3B42V6. This clearly shows the improvements of the new Version-7 algorithm upon its predecessor V6 products both statistically and now hydrologically.

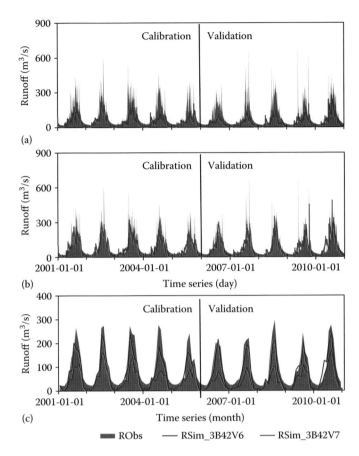

FIGURE 6.7 Comparison of CREST simulated sreamflow from 3B42V6 (blue line) and 3B42V7 (red line) with gauge calibrated parameters and observed streamflow in both calibration (January 1, 2001–December 31, 2005) and validation (January 1, 2006–December 31, 2010) period. (a) Daily time series from 3B42V6, (b) daily time series from 3B42V7, and (c) monthly time series from 3B42V6 and 3B42V7.

6.3.2.2 Scenario II: CREST Calibrated by Individual TMPA Products

To further assess the effects of TMPA 3B42 (V6 and V7) products on streamflow, the CREST model is recalibrated and validated with 3B42V6 and 3B42V7 for the same period as Scenario I. This scenario is often used as an alternative strategy for remote sensing precipitation over ungauged basins. As shown in Figure 6.8, all simulations are significantly improved after the recalibration, and they capture most of the daily and monthly peak flows. Comparatively, the CREST model simulations based on 3B42V7 inputs have better skill than those based on 3B42V6. As summarized in Tables 6.3 and 6.4, simulations have good statistical agreement with observed streamflow at daily and monthly scale.

The statistical indices of daily NSCE, Bias, and CC in Table 6.3 were selected for visual comparison of the two modeling scenarios. Figure 6.9 indicates that the

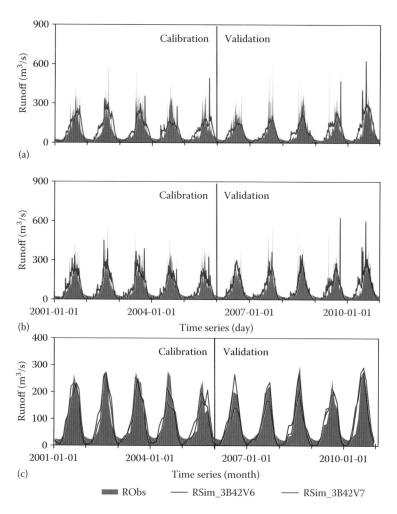

(a)

(b)

(c)

FIGURE 6.8 As in Figure 6.9, but the parameters was recalibrated using 3B42V6 and 3B42V7, respectively.

product-specific recalibration in Scenario II has obviously improved the NSCE and CC values and reduced the Bias values for both the calibration and validation periods. It is noted that the recalibration forcing with 3B42V7 in Scenario II has much higher NSCE and smaller Bias than 3B42V6, and very comparable CC values, all of which improved upon the rain gauge-benchmarked model.

6.3.3 DISCUSSION OF PARAMETER COMPENSATION EFFECT FROM SCENARIO II

Table 6.5 shows the optimum parameter sets forced by 3B42V6 and 3B42V7, relative to the gauge forcing, for the calibration period from 2001 to 2005 using the SCE-UA algorithm. Note that the parameter values of Ksat and WM are spatially

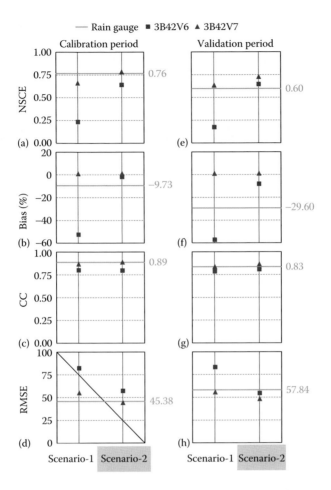

FIGURE 6.9 Comparison of the streamflow performance statistics of the TMPA 3B42V6 and 3B42V7 precipitation for the two simulation scenarios in both calibration period (a–d) and validation period (e–h).

distributed but have been basin-averaged and summarized in Table 6.4. It shows that 3B42V7-calibrated parameters have less deviation from the gauge-calibrated parameter values than 3B42V6. For example, RainFact is the adjustment factor of the precipitation either due to canopy interception or under/over estimation. Table 6.5 shows that 3B42V6 increases the RainFact value from 0.87 to 1.34, to compensate its underestimation as shown in Figures 6.3 and 6.4, while 3B42V7's estimated value (0.98) is closer to 1 and the Gauge value (0.87). Another example is KE, the ratio of the PET to actual evapotranspiration. Table 6.5 reveals that 3B42V6 demands a reduced KE value from 0.10 to 0.05 to partition more precipitation into runoff while 3B42V7 only slightly increases it from 0.10 to 0.13, possibly to partially offset the above RainFact increase, amongst other parametric interactions. The third example is Ksat, the soil saturated hydraulic conductivity. Table 6.5 shows that the Ksat of

TABLE 6.5

CREST Model Parameter Values Calibrated with Different
Precipitation Data for the Calibration Period of January
2001–December 2005

Parameters	Gauge	3B42V6	3B42V7
RainFact	0.87	1.34	0.98
Ksat	56.90	33.09	52.73
WM	166.50	142.71	166.52
B	1.48	1.48	1.48
IM	0.20	0.20	0.20
KE	0.10	0.05	0.13
coeM	88.05	63.67	67.95
coeR	2.68	1.33	1.44
coeS	0.43	0.47	0.67
KS	0.99	0.71	0.78
KI	0.20	0.13	0.14

3B42V6 reduced from 56.90 to 33.09 while V7 only changed slightly from 56.90 to 52.73. Regarding the mean water capacity, WM, 3B42V6 decreased from 166.50 to 142.71 to hold less water in the soils while 3B42V7 did not change much from the gauge-calibrated value, which is presumably closer to the truth. It also shows the overland flow coefficient, coeM, the average channel flow speed, coeR, the overland flow recession coefficient, KS, and the interflow recession coefficient, KI, all had reduced values to retain more water in the river basin after recalibrating the parameters to both of the satellite products.

Not surprisingly, Table 6.4 also shows some opposite changes of values such as KE for 3B42V7 and coeS, the surface-interflow conversion factor, for both 3B42V6 and 3B42V7, resulting in a slight decrease in streamflow. However, the overall effect of the recalibrated parameter sets is to largely compensate for rainfall underestimation in 3B42V6 while less so for 3B42V7. The effect of arriving at a very similar simulation with different combinations of parameter settings has been called "Equifinality" of the hydrologic model (Aronica et al., 1998; Zak and Beven, 1999; Beven and Freer, 2001). This study clearly shows how different parameter settings can compensate for errors in the satellite rainfall forcing and can thus improve model prediction of streamflow. It is possible that model structural deficiencies, such as not accounting for snowmelt contribution to total runoff, also benefits from model calibration. However, this parameter compensation effect comes with the price of having a locally optimized model with parameter values unrepresentative of reality. This might limit the model's predictive capability at internal sub-basins, or under different initial conditions. This is particularly concerning under scenarios involving climate change. In any case, the recalibration strategy could be especially problematic for 3B42V6 (Bitew and Gebremichael, 2011; Jiang et al., 2012); however, the 3B42V7 product gives higher confidence for use in ungauged basins even without the need for recalibration.

6.4   SUMMARY AND CONCLUSIONS

Satellite precipitation products are very important for regional and global hydrologic studies, particularly for remote regions and developing countries. This study first focuses on statistically assessing the accuracy of the TMPA 3B42V6 product versus its latest successive version 3B42V7, and then hydrologically evaluates their streamflow prediction utility using the CREST distributed hydrologic model in the mountainous Wangchu basin of Bhutan.

The two versions of TMPA satellite products are statistically compared with a decade-long (2001–2010) rain gauge dataset at daily and monthly scales. In general, 3B42V7 consistently improves upon 3B42V6's underestimation both for the whole basin (bias improved from −41.15% to −8.38%) and for a 0.25° × 0.25° grid cell with high-density gauges (bias improved from −40.25% to 0.04%), though with modest enhancement of CC (from 0.36 to 0.40 for entire basin and from 0.37 to 0.41 for the grid cell). 3B42V7 also improves upon 3B42V6 in terms of occurrence frequency across the rain intensity spectrum. Apparently, the results show that the new algorithm 3B42V7 has much improved accuracy upon 3B42V6, in addition to assimilating more satellite sensors (such as SSMIS, MHS, and MetOp) (Huffman et al., 2011).

For the hydrologic evaluation, two scenario-based calibration and validation experiments are conducted over the same 10-year time span. Scenario I, in situ gauge benchmarking, is widely used by the hydrologic community especially over gauged basins, while Scenario II, input-specific recalibration, is arguably deemed as an alternative for application to ungauged basins where only remote sensing rainfall data may be available for use. In Scenario I, the 3B42V6-based simulation shows lower hydrologic prediction skill in terms of NSCE (0.23 at daily scale and 0.25 at monthly scale) while 3B42V7 performs fairly well (0.66 at daily scale and 0.77 at monthly scale), a comparable skill score with the simulations gauge benchmark. For the precipitation input-specific calibration in Scenario II, significant improvements are observed for 3B42V6 across all statistics. These enhancements are not as obvious for the already-well-performing 3B42V7-calibrated model, except for some reduction in bias (from −26.98% to −4.81%). This behavior is consistent with previous studies (Bitew and Gebremichael, 2011; Bitew et al., 2011; Jiang et al., 2012). This study offers unique insights into 3B42V6 and 3B42V7 products in a mountainous South Asian basin.

In conclusion, the latest 3B42V7 algorithm presents a significant upgrade from 3B42V6 both in terms of accuracy (i.e., correcting the underestimation) and in its promising hydrologic utility in the study region, even without the requirement of recalibrating the hydrologic model. The parameter compensation effect is often recognized but still used by the hydrology community. This approach has been noted to be problematic due to unrealistic parameter settings that may ultimately limit the model's predictive capability under conditions of climate change and differing initial conditions.

ACKNOWLEDGMENTS

The current study was supported by the NASA/Marshall Space Flight Center Grants NNM11AB34P and NNMi2428088Q to the University of Oklahoma.

REFERENCES

Aronica, G., Hankin, B., and Beven, K.J., 1998. Uncertainty and equifinality in calibrating distributed roughness coefficients in a flood propagation model with limited data. *Advances in Water Resources*, 22(4):349–365.

Beven, K.J. and Freer, J., 2001. Equifinality, data assimilation, and uncertainty estimation in mechanistic modelling of complex environmental systems using the GLUE methodology. *Journal of Hydrology*, 249(1–4):11–29.

Bitew, M.M. and Gebremichael, M., 2011. Evaluation of satellite rainfall products through hydrologic simulation in a fully distributed hydrologic model. *Water Resources Research*, 47(6):W06526.

Bitew, M.M., Gebremichael, M., Ghebremichael, L.T., and Bayissa, Y.A., 2011. Evaluation of high-resolution satellite rainfall products through streamflow simulation in a hydrological modeling of a small mountainous watershed in Ethiopia. *Journal of Hydrometeorology*, 13(1):338–350.

Bookhagen, B. and Burbank, D.W., 2010. Toward a complete Himalayan hydrological budget: Spatiotemporal distribution of snowmelt and rainfall and their impact on river discharge. *Journal of Geophysical Research*, 115(F3):F03019.

Chokngamwong, R. and Chiu, L.S., 2008. Thailand daily rainfall and comparison with TRMM products. *Journal of Hydrometeorology*, 9(2):256–266.

Duan, Q., Sorooshian, S., and Gupta, V., 1992. Effective and efficient global optimization for conceptual rainfall-runoff models. *Water Resources Research*, 28(4):1015–1031.

Duan, Q.Y., Gupta, V.K., and Sorooshian, S., 1993. Shuffled complex evolution approach for effective and efficient global minimization. *Journal of Optimization Theory and Applications*, 76(3):501–521.

FAO/IIASA/ISRIC/ISSCAS/JRC, 2009. *Harmonized World Soil Database* (version 1.1), FAO, Rome, Italy.

Hansen, M.C., Defries, R.S., Townshend, J.R.G., and Sohlberg, R., 2000. Global land cover classification at 1 km spatial resolution using a classification tree approach. *International Journal of Remote Sensing*, 21(6):1331–1364.

Hong, Y., Hsu, K.-L., Sorooshian, S., and Gao, X., 2004. Precipitation estimation from remotely sensed imagery using an artificial neural network cloud classification system. *Journal of Applied Meteorology*, 43(12):1834–1853.

Huffman, G.J., Bolvin, D.T., Nelkin, E., and Adler, R.F., 2011. Highlights of version 7 TRMM multi-satellite precipitation analysis (TMPA). In: Klepp, C. and Huffman, G.J., (Eds.), *5th International Precipitation Working Group Workshop, Workshop Program and Proceedings*. Reports on Earth System Science, 100/2011, Max-Planck-Institut für Meteorologie, Hamburg, Germany.

Huffman, G.J. et al., 2007. The TRMM multisatellite precipitation analysis (TMPA): Quasi-global, multiyear, combined-sensor precipitation estimates at fine scales. *Journal of Hydrometeorology*, 8(1):38–55.

Islam, M.N. and Uyeda, H., 2007. Use of TRMM in determining the climatic characteristics of rainfall over Bangladesh. *Remote Sensing of Environment*, 108(3):264–276.

Jamandre, C.A. and Narisma, G.T., 2013. Spatio-temporal validation of satellite-based rainfall estimates in the Philippines. *Atmospheric Research*, 122:599–608.

Jiang, S. et al., 2012. Comprehensive evaluation of multi-satellite precipitation products with a dense rain gauge network and optimally merging their simulated hydrological flows using the Bayesian model averaging method. *Journal of Hydrology*, 452–453:213–225.

Joyce, R.J., Janowiak, J.E., Arkin, P.A., and Xie, P., 2004. CMORPH: A method that produces global precipitation estimates from passive microwave and infrared data at high spatial and temporal resolution. *Journal of Hydrometeorology*, 5(3):487–503.

Khan, S.I. et al., 2011a. Hydroclimatology of Lake Victoria region using hydrologic model and satellite remote sensing data. *Hydrology and Earth System Sciences*, 15(1):107–117.

Khan, S.I. et al., 2011b. Satellite remote sensing and hydrologic modeling for flood inundation mapping in Lake Victoria basin: Implications for hydrologic prediction in ungauged basins. *IEEE Transactions on Geoscience and Remote Sensing*, 49(1):85–95.

Li, X.-H., Zhang, Q., and Xu, C.-Y., 2012. Suitability of the TRMM satellite rainfalls in driving a distributed hydrological model for water balance computations in Xinjiang catchment, Poyang lake basin. *Journal of Hydrology*, 426–427:28–38.

Liang, X., Lettenmaier, D.P., Wood, E.F., and Burges, S.J., 1994. A simple hydrologically based model of land surface water and energy fluxes for general circulation models. *Journal of Geophysical Research*, 99(D7):14415–14428.

Liang, X., Wood, E.F., and Lettenmaier, D.P., 1996. Surface soil moisture parameterization of the VIC-2L model: Evaluation and modification. *Global and Planetary Change*, 13(1–4):195–206.

Mishra, A., Gairola, R.M., Varma, A.K., and Agarwal, V.K., 2010. Remote sensing of precipitation over Indian land and oceanic regions by synergistic use of multisatellite sensors. *Journal of Geophysical Research*, 115(D8):D08106.

Moriasi, D.N. et al., 2007. Model evaluation guidelines for systematic quantification of accuracy in watershed simulations. *ASABE*, 50(3):885–900.

Sorooshian, S. et al., 2000. Evaluation of PERSIANN system satellite–based estimates of tropical rainfall. *Bulletin of the American Meteorological Society*, 81(9):2035–2046.

Stisen, S. and Sandholt, I., 2010. Evaluation of remote-sensing-based rainfall products through predictive capability in hydrological runoff modelling. *Hydrological Processes*, 24(7):879–891.

Su, F., Hong, Y., and Lettenmaier, D.P., 2008. Evaluation of TRMM multisatellite precipitation analysis (TMPA) and its utility in hydrologic prediction in the La Plata basin. *Journal of Hydrometeorology*, 9(4):622–640.

Wang, J. et al., 2011. The coupled routing and excess storage (CREST) distributed hydrological model. *Hydrological Sciences Journal*, 56(1):84–98.

Wu, H., Adler, R.F., Hong, Y., Tian, Y., and Policelli, F., 2012. Evaluation of global flood detection using satellite-based rainfall and a hydrologic model. *Journal of Hydrometeorology*, 13(4):1268–1284.

Xue, X. et al., 2013. Statistical and hydrological evaluation of TRMM-based multi-satellite precipitation analysis over the Wangchu basin of Bhutan: Are the latest satellite precipitation products 3B42V7 ready for use in ungauged basins? *Journal of Hydrology*, 499:91–99.

Yong, B. et al., 2010. Hydrologic evaluation of multisatellite precipitation analysis standard precipitation products in basins beyond its inclined latitude band: A case study in Laohahe basin, China. *Water Resources Research*, 46(7):W07542.

Yong, B. et al., 2012. Assessment of evolving TRMM-based multisatellite real-time precipitation estimation methods and their impacts on hydrologic prediction in a high latitude basin. *Journal of Geophysical Research*, 117:D09108.

Zak, S.K. and Beven, K.J., 1999. Equifinality, sensitivity and predictive uncertainty in the estimation of critical loads. *The Science of The Total Environment*, 236(1–3):191–214.

Zhao, R.J., 1992. The Xinanjiang model applied in China. *Journal of Hydrology*, 135(1–4):371–381.

Zhao, R.J., Zhuang, Y.G., Fang, L.R., Liu, X.R., and Zhang, Q.S., 1980. *The Xinanjiang model, hydrological forecasting proceedings Oxford symposium*. IAHS Publication, Oxford University, pp. 351–356.

7 An Advanced Distributed Hydrologic Framework
The Development of CREST

*Xinyi Shen, Yang Hong, Emmanouil
N. Anagnostou, Ke Zhang, and Zengchao Hao*

CONTENTS

7.1 BACKGROUND

Water cycle has been extensively yet not thoroughly explored because it is closely related to multidisciplines including atmospheric science, meteorology, hydrology, and remote sensing and perplexed by the complexity of the land surface condition and the uncertainty of multisource dataset. With the aid of advanced computing, remote sensing, and Geographic Information System technologies, and available meteorological products with high spatiotemporal resolution, distributed hydrologic models have been rapidly developed and widely used in real-time flood forecasting, near/long-term drought prediction, and general water resources management. In the past, a hydrologic model focuses primarily on the rainfall infiltration and runoff routing components with a separated land surface model giving the vertical water input. At present, with more frequent interactions among earth sciences, modern distributed hydrologic models usually appear in a more integrated form. Not only a routing model is coupled with a land surface one, ground water and human interference (e.g., irrigation) modules are also introduced to some hydrologic models. In this way, modern hydrologic models play an engine role for many applications in geosciences. In this chapter, we review the evolution of the Coupled Routing and Excess water Storage (CREST) model as an example of a modern integrated hydrologic model.

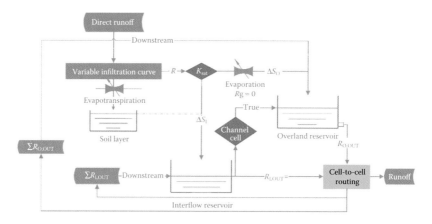

FIGURE 7.1   Model structure of the CREST v1.5.

The CREST (Wang et al., 2011) model is among the early works that fully couple the land surface and the routing modules. Figure 7.1 shows that the routing module not only takes the output of the land surface module, but also feeds back the land surface module by means of adding the transported water as the bare rainfall to the arrival downstream cell. Moreover, the land surface and routing modules are calibrated as a whole as well.

Besides the coupling benchmark, developers of CREST also tried to develop more efficient computational tool in response to the high demands of regional and global hydrologic model at hyper-resolution (Wood et al., 2011). The CREST model was initially developed in the University of Oklahoma and supported jointly by National Aeronautics and Space Administration. CREST has been deployed as a kernel model in several operational systems including the FLASH FLOOD (http://www.nssl.noaa.gov/projects/flash/) and Near Real-time Global Hydrologic Simulation and Flood Monitoring Demonstration System (http://eos.ou.edu/).

7.2   MODEL DEVELOPMENT

7.2.1   CREST VERSIONS 1.0–1.5

Wang et al. (2011) developed the first versions of CREST featured by its cell-to-cell routing framework at fine resolution, typically ~1 km × 1 km. The framework is conceptualized in Figure 7.2. Following the flow direction map, which is globally available from 90 m (Lehner et al., 2006) to 0.5° (Wu et al., 2011), the routing module computes the distance water could travel from every grid-cell during one model time step, Δt. Figure 7.2a illustrates three examples of such a travel, that is, cell A to cell D, cell B to cell E, and cell C to cell F. Because of the finite discretization of basin grids and the time span, the travel distance has zero probability of passing an integer number of grids during Δt. Converting a two-dimensional travel path to one dimensional, as illustrated in Figure 7.2b, and supposing the

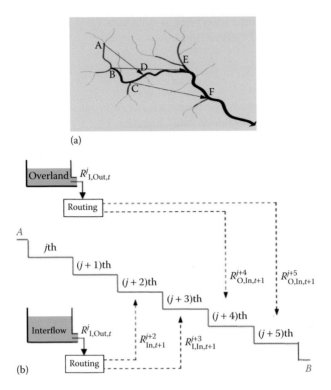

FIGURE 7.2 The cell-to-cell routing framework. (a) Water transportation within one time step. (b) Two conceptual linear reservoirs that account for the surface flow and interflow, respectively.

travel distance is $4x$ grids, then $(1-x)/10$ of the total amount of the transported water starting from the jth is redistributed to the $(j+4)$th cell and $x/10$ to the $(j+5)$th cell. In addition, two virtual linear reservoirs, the overland flow and interflow reservoirs, at each grid cell are employed to separate the floodplain storage, groundwater flow, storage by lakes and wetlands, and the simulation of operations of real reservoirs, in order to simulate subgrid cell routing in a cell-to-cell routing scheme (Wang et al., 2011).

The outflows of reservoirs are proportional to the water storage of the source cell, as given in the following equation,

$$R(\text{out})_{S,I}^t = k_{S,I} S_{S,I}^{t*} \tag{7.1}$$

where:
S stands for the water depth (in L) of the reservoirs
$R(\text{out})$ stands for the outflow
k is the storage-flow coefficient
the subscripts S and I refer to surface flow and interflow, respectively

The water storages in linear reservoirs at time step $t + 1$ are further updated by accounting for inflow and outflow (in L):

$$S_{S,I}^{t+1} = S_{S,I}^{t*} + \sum R(in)_{S,I}^{t} - R(out)_{S,I}^{t} \tag{7.2}$$

At each time step t, the discharge from a given cell is calculated by converting the runoff depth, $R(out)_{S,I}^{t}$, to discharge using Equation 7.4

$$\text{runoff} = \left[R(out)_S + R(out)_I \right] A_g / \Delta t \tag{7.3}$$

Besides the light computation from using linear reservoirs, developers of CREST v1.5 also simplified the computation of the land surface process to make the model efficient enough to run at 1 km resolution yet in large basins, that is, larger than 100,000 km². CREST employs the Xin'an Jiang model's conception of a variable infiltration curve (Ren-Jun, 1992; Zaitarian, 1992; Zhao et al., 1995) to account for the water vertical redistribution from precipitation to evaporation, soil moisture, and excessive water storages at every grid.

7.2.2 CREST VERSION 2.0

As one of the initial versions, CREST version 1.5 emphasizes on flow simulation, especially on flash flood. It takes no consideration of the vegetation transpiration effects and simply assumes bare soil evaporation at a resisted rate calculated by the ratio of available soil moisture versus the field capacity and the available energy given by the potential evapotranspiration (PET) forcing data. Furthermore, soil is conceptualized by a single layer. Therefore, soil moisture variability at the vertical direction could not be accounted for. The inaccuracy introduced by such a simplified land surface scheme is mitigated by the adjustment of 12 parameters including scaling factors for precipitation and PET (Xue et al., 2013). These scaling factors partially account for under- or overestimations of precipitation and ET and partially for the water-interception effects of vegetation in Equations 7.4 and 7.5

$$P = R_P \hat{P} \tag{7.4}$$

$$\text{PET} = R_{PET} \widehat{\text{PET}} \tag{7.5}$$

where:

$\hat{P}$ indicates precipitation
$\widehat{\text{PET}}$ refers to PET values read from the forcing data
the subscripts P and PET are the rescaled ones used in the model

Both R_P and R_{PET} are static model parameters to be calibrated. Additionally, the Shuffled Complex Evolution Algorithm was adopted in CREST v2.0 and therefore effectively reduced the calibration time.

Shen et al. (2015) noted that both CREST v1.5 and v2.0 gave significant underestimation of streamflow at the outlet of basins larger than 40,000 km² and a discontinuous flow pattern along the river network at very small scales. Such misrepresentation is caused by using the same procedure to accumulating water storage and streamflow. Taking cell C as our exemplified observing point, from Equation 7.3, the discharge of cell C is written in Equation 7.6:

$$\text{runoff}_C = \left[R_C(\text{out})_S + R_C(\text{out})_I \right] A_g(C) / \Delta t \tag{7.6}$$

From Equation 7.6, we can observe that the water from C to F during Δt is the sole contributor to runoff_C, while runoff_A and runoff_B that pass through cell C during the same period are not taken into account. We hence propose a fully distributed linear reservoir routing method that redefines the relationship between the runoff and the transported water by rewriting Equation 7.3 as

$$\text{runoff} = \frac{\left[R(\text{out})_S + R(\text{out})_I \right] A_g + \sum R(\text{via})_S A_g(\text{depart})_S + \sum R(\text{via})_I A_g(\text{depart})_I}{\Delta t} \tag{7.7}$$

where:
$R(\text{out})_S$ and A_g refer to the outgoing flow and grid area of a given cell, respectively
$R(\text{via})_{S,I}$ stand for the overland and interflow flow that pass through the given cell to any downstream receptor grid cells
$A_g(\text{depart})_{S,I}$ represent the grid areas of the donor cells of $R(\text{via})_{S,I}$

For comparison, the revised cell-to-cell routing is conceptualized in Figure 7.3 by colored arrows where all these three terms, that is, water from cell A to cell D, cell B to cell E, and cell C to cell F contribute to the runoff at cell C because it either sets off from or passed via cell C.

The original and revised routing methods have been tested by means of the hydrograph of outlets and spatial pattern along the river network in three basins characterized by different drainage area, location, climate condition, and topography in the updated CREST v2.1 by Shen et al. (2015). They are the Tar River in North Carolina, USA, Kankakee River in Illinois, USA, and Gan River in Jiangxi, China. Figure 7.4a–c gives the validation of simulated streamflow against the observed one.

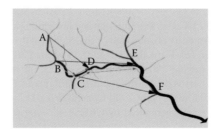

FIGURE 7.3 Routing scheme of the CREST v2.1.

CREST v2.1 catches the discharge patterns more precisely in all basins than CREST v2.0, with reduced underestimation. Comparing Figure 7.4b and d, the performances of CREST v2.1 at different spatial scales in the Kankakee River are almost identical, which again verifies the stability of the revised routing scheme.

In Figure 7.5a, it is clear that CREST v2.0 produces bumpy and discontinuous daily runoff values along the mainstream because of the previously mentioned water jumping described in Section 7.2, which is unnatural. In contrast, Figure 7.5b shows

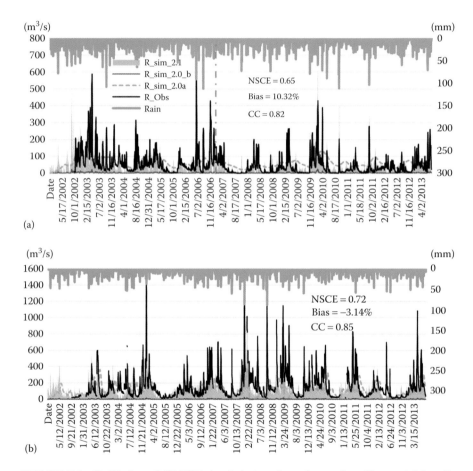

FIGURE 7.4 Validation of CREST simulation against observation. The model is calibrated and validated against daily observation of streamflow at the outlet of the three basins which are the (a) Tar River, (b), Kankakee River. In (a–c) are CREST runs at 1 km resolution while in (d) CREST runs at 250 m resolution. Besides the hydrographs produced by CREST v2.1 in the area chart style, hydrographs of calibrated CREST v2.0 has have also been plotted as a gray dash line (R_sim_2.0a) from (a) to (c). Using the same parameter set calibrated by v2.1, hydrographs generated by CREST v2.0 are plotted in a red solid line (R_sim_2.0b) from (a) to (c).   *(Continued)*

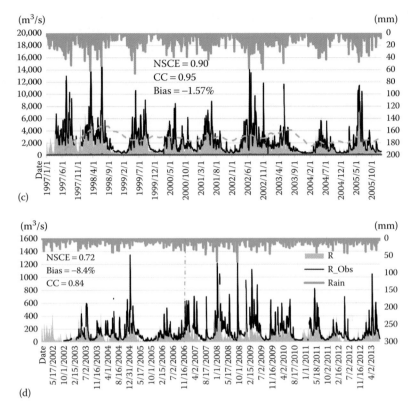

(c)

(d)

FIGURE 7.4 (Continued)   Validation of CREST simulation against observation. The model is calibrated and validated against daily observation of streamflow at the outlet of the three basins which are the (c) Gan River, and (d) Kankakee River. In (a–c) are CREST runs at 1 km resolution while in (d) CREST runs at 250 m resolution. Besides the hydrographs produced by CREST v2.1 in the area chart style, hydrographs of calibrated CREST v2.0 has have also been plotted as a gray dash line (R_sim_2.0a) from (a) to (c). Using the same parameter set calibrated by v2.1, hydrographs generated by CREST v2.0 are plotted in a red solid line (R_sim_2.0b) from (a) to (c).

that CREST v2.1 produces naturally continuous and basically monotonic daily runoff values from the upstream to the downstream as a result of taking the contribution from all the water that passes through the observing cell into account. From this point, we were confident to confirm that significant performance improvement has been observed over all three basins after using the revised routing scheme.

Nevertheless, both precipitation error and vegetation interception are time dynamic. Therefore, the adjustments in Equations 7.4 and 7.5 are insufficient physically and break water balance. It causes downgrade of the model performance over a longer validation period. The model is not applicable to cold regions because a snow process of energy balance is not represented within the model.

7.2.3 THE INCOMING VERSION OF CREST

To date, in the University of Connecticut (UCONN), supported by a 35-year flood frequency analysis project funded by Connecticut Institute for Resilience & Climate Adaptation, Shen and Anagnostou (2015) developed a new version of CREST equipped with a snow ablation process by replacing its previous simplified scheme of a land surface process coupled with water and energy balances. This new version explicitly computes vegetation's evapotranspiration (ET) based on their type, root, and growth. It integrates the snow accumulation and ablation model, SNOW17 (Anderson, 2006). Similar to (Andreadis et al., 2009), the new CREST divides the snow pack layer into two thermal layers in order to take the insulator effect of snow

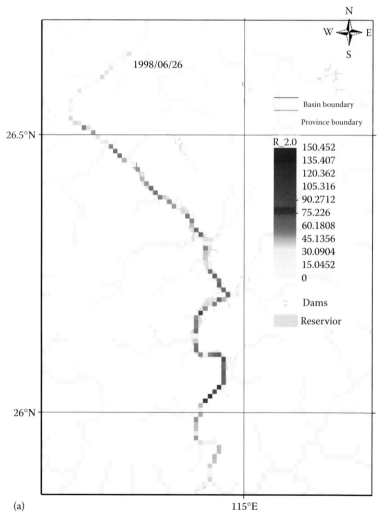

(a)   115°E

FIGURE 7.5   River response to flash flood event. Channel flow of the river network by (a) CREST v2.0.   *(Continued)*

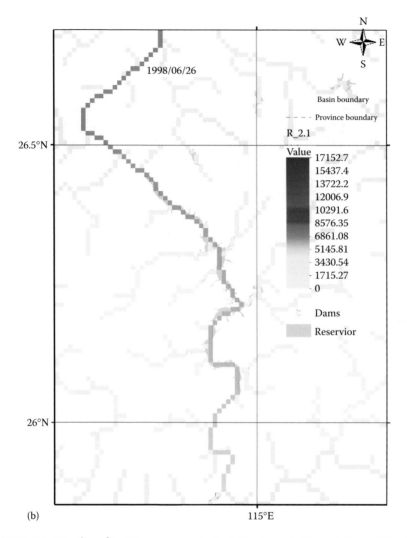

FIGURE 7.5 (Continued) River response to flash flood event. Channel flow of the river network by (b) CREST v2.1.

into consideration. A multilayer soil structure is implemented in the model to account for both thermal and moist variability vertically.

The new CREST was tested over 17 river basins trespassing Connecticut and exhibits perfect adaption in cold regions in terms of simulated ET, snow pack, and streamflow. Figure 7.6 shows its daily ET graph validation and error distribution over Connecticut basins. It should be noted that all parameters of the land surface module in this version is physically derived thus do not require calibration. Using the physically based land surface scheme in the new CREST, it is not necessary to adjust the precipitation ratio, resulting in possible water imbalance. Consequently,

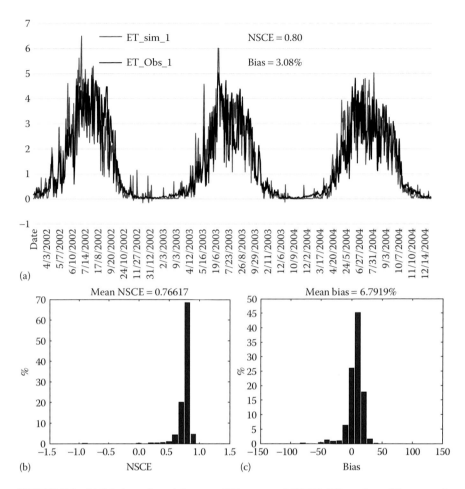

(a)

(b)

(c)

FIGURE 7.6 Validation of model output ET versus a MODIS ET product (Zhang et al., 2010) with (a) daily ET graph of one randomly selected cell of 8 km by 8 km. (b) NSCE distribution of all cells and (c) relative bias distribution of the simulation.

the simulation of streamflow even in the cold regions is more accurate. Figure 7.7 shows hydrographs of the Connecticut and Housatonic Rivers with less than 7% relative bias.

7.2.4 EFFICIENCY AND CONVERGENCE CONSIDERATIONS

In addition to the revised routing scheme in the new version, the computational efficiency of CREST v2.1 is further improved by the adoption of full vectorization (Van Der Walt et al., 2011) in its Matrix implementation. This computational style effectively avoids looping over grid cells and thus reducing runtime of any grid-based models significantly. In the latest version, closed energy balance requires to iteratively solve a nonlinear equation set over all grids by all time steps. Therefore, the computational load of the land surface process is considerable. To solve this

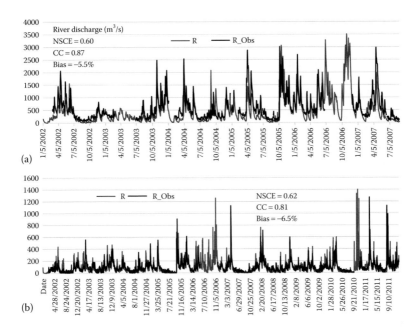

FIGURE 7.7 Streamflow validation versus USGS observation at (a) Thompsonville, Connecticut River, and (b) Stevenson, Housatonic River.

issue, the land surface module is implemented in a vectorized and parallel style. Basin cells are evenly loaded into different processors while within each processor, vectorized computation is carried out for all computational steps. In this way, the runtime of simulating the land surface process in the Connecticut River that consists of 260,000 grid cells at 500 m resolution of 1-year/hourly time span is reduced to approximately 1 h using 90 cores in UCONN's HORNET cluster. Compared with most other models in earth sciences, CREST is among the very few that utilize both vectorized and parallel computation and is thus among the topmost efficient global hydrologic models.

To reach the energy balance within the model, all thermal equations at all cells within the timeframe have to converge. The Variable Infiltration Capacity (VIC) model (Liang et al., 1994) employs the fixed point approach to solve only one equation of the nonlinear system each time. However, this method hardly guarantees the convergence. Judged by the example Stehekin basin downloadable from http://vic.readthedocs.org/en/vic.4.2.c/, the closed energy option is difficult to achieve. In practice, users of VIC tend to switch off the closed energy option and most of them even turn off the entire energy balance option for efficiency consideration. Either way, its energy budget is not balanced. CREST instead uses the Broyden nonlinear system solver to solve all equations in the nonlinear thermal equilibrium system simultaneously and effectively obtain convergence in all seasons and land covers. As a result, the users of the new CREST version do not have to choose one of the following options: closed energy, energy balance, or water balance modes, and thus can model with energy and mass balance simultaneously.

7.3 CONCLUSION

CREST has been under rapid development since its debut in 2010. At present, it has been successfully applied to basins of various drainage sizes and climate conditions in Africa, Asia, Europe, and America. It enjoyed streamflow simulation efficiency while maintaining hydrologic fundamentals in the beginning. The CREST version 2.1 proposed a new cell-to-cell routing continuum scheme and significantly improved the flash flood streamflow simulation. In its most recent version, both physics of the land surface process and the efficiency of the model are redeemed and the model therefore becomes more physically based and widely applicable.

REFERENCES

Anderson, E. A. (1976). A point energy and mass balance model of a snow cover. NOAA Technical Report NWS 19, Office of Hydrology, National Weather Service, Silver Spring, MD.

Andreadis, K. M., P. Storck, and D. P. Lettenmaier (2009), Modeling snow accumulation and ablation processes in forested environments. *Water Resour. Res.*, 45(5), W05429.

Lehner, B., K. Verdin, and A. Jarvis (2006), *HydroSHEDS Technical Documentation, Version 1.0*. World Wildlife Fund US, Washington, DC, pp. 1–27.

Liang, X., D. P. Lettenmaier, E. F. Wood, and S. J. Burges (1994), A simple hydrologically based model of land surface water and energy fluxes for general circulation models. *J. Geophys. Res.Atmos*, 99(D7), 14415–14428.

Ren-Jun, Z. (1992), The Xinanjiang model applied in China. *J. Hydrol.*, 135(1), 371–381.

Shen, X. and E. N. Anagnostou (2015), An improved distributed hydrologic simulation framework—Integrating energy/water balances with snow processes. *J. Hydrol.*, (in press).

Shen, X., Y. Hong, K. Zhang, and H. Zengchao (2015), Refine a distributed linear reservoir routing method to improve performance of the CREST model. *J. Hydrol. Eng.*, (in press).

Van Der Walt, S., S. C. Colbert, and G. Varoquaux (2011), The NumPy array: A structure for efficient numerical computation. *Comput. Sci. Eng.*, 13(2), 22–30.

Wang, J., Y. Hong, L. Li, J. J. Gourley, S. I. Khan, K. K. Yilmaz, R. F. Adler, F. S. Policelli, S. Habib, and D. Irwn (2011), The coupled routing and excess storage (CREST) distributed hydrological model. *Hydrolog. Sci. J.*, 56(1), 84–98.

Wood, E. F., J. K. Roundy, T. J. Troy, L. Van Beek, M. F. Bierkens, E. Blyth, A. de Roo, P. Döll, M. Ek, and J. Famiglietti (2011), Hyperresolution global land surface modeling: Meeting a grand challenge for monitoring Earth's terrestrial water. *Water Resour. Res.*, 47(5), W05301.

Wu, H., J. S. Kimball, N. Mantua, and J. Stanford (2011), Automated upscaling of river networks for macroscale hydrological modelling. *Water Resour. Res.*, 47(3), W03517.

Xue, X., Y. Hong, A. S. Limaye, J. J. Gourley, G. J. Huffman, S. I. Khan, C. Dorji, and S. Chen (2013), Statistical and hydrological evaluation of TRMM-based multi-satellite precipitation analysis over the Wangchu basin of Bhutan: Are the latest satellite precipitation products 3B42V7 ready for use in ungauged basins? *J. Hydrol.*, 499(30), 91–99.

Zaitarian, V. (1992), A land-surface hydrology parameterization with sub-grid variability for general circulation models. *J. Geophys. Res.*, 97(D3), 2717–2728.

Zhang, K., J. S. Kimball, R. R. Nemani, and S. W. Running (2010), A continuous satellite-derived global record of land surface evapotranspiration from 1983 to 2006. *Water Resour. Res.*, 46(9), W09522.

Zhao, R.-J., X. Liu, and V. Singh (1995), The Xinanjiang model. In *Computer Models of Watershed Hydrology*, Singh, V. P. (ed.), Water Resources Publications, CO, pp. 215–232.

8 Assimilation of Remotely Sensed Streamflow Data to Improve Flood Forecasting in Ungauged River Basin in Africa

Yu Zhang and Yang Hong

CONTENTS

8.1 INTRODUCTION

Flooding, which is considered as one of the most hazardous and frequently occurring disasters in both rural and urban areas, accounts for around one-third of all global geophysical hazards. Every year there are hundreds and thousands of flooding events around the world that cause a significant number of fatalities and economic loss (Adhikari et al., 2010; Hong et al., 2007a). With strong impacts from human activities and changing climate, it is reasonably anticipated that the flood risk will not decrease, but rather become more severe and frequent, thus threatening more regions around the world in the future (McCarthy, 2001). The increasing adverse worldwide impact from floods indicates that flooding is not only a regional or national-level issue, but is instead a global problem that greatly motivates a global flood monitoring and forecasting system coordinated among worldwide research institutions and government decision makers, and this is particularly essential for developing regions with sparsely hydrologic gauges or even without gauge observations. However, insufficient reliable observations have been a barrier in hydrologic predictions, especially in those developing regions.

The development of the remote sensing technique explores the possibility of hydrologic predictions for ungauged regions as well as for large-scale or even global-scale predictions by utilizing the remote sensing data for model setup, parameterization, forcing, and so on. The topography data such as the digital elevation model can be observed from the freely available Advanced Spaceborne Thermal Emission and Reflection Radiometer (Abrams, 2000) with a coverage from 83° N to 83° S at 30 m spatial resolution, and the shuttle radar and topography mission (Farr and Kobrick, 2000; Farr et al., 2007; Rabus et al., 2003; Van Zyl, 2001) with a coverage of 60° N to 56° S at 90 m spatial resolution from NASA can be applied to calculate the slope and extract the river net for model setup. Worldwide, both the advanced very high-resolution radiometer (Goward et al., 1985, 1991) and the MODerate resolution imaging spectroradiometer (MODIS, Pagano and Durham, 1993) provide global land cover information which estimates the hydrologic physical parameters, such as the saturated hydraulic conductivity and the maximum soil water capacity. Besides contributing to the hydrologic model setup and parameterization, remote sensing data also provide information for hydrologic forcing (e.g., precipitation) and observation (e.g., soil moisture, streamflow, and reservoir) which can be assimilated into the hydrologic system to improve the prediction skills. The global-wide precipitation can be estimated from multiple satellite missions and sensors such as TRMM—Tropical Rainfall Measurement Mission from NASA (Huffman et al., 2007, 2010), CMORPH—CPC morphing technique from the Climate Prediction Center at National Oceanic and Atmospheric Administration (NOAA, Joyce et al., 2004), PERSIANN—precipitation estimation from remotely sensed information using artificial neural networks that is operational at the University of Arizona (Hong et al., 2007b; Sorooshian et al., 2000), and so on. Soil moisture, another major hydrologic component that controls how much of the excess rainfall is infiltrated into the soil and how much of the excess rainfall is generated as the overland flow, can be retrieved from either active microwave sensors such as the Advanced Scatterometer from NOAA (Bartalis et al., 2007) or passive microwave sensors such as the Advanced Microwave Scanning Radiometer–Earth Observing System (AMSR-E; Kawanishi et al., 2003), from NASA and the Soil Moisture and Ocean Salinity (SMOS; Kerr et al., 2001) from European Space Agency (ESA), or from both active and passive such as Soil Moisture Active and Passive (SMAP; Entekhabi et al., 2010) from NASA. In addition to estimating physical parameters for the hydrologic model, MODIS can also be applied to estimate surface water such as lakes and reservoirs (Gao, 2015; Gao et al., 2012; Zhang et al., 2014a). Moreover, the AMSR-E and MODIS were recently used to estimate the streamflow at ungauged basins; this can be either directly applied to estimate the streamflow for flood detection (Brakenridge et al., 2007) or be utilized as the data assimilation source to update the internal states of the hydrologic model, thus improving the hydrologic predictability (Zhang et al., 2013, 2014b).

Though satellite remote sensing bridges the gap between ungauged regions and regions with sufficient in situ observations, and makes it feasible to model the hydrologic processes over the globe, it still has large uncertainties especially in regions with limited in situ observations for the development of the satellite retrieval algo-rithms. For example Yong et al. (2012) pointed out that TRMM has the key limita-tion in the underestimation of the precipitation in higher latitude and in the regions of intense convection over land. Ensemble data assimilation, which was firstly used in engineering and aerospace applications dating back to the 1960s, can further fuse reliable observations, no matter in situ or satellite remote sensing, into the dynamical models to improve the quality and accuracy of the estimates (Robinson et al., 1998). The ensemble Kalman filter (EnKF) has broadly been used in dynamic meteorol-ogy as well as numerical predictions (Anderson et al., 2005; Li et al., 2012; Wang, 2011; Wang and Bishop, 2003; Wang et al., 2007), and was proved for the potential of enhancing forecasting skills. Meanwhile, ensemble data assimilation has increas-ingly been applied in hydrology modeling for assimilating soil moisture (Aubert et al., 2003; Chen et al., 2011; Crow and Ryu, 2009; Crow et al., 2005; Gao et al., 2007; Pauwels et al., 2002) or streamflow information (Aubert et al., 2003; Chen et al., 2011; Clark et al., 2008; Pauwels and De Lannoy, 2006), thus increasing the predictability of hydrologic events. However, the degree of improvement in forecast skill is contingent on the model structure, model parameterizations, and the quality of the observed data that are assimilated into the model. For example, in the study by Chen et al. (2011), the failure of improving streamflow prediction via assimilating soil moisture into the Surface Water & Ocean Topography (SWOT) model was due to the deficiency of the model structure.

Building on Zhang et al. (2013, 2014b), this chapter introduces the applicability of utilizing satellite remote-sensed TRMM precipitation as forcing and assimilating spaceborne AMSR-E streamflow signals via Ensemble Square Root Filter (EnSRF) to improve the streamflow prediction in a sparsely gauged basin—Okavango, Africa. The application of AMSR-E streamflow signals for hydrologic perspective, which are impacted by factors including width of the river, channel geometry, water tempera-ture relative to land, and measurement pixel resolution, and this will be introduced in detail in Section 8.2. Section 8.3 talks about the study region, data used, methodol-ogy and results, followed by a summary and future perspective in Section 8.4.

8.2   STREAMFLOW ESTIMATED FROM REMOTE SENSING

Traditional in situ streamflow observations, especially over most of the developing countries and emerging regions, still require human labors, and the data information bears time delay, though some of them can be automatically gathered and transmit-ted to a central data center and processed over those developed countries or regions. Satellite remote sensing streamflow observations, with more comprehensive cover-age over the globe, are much more efficient and involve less human labor, and also, are capable of providing near-real-time streamflow observations. The Global Flood Detection System under Global Disaster Alert and Coordination System (GDACS-GFDS, http://www.gdacs.org/flooddetection/overview.aspx) and Dartmouth Flood Observatory (DFO, http://www.dartmouth.edu/~floods/index.html) use near-real-time satellite remote sensing, which is basically based on a passive microwave sensor, AMSR-E (June 1, 2002, to October 4, 2011), together with a TRMM TMI (TRMM microwave imager) sensor (January 1, 1998, to present), to measure the surface

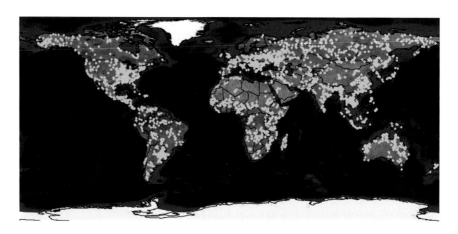

FIGURE 8.1 Spatial distribution of the satellite remote sensing streamflow observation gauges over the globe from Dartmouth Flood Observatory (DFO). (Data from http://www .dartmouth.edu/~floods/AMSR-E%20Gaging%20Reaches/Summary3.htm.)

brightness temperature which can be converted into streamflow using the algorithm developed by Brakenridge et al. (2007) over 2000 selected streamflow sites to monitor floods over the globe (Figure 8.1). This is realized by measuring the brightness temperature at 36.5 GHz, descending orbit with horizontal polarization, which responds to surface wetness and, thus, flooding (Brakenridge et al., 2007). In addition to Brakenridge et al. (2007), previous studies such as Salvia et al. (2011) and Temimi et al. (2007, 2011) also explored the potential of using AMSR-E brightness temperature in estimating discharge, thus, monitoring flood.

AMSR-E measures horizontally and vertically polarized brightness temperature at six different frequencies from 6.9 to 89.0 GHz, among which 36.5 GHz at horizontal polarization is approved to be the most sensitive and appropriate to measure the variability of river discharge (Brakenridge et al., 2007). The M/C ratio signal is defined as the ratio of the brightness temperature of the measurement (M) wet pixel (usually over the surface of a river, Figure 8.2) over the brightness temperature of the calibration (C) dry pixel (usually over the land near the wet pixel, Figure 8.2):

$$\text{M/C Ratio} = \frac{Tb_m}{Tb_c} \tag{8.1}$$

The M/C ratio signal data can be downloaded from GDACS-GFDS. Some technical details for selecting the wet and dry pixels should be noted: (1) The calibrated dry pixel C should be close to the measurement wet pixel M so that changes such as vegetation, soil texture, and so on at those locations are more likely to be correlated. In other words, those two locations are more likely under similar conditions (e.g., vegetation and soil texture); meanwhile, the short distance between the dry pixel C and the wet pixel M can also guarantee that the measurements acquired by AMSR-E are effectively contemporaneous; (2) M should be selected to have the largest change in water surface area and relatively high sensitivity; (3) C should be selected to be close to M but is located far

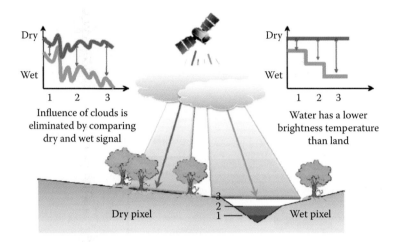

FIGURE 8.2 Illustration of the M/C ratio methodology of the Global Flood Detection System (GFDS). (From Kugler, Z. and T. De Groeve, *The Global Flood Detection System*, Office for Official Publications of the European Communities, Luxembourg, Europe, pp. 43, 2007 [Figure 3].)

enough to not be affected by flood inundation; (4) MODIS is applied to assist selecting M where flow area expansions occur (Brakenridge et al., 2007; Kugler and De Groeve, 2007). In contrast with the visible or infrared sensors which have limited penetration through clouds, the major merit of the AMSR-E passive microwave sensor onboard the NASA EOS Aqua satellite is not restricted by cloud cover and can provide data for daily flood monitoring over the globe. However, because radiation during night-time is more stable than during the day, only the H-polarization of 36 GHz band of descending (nightly) swaths with a footprint size of approximately 8 × 12 km is used.

The radiance signal is highly correlated with streamflow observation as shown in Figure 8.3 in Zhang et al. (2014b). Following Brakenridge et al. (2007), the in situ observed streamflow was used to calibrate the orbital gauging measurements (the radiance signal) into streamflow with a unit of m³/s for a variety of polynomial regressions (Zhang et al., 2013). The nonlinear quadratic polynomial regression, which has the best goodness-of-fit among all regressions tested in Zhang et al. (2013), was selected to convert the satellite radiance signal into streamflow. The AMSR-E observation (Figure 8.4), which is the streamflow converted from the satellite-observed radiance signal using the nonlinear quadratic polynomial regression in Zhang et al. (2013), is lined well with the in situ observation, especially during the peak flows. However, AMSR-E sensors are not sensitive to low flows because the converted streamflow from AMSR-E radiance signal shows overestimations during the low flows. The limitation of this approach in estimating streamflow using satellite remote sensing is that it is only applicable to medium- to large-sized basins as factors such as the width of the river, channel geometry, water temperature relative to land, and measurement pixel resolution would impact the streamflow conversion results. The accuracy of the AMSR-E signals for basins with less than 50,000 km² drainage areas needs further investigation (Khan et al., 2012).

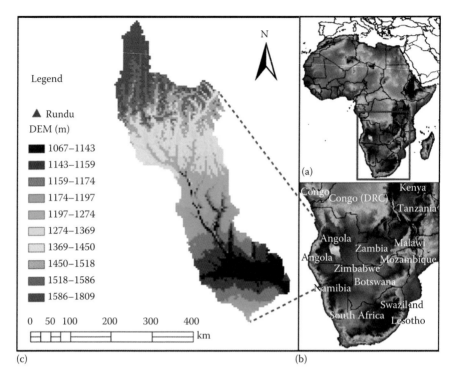

FIGURE 8.3 Location of the Okavango River basin. (From Zhang, Y. et al., *IEEE J. Select. Top. Appl. Earth Observ. Remote Sens.,* 6, 2375–2390, 2013 [Figure 1].)

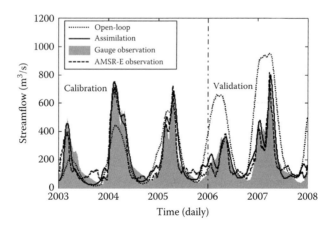

FIGURE 8.4 Comparison of streamflow predictions before (open loop) and after (assimilation) data assimilation against gauge observation and AMSR-E observation.

8.3 ASSIMILATION OF REMOTELY SENSED INFORMATION FOR IMPROVING HYDROLOGIC PREDICTIONS

8.3.1 STUDY REGION

The Okavango River, with a drainage area around 413,000 km^2, runs through central Angola and flows through Namibia and Botswana (Figure 8.3). The precipitation of the Okavango River basin decreases from the north upper stream subtropical climate region (~1300 mm/year) to the south downstream Kalahari Desert (~450 mm/year; Hughes et al., 2006; Milzow et al., 2009b). Okavango River basin is one of the most important economic and water resources in southern Africa; thus, a number of studies have been carried out over this basin for hydrologic analysis enabled by models, remote sensing for understanding the water resources availability, flooding patterns, and the interactions between hydrologic cycle and climate change (Andersson et al., 2006; Dinçer et al., 1987; Hughes et al., 2006, 2011; McCarthy et al., 1998, 2003; Milzow et al., 2009a; Wilk et al., 2006; Wolski et al., 2006). Figure 8.3 shows the location of this river basin. The Rundu gauge station is the outlet with both the in situ streamflow and the remote sensing discharge estimates (i.e., AMSR-E M/C ratio signal) available.

8.3.2 DATA USED

The real-time 3B42-level product from TRMM multisatellite precipitation analysis is the near-real-time observation that was calibrated by the TRMM-combined instrument dataset. TRMM 3B42 RT, with quasi-global coverage from 50°N to 50°S latitude at 0.25° (approximate to 25 km in the tropical area) and 3 hourly, is processed into daily accumulation as well as basin-average and applied as the forcing data to drive the hydrologic model. Potential evapotranspiration comes from the Famine Early Warning Systems Network (http://igskmncnwb015.cr.usgs.gov/Global/) with a spatial resolution of 0.25°, and is likewise processed into daily and basin-average as additional force to the model.

The AMSR-E signal converted streamflow was applied to calibrate the model and then was also assimilated into the model to correct the model internal states for each assimilation cycle, without in situ observations involved. The in situ observation here is only used for validation.

8.3.3 METHODOLOGY

First, the AMSR-E/TMI passive microwave streamflow signals are converted into actual streamflow domain with the unit of m^3/s by adopting the algorithm from Brakenridge et al. (2007) and Zhang et al. (2013); then it was used to calibrate the hydrologic model (HyMOD; please refer to Zhang et al., 2013, for model details) for 2003–2005; last, the model was coupled with EnSRF to account for uncertainty in both forcing data and model initial conditions and, thus, improve the accuracy of hydrologic simulation by assimilating the signal converted streamflow, in comparison to the in situ streamflow observations.

EnSRF, which is a sequential data assimilation technique and also referred to as EnKF without perturbing observations, is applied to assimilate different AMSR-E signal converted streamflow into HyMOD. Compared to the traditional EnKF which requires perturbing both forcing data and observations, EnSRF only perturbs the forcing, which saves computation especially for large-scale and high-resolution assimilation. Whitaker and Hamill (2002) found out that there is no additional computational cost by EnSRF relative to EnKF, and EnSRF performs more accurately than EnKF for the same ensemble size. However, it is still an open research topic to compare the accuracy and efficiency of different sequential data assimilation approaches (e.g., EnKF and EnSRF).

Let X^b denote the background model forecast, which is also called the first guess in data assimilation ($n \times 1$ dimension and n is the total number of ensemble member); let y denote the observation ($p \times 1$ dimension and p is the number of observations), which is the AMSR-E signal converted streamflow in this study; let H denote the observation operator that converts the states in the model into observation space ($p \times n$ dimension); the estimate of the analysis X^a (which is updated after EnSRF data assimilation with $n \times 1$ dimension) can be calculated as

$$X^a = X^b + \widehat{K}(y - H(X^b)). \tag{8.2}$$

In Equation 8.2, $\widehat{K}$ denotes the traditional Kalman gain. The ensemble X^b is denoted as

$$X^b = \left(x_1^b, x_2^b, \ldots, x_n^b \right). \tag{8.3}$$

The ensemble mean is then defined as

$$\overline{X^b} = \frac{1}{n} \sum_{i=1}^{n} x_i^b. \tag{8.4}$$

And the perturbation from the mean for the ith member is defined as

$$x_i'^b = x_i^b - \overline{x^b}. \tag{8.5}$$

Then X'^b is defined as a matrix formed from the ensemble of perturbations:

$$X'^b = \left(x_1'^b, x_2'^b, \ldots, x_n'^b \right). \tag{8.6}$$

The background error covariance is calculated as

$$\widehat{P}^b = \frac{1}{n-1} X'^b (X'^b)^T. \tag{8.7}$$

In practice, we do not calculate $\widehat{P}^b$, but rather calculate $\widehat{P}^b H^T$ and $H\widehat{P}^b H^T$ via the following equations:

$$\widehat{P}^b H^T = \frac{1}{m-1} \sum_{i=1}^{m} \left(X_i^b - \overline{X}^b \right) \left(H\left(X_i^b - \overline{H\left(X^b\right)} \right) \right)^T, \tag{8.8}$$

$$HP\hat{}^bH^T = \frac{1}{m-1}\sum_{i=1}^{m}\left(H\left(X_i^b\right)-H\left(\overline{X}^b\right)\right)\left(H\left(X_i^b-\overline{H\left(X^b\right)}\right)\right)^T. \qquad (8.9)$$

Here, m is the ensemble size. Then the traditional Kalman gain $\hat{K}$ can be calculated by Equation 8.10,

$$\hat{K} = \hat{P}^b H^T \left(H\hat{P}^b H^T + R \right)^{-1}. \qquad (8.10)$$

R is the observation error covariance with a dimension of $p \times p$.

The reduced Kalman gain $\tilde{K}$, which is used to update the deviation from the ensemble mean, is estimated by the following equation:

$$\tilde{K} = \left(1 + \sqrt{\frac{R}{H\hat{P}^b H^T + R}} \right)^{-1} \hat{K}. \qquad (8.11)$$

The ensemble mean can be updated by

$$\overline{X}_i^a = \overline{X}_i^b + \hat{K}\left(y - H\left(\overline{X}_i^b \right) \right). \qquad (8.12)$$

The perturbation (deviation of ensemble mean) can be updated by

$$X_i'^a = X_i'^b - \hat{K}H\left(X_i'^b \right). \qquad (8.13)$$

The final analysis is

$$X_i^a = \overline{X}_i^a + X_i'^a. \qquad (8.14)$$

The forcing data (precipitation in this study) is perturbed using Equation 8.15:

$$P_i = P + \varepsilon_i, \qquad (8.15)$$

where ε_i is a random noise factor drawn from a Gaussian distribution,

$$\varepsilon_i \sim N\left(0, R \right). \qquad (8.16)$$

As discussed in Zhang et al. (2013), no spatial error correlation is computed in the generation of the precipitation perturbation due to the feature of the lumped model applied in this study. No temporal error correlations are counted as well. At each time step, an independent precipitation error is generated by Gaussian distribution (Equations 8.15 and 8.16) and added up to the original basin-average precipitation.

8.3.4 STREAMFLOW ASSIMILATION AND PREDICTIONS

Zhang et al. (2013) performed three experiments and demonstrated the apparent capability to use the AMSR-E signal converted streamflow to calibrate a hydrologic model because it is capable to achieve nearly the same degree of high skill as using in situ gauge observations. The calibration (2003–2005) and validation (2006–2007) results are shown in Figure 8.4, which demonstrates the great potential of the AMSR-E streamflow to be used in tandem with remotely sensed precipitation data and PET for providing real-time flood detection and forecasts in sparsely gauged or ungauged basins.

EnSRF is then used to assimilate the AMSR-E streamflow observations into the hydrologic model and to estimate all the internal states, thus improving the streamflow predictions. Based on the sensitivity test from Zhang et al. (2013), an ensemble size of 50, a spread of precipitation of 150%, and an observation error (i.e., AMSR-E streamflow in this study) of 10% were assumed. The precipitation was perturbed by adding Gaussian white noise through multiplying the basin-averaged TRMM RT daily data by a multiplier of which the mean is 1.0 and the standard deviation is 150%.

- Impact of data assimilation during the calibration period.
  Compared to streamflow from "Open Loop" before data assimilation (dotted line), the "Assimilation" streamflow ensemble mean after data assimilation (solid line) is much closer to the ground truth—gauge observations (gray shaded area). This result reflects the effectiveness of the EnSRF.
- Impact of data assimilation during the validation period.
  During the validation period of 2006–2007, the modeling performance without streamflow assimilation has deteriorated at a significant level compared to the calibration period. The simplicity of the model structure, as well as the inter-annual uncertainties in the remotely sensed TRMM RT precipitation, might have contributed to this deterioration. However, the application of EnSRF for assimilating the AMSR-E improves the one-day streamflow prediction. The modeling results have been remarkably enhanced for the "Assimilation" component compared to the "Open Loop" during the validation period. And this result clearly highlights the potential of using the remote sensing data as a proxy for streamflow with application for flood early warning in sparsely gauged or ungauged basins. The results above demonstrate that even using a simple hydrologic model, when coupled with the EnSRF data assimilation approach, together with large perturbations of precipitation to compensate for the model structural deficiencies, a satisfactory modeling performance can be produced for streamflow forecasting.

8.4   SUMMARY AND FUTURE PROSPECTIVE

Though data scarcity still remains a big challenge in understanding the hydrologic system, remote sensing data bears the mission in providing a promising perspective on advancements in bridging the gap between ungauged or sparsely gauged regions and those regions with sufficient in situ observations. In addition, data assimilation techniques, together with hydrologic models or land surface models, can incorporate the uncertainties from both the forcing data and model internal states, and also have the potential to enhance the hydrologic modeling performance. To summarize, this study coupled the EnSRF with a widely used conceptual rainfall–runoff model to assimilate streamflow data from satellite remote sensing and update all the internal states in the model at daily steps, thus providing one-day streamflow predictions. The following key conclusions are reached in this chapter:

- AMSR-E streamflow signals, with high correlation to in situ observations, can be successfully used to estimate streamflow, which is highly consistent with the in situ observation. The signal converted streamflow particularly matches well with the observation over peak flows due to the high sensitivity of AMSR-E signals to land surface wetness.
- The hydrologic modeling performance is contingent on the uncertainties in the forcing data, parameterization, internal states, and model structure. The general poor performance of "Open Loop" in this study can be attributed to the weakness of the traditional calibration techniques that are normally constrained or limited from the inaccuracy of input remote sensing precipitation data and the simplification of the model structure. Data assimilation can account for both the uncertainties in the input data and the model structure by updating the internal model states; so, it is a promising tool in improving hydrologic modeling performance with reliable observations, especially for the applications of real-time forecasts for decision makers.
- By assimilating the AMSR-E signal converted to streamflow into the hydrologic model by EnSRF, the difference between the streamflow simulation and observation can be reduced, and this is particularly prominent during the period without model calibration. This demonstrates the effectiveness and efficiency of EnSRF in improving modeling performance by assimilating reliable AMSR-E signal converted streamflow data. In this study, only the remote sensing streamflow signals were used first to calibrate the model and then assimilated into the model without in situ streamflow data, thus demonstrating the potential usefulness of the AMSR-E signal data to benchmark and improve hydrologic predictions in ungauged or under-gauged basins. And this technique can be transferred to global application in theory, but with a basin size requirement ($>50,000$ km^2).
- Previous studies on hydrologic data assimilation commonly assimilate in situ observations such as soil moisture (Aubert et al., 2003; Chen et al., 2011) or streamflow (Aubert et al., 2003; Clark et al., 2008; Pauwels and De Lannoy, 2006). Somehow, for global wise or regions without sufficient in situ observations, satellite remote sensing can provide assimilation data sources such as soil moisture to improve the streamflow prediction from hydrologic modeling (Crow and Ryu, 2009; Crow et al., 2005; Gao et al., 2007; Pauwels et al., 2002). To the authors' knowledge, there is very limited assimilating remote sensing streamflow information for hydrologic predictions. Back to 2001, Wagener et al. (2001) mentioned that the river discharge cannot be directly measured by satellite sensors. However, with the development of remote sensing techniques, the passive microwave sensors—AMSR-E together with TRMM TMI—have been recently used to detect river

discharge changes, and that information can be converted into streamflow by using the algorithm mentioned in Brakenridge et al. (2007). Recent studies about the assimilation of the AMSR-E/TMI streamflow information into hydrologic models include Zhang et al. (2013, 2014b). Zhang et al. (2013) is the "first attempt" to exploit and demonstrate the applicability of assimilating spaceborne passive microwave streamflow signals to improve flood prediction in the sparsely gauged Okavango River basin in Africa. Compared to the closest previous publication (Khan et al., 2012) which has also investigated the applicability of the AMSR-E signals in hydrologic modeling in the same research region, Zhang et al. (2013) used a simple yet robust model and conducted competitive results by the assimilation of satellite remote sensing streamflow information via EnSRF. When combined with the EnSRF data assimilation approach, even the simple-structured HyMOD is capable to achieve similar results compared to a complex, distributed CREST hydrologic model applied in Khan et al. (2012).

To summarize, Zhang et al. (2013, 2014b) firstly exploited and demonstrated the applicability of assimilating spaceborne AMSR-E/TMI streamflow signals to improve flood prediction over the Okavango River basin, in actual streamflow domain and probability domain, respectively. Building on these two studies, this chapter shows that opportunities and challenges exist for an integrated application of a suite of satellite data to flood prediction by careful fusing satellite remote sensing and in situ observations and further effective assimilation of information into hydrologic models. Given the global availability of satellite-based precipitation (e.g., TRMM and GPM—Global Precipitation Mission) and AMSR-E/TMI signal information in near real time, we argue that this work will also contribute to the decadal initiative of prediction in ungauged basins: a paradigm shift in the streamflow prediction methods away from traditional methods reliant on statistical analysis and calibrated models, and towards new techniques and new kind of observations, particularly imperative for the vast ungauged or sparsely gauged basins around the world. More promising, data assimilation of remote sensing information for improving hydrologic prediction can be increasingly appreciated and supported by multiple satellite missions: the current TRMM and GPM together with the current SMAP, and future SWOT (to be launched in 2020 and is introduced in details in the last section of Chapter 4), together with more physical processed hydrologic or land surface models.

Nevertheless, it would be cautious to apply this approach to smaller basins with a basin area of less than 50,000 km^2. And the AMSR-E/TMI streamflow signal record, which only goes back to year 2002, limits its application to look at the long-term hydrologic shift under climate change. But the record of 10+ years' streamflow information should be capable to carry out the hydrologic model calibration for all the basins (with sizes larger than 50,000 km^2) where the AMSR-E record is available over the globe, thus providing real-time hydrologic prediction at multiple scales from days to seasons, for example, by implementing numerical weather forecast (e.g., Global Forecasting System) and seasonal forecast (e.g., Climate Forecast System version 2).

REFERENCES

Adhikari, P., Y. Hong, K. R. Douglas, D. B. Kirschbaum, J. Gourley, R. Adler, and G. R. Brakenridge, 2010: A digitized global flood inventory (1998–2008): Compilation and preliminary results. *Natural Hazards*, **55**, 405–422.

Anderson, J. L., B. Wyman, S. Zhang, and T. Hoar, 2005: Assimilation of surface pressure observations using an ensemble filter in an idealized global atmospheric prediction system. *Journal of the atmospheric sciences*, **62**, 2925–2938.

Andersson, L. and Coauthors, 2006: Impact of climate change and development scenarios on flow patterns in the Okavango River. *Journal of Hydrology*, **331**, 43–57.

Abrams, M. 2000: The advanced spaceborne thermal emission and reflection radiometer (ASTER): Data products for the high spatial resolution imager on NASA's Terra platform. *International Journal of Remote sensing*, **21**(5), 847–859.

Aster, N., 2006: Advanced Spaceborne Thermal Emission and Reflection Radiometer.

Aubert, D., C. Loumagne and L. Oudin, 2003: Sequential assimilation of soil moisture and streamflow data in a conceptual rainfall–runoff model. *Journal of Hydrology*, **280**, 145–161.

Bartalis, Z. and Coauthors, 2007: Initial soil moisture retrievals from the METOP-A advanced scatterometer (ASCAT). *Geophysical Research Letters*, **34**, L20401.

Brakenridge, G. R., S. V. Nghiem, E. Anderson, and R. Mic, 2007: Orbital microwave measurement of river discharge and ice status. *Water Resources Research*, **43**, W04405.

Chen, F., W. T. Crow, P. J. Starks, and D. N. Moriasi, 2011: Improving hydrologic predictions of a catchment model via assimilation of surface soil moisture. *Advances in Water Resources*, **34**, 526–536.

Clark, M. P. and Coauthors, 2008: Hydrological data assimilation with the ensemble Kalman filter: Use of streamflow observations to update states in a distributed hydrological model. *Advances in water resources*, **31**, 1309–1324.

Crow, W. T. and D. Ryu, 2009: A new data assimilation approach for improving runoff prediction using remotely-sensed soil moisture retrievals. *Hydrology and Earth System Sciences*, **13**, 1–16.

Crow, W. T., R. Bindlish, and T. J. Jackson, 2005: The added value of spaceborne passive microwave soil moisture retrievals for forecasting rainfall-runoff partitioning. *Geophysical Research Letters*, **32**, L18401.

Dartmouth Flood Observatory (DFO). http://www.dartmouth.edu/~floods/AMSR-E%20Gaging%20Reaches/Summary3.htm.

Dinçer, T., S. Child, and B. Khupe, 1987: A simple mathematical model of a complex hydrologic system—Okavango Swamp, Botswana. *Journal of Hydrology*, **93**, 41–65.

Entekhabi, D. and Coauthors, 2010: The soil moisture active passive (SMAP) mission. *Proceedings of the IEEE*, **98**, 704–716.

Farr, T. G. and M. Kobrick, 2000: Shuttle radar topography mission produces a wealth of data. *Eos, Transactions American Geophysical Union*, **81**, 583–585.

Farr, T. G. and Coauthors, 2007: The shuttle radar topography mission. *Reviews of geophysics*, **45**(2).

Gao, H., 2015: Satellite remote sensing of large lakes and reservoirs: from elevation and area to storage. *Wiley Interdisciplinary Reviews: Water*, **2**, 147–157.

Gao, H., C. Birkett, and D. P. Lettenmaier, 2012: Global monitoring of large reservoir storage from satellite remote sensing. *Water Resources Research*, **48**, W09504.

Gao, H., E. F. Wood, M. Drusch, and M. F. McCabe, 2007: Copula-derived observation operators for assimilating TMI and AMSR-E retrieved soil moisture into land surface models. *Journal of Hydrometeorology*, **8**, 413–429.

Goward, S. N., B. Markham, D. G. Dye, W. Dulaney, and J. Yang, 1991: Normalized difference vegetation index measurements from the advanced very high resolution radiometer. *Remote sensing of environment*, **35**, 257–277.

Goward, S. N., C. J. Tucker, and D. G. Dye, 1985: North American vegetation patterns observed with the NOAA-7 advanced very high resolution radiometer. *Vegetatio*, **64**, 3–14.

Hong, Y., R. F. Adler, A. Negri, and G. J. Huffman, 2007a: Flood and landslide applications of near real-time satellite rainfall products. *Natural Hazards*, **43**, 285–294.

Hong, Y., D. Gochis, J.-T. Cheng, K.-L. Hsu, and S. Sorooshian, 2007b: Evaluation of PERSIANN-CCS rainfall measurement using the NAME event rain gauge network. *Journal of Hydrometeorology*, **8**, 469–482.

Huffman, G. J., R. F. Adler, D. T. Bolvin, and E. J. Nelkin, 2010: The TRMM multi-satellite precipitation analysis (TMPA). In *Satellite rainfall applications for surface hydrology*, Hossain, F. and M. Gebremichael (eds.), Springer, the Netherlands, pp. 3–22.

Huffman, G. J. and Coauthors, 2007: The TRMM multisatellite precipitation analysis (TMPA): Quasi-global, multiyear, combined-sensor precipitation estimates at fine scales. *Journal of Hydrometeorology*, **8**, 38–55.

Hughes, D. A., D. G. Kingston, and M. C. Todd, 2011: Uncertainty in water resources availability in the Okavango River basin as a result of climate change. *Hydrology and Earth System Sciences*, **15**, 931–941.

Hughes, D. A., L. Andersson, J. Wilk, and H. H. G. Savenije, 2006: Regional calibration of the Pitman model for the Okavango River. *Journal of Hydrology*, **331**, 30–42.

Joyce, R. J., J. E. Janowiak, P. A. Arkin, and P. Xie, 2004: CMORPH: A method that produces global precipitation estimates from passive microwave and infrared data at high spatial and temporal resolution. *Journal of Hydrometeorology*, **5**, 487–503.

Kawanishi, T. and Coauthors, 2003: The advanced microwave scanning radiometer for the earth observing system (AMSR-E), NASDA's contribution to the EOS for global energy and water cycle studies. *IEEE Transactions on Geoscience and Remote Sensing*, **41**, 184–194.

Kerr, Y. H., P. Waldteufel, J.-P. Wigneron, J. -M. Martinuzzi, J. Font, and M. Berger, 2001: Soil moisture retrieval from space: The soil moisture and ocean salinity (SMOS) mission. *IEEE Transactions on Geoscience and Remote Sensing*, **39**, 1729–1735.

Khan, S. and Coauthors, 2012: Microwave satellite data for hydrologic modeling in ungauged basins. *IEEE Geoscience and Remote Sensing Letters*, **9**, 663–667.

Kugler, Z. and T. De Groeve, 2007: *The Global Flood Detection System*. Office for Official Publications of the European Communities, Luxembourg, Europe, pp. 43.

Li, Y., X. Wang, and M. Xue, 2012: Assimilation of radar radial velocity data with the WRF hybrid ensemble–3DVAR system for the prediction of Hurricane Ike (2008). *Monthly Weather Review*, **140**, 3507–3524.

McCarthy, J. J., 2001: *Climate change 2001: Impacts, adaptation, and vulnerability: Contribution of Working Group II to the third assessment report of the Intergovernmental Panel on Climate Change*. Cambridge University Press, Cambridge.

McCarthy, J. M., T. Gumbricht, T. McCarthy, P. Frost, K. Wessels, and F. Seidel, 2003: Flooding patterns of the Okavango wetland in Botswana between 1972 and 2000. *AMBIO: A Journal of the Human Environment*, **32**, 453–457.

McCarthy, T. S., A. Bloem, and P. A. Larkin, 1998: Observations on the hydrology and geohydrology of the Okavango Delta. *South African Journal of Geology*, **101**, 101–117.

Milzow, C., L. Kgotlhang, P. Bauer-Gottwein, P. Meier, and W. Kinzelbach, 2009b: Regional review: the hydrology of the Okavango Delta, Botswana—processes, data, and modelling. *Hydrogeology Journal*, **17**, 1297–1328.

Milzow, C., L. Kgotlhang, W. Kinzelbach, P. Meier, and P. Bauer-Gottwein, 2009a: The role of remote sensing in hydrological modelling of the Okavango Delta, Botswana. *Journal of environmental management*, **90**, 2252–2260.

Pagano, T. S. and R. M. Durham, 1993: Moderate resolution imaging spectroradiometer (MODIS). In *Optical Engineering and Photonics in Aerospace Sensing*, International Society for Optics and Photonics, Orlando, FL, pp. 2–17.

Pauwels, V., R. Hoeben, N. E. C. Verhoest, F. P. De Troch, and P. A. Troch, 2002: Improvement of TOPLATS-based discharge predictions through assimilation of ERS-based remotely sensed soil moisture values. *Hydrological Processes*, **16**, 995–1013.

Pauwels, V. R. N. and G. J. M. De Lannoy, 2006: Improvement of modeled soil wetness conditions and turbulent fluxes through the assimilation of observed discharge. *Journal of Hydrometeorology*, **7**, 458–477.

Rabus, B., M. Eineder, A. Roth, and R. Bamler, 2003: The shuttle radar topography mission—a new class of digital elevation models acquired by spaceborne radar. *ISPRS Journal of Photogrammetry and Remote Sensing*, **57**, 241–262.

Robinson, A. R., P. F. J. Lermusiaux, and N. Q. Sloan, 1998: Data assimilation. *The sea*, **10**, 541–594.

Salvia, M. and Coauthors, 2011: Estimating flooded area and mean water level using active and passive microwaves: the example of Paraná River Delta floodplain. *Hydrology and Earth System Sciences*, **15**, 2679–2692.

Sorooshian, S., K.-L. Hsu, X. Gao, H. V. Gupta, B. Imam, and D. Braithwaite, 2000: Evaluation of PERSIANN system satellite-based estimates of tropical rainfall. *Bulletin of the American Meteorological Society*, **81**, 2035–2046.

Temimi, M., R. Leconte, F. Brissette, and N. Chaouch, 2007: Flood and soil wetness monitoring over the Mackenzie River Basin using AMSR-E 37GHz brightness temperature. *Journal of Hydrology*, **333**, 317–328.

Temimi, M., T. Lacava, T. Lakhankar, V. Tramutoli, H. Ghedira, R. Ata, and R. Khanbilvardi, 2011: A multi-temporal analysis of AMSR-E data for flood and discharge monitoring during the 2008 flood in Iowa. *Hydrological Processes*, **25**, 2623–2634.

Van Zyl, J. J., 2001: The shuttle radar topography mission (SRTM): A breakthrough in remote sensing of topography. *Acta Astronautica*, **48**, 559–565.

Wagener, T., D. P. Boyle, M. J. Lees, H. S. Wheater, H. V. Gupta, and S. Sorooshian, 2001: A framework for development and application of hydrological models. *Hydrology and Earth System Sciences Discussions*, **5**, 13–26.

Wang, X., 2011: Application of the WRF hybrid ETKF-3DVAR data assimilation system for hurricane track forecasts. *Weather and Forecasting*, **26**, 868–884.

Wang, X. and C. H. Bishop, 2003: A comparison of breeding and ensemble transform Kalman filter ensemble forecast schemes. *Journal of the atmospheric sciences*, **60**, 1140–1158.

Wang, X., C. Snyder, and T. M. Hamill, 2007: On the theoretical equivalence of differently proposed ensemble-3DVAR hybrid analysis schemes. *Monthly weather review*, **135**, 222–227.

Whitaker, J. S. and T. M. Hamill, 2002: Ensemble data assimilation without perturbed observations. *Monthly Weather Review*, **130**, 1913–1924.

Wilk, J. and Coauthors, 2006: Estimating rainfall and water balance over the Okavango River Basin for hydrological applications. *Journal of Hydrology*, **331**, 18–29.

Wolski, P., H. H. G. Savenije, M. Murray-Hudson, and T. Gumbricht, 2006: Modelling of the flooding in the Okavango Delta, Botswana, using a hybrid reservoir-GIS model. *Journal of Hydrology*, **331**, 58–72.

Yong, B. and Coauthors, 2012: Assessment of evolving TRMM-based multisatellite real-time precipitation estimation methods and their impacts on hydrologic prediction in a high latitude basin. *Journal of Geophysical Research: Atmospheres (1984–2012)*, **117**, D09108.

Zhang, S., H. Gao, and B. S. Naz, 2014a: Monitoring reservoir storage in South Asia from multisatellite remote sensing. *Water Resources Research*, **50**, 8927–8943.

Zhang, Y., Y. Hong, J. J. Gourley, X. Wang, G. R. Brakenridge, T. De Groeve, and H. Vergara, 2014b: Impact of assimilating spaceborne microwave signals for improving hydrological prediction in ungauged basins. *Remote Sensing of the Terrestrial Water Cycle*, **206**, 439.

Zhang, Y., Y. Hong, X. Wang, J. J. Gourley, J. Gao, H. J. Vergara, and B. Yong, 2013: Assimilation of passive microwave streamflow signals for improving flood forecasting: A first study in Cubango river basin, Africa. *IEEE Journal of Selected Topics in Applied Earth Observations and Remote Sensing*, **6**, 2375–2390.

9 Multi-Sensor Geospatial Data for Flood Monitoring along Indus River, Pakistan

Sadiq Ibrahim Khan and Zachary L. Flamig

CONTENTS

9.1 INTRODUCTION

Hydrometeorological hazards are among the most recurring and devastating natural hazards, impacting human lives and causing severe economic damage throughout the world. It is understood that flood risks will not subside in the future and with the onset of climate change, flood intensity and frequency will threaten many regions of the world (Jonkman, 2005; McCarthy, 2001). The current trend and future scenarios of flood risks demand accurate spatial and temporal information on the flood hazards and risks, particularly in emerging economies. Natural hazards commonly cause loss of life and bring about extensive economic losses and social disruptions. Floods cannot be completely controlled by structural means, but the impacts and aftereffects can be managed by developing effective risk reduction strategies through application of geospatial tools and decision-support systems. Implementation of a flood prediction system can potentially help mitigate flood-induced hazards. Such a system typically requires implementation and calibration of a hydrologic model using in situ observations (e.g., rain gauges and stream gauges).

In South Asia, Pakistan is among the top five countries with the highest annual average number of people physically exposed to floods and precipitation-triggered landslides. These events occur normally due to storm systems during the monsoon from July to September. During 2010, floods particularly hit Punjab, Khyber Pakhtunkhwa (KP), and Sindh, while torrents tend to affect the hilly areas of Balochistan and the northern and western federally administered areas. The Indus River basin is one such case where river flooding has historically been a very significant problem to socioeconomics. The flood events of 1950, 1992, and 1998 resulted in devastating loss of life and colossal damage to the national economy. Repeated and increasingly frequent flooding every year somehow threatens communities in all the five provinces of Pakistan, as seen during the 2010 floods (Figure 9.1), and this enhances the need for effective flood risk management. Robust techniques that integrate both in situ and remotely sensed data can improve the flood monitoring system in the country.

The advancement of satellite sensors has provided better measurement capability to estimate precipitation at high spatial and temporal resolution at global scale. In the last three decades, numerous methods have been developed to estimate precipitation through satellite and in situ remote sensing techniques, which include both spaceborne and ground-based radar systems. Recently, satellite remote sensing provides information of the spatial distribution and dynamics of hydrologic phenomena, often unattainable by conventional ground-based methods. Multi-sensor data have been used to provide timely and cost-effective hydrologic monitoring in sparsely gauged basins, irrespective of the political boundaries and other geophysical barriers.

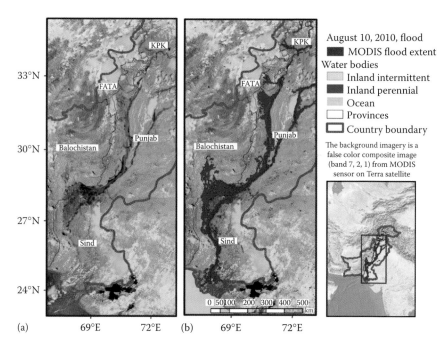

FIGURE 9.1 (a) MODIS false-color compsote (bands 7, 2, 1) for August 10, 2010, and (b) Indus River flood inundation delinated from the false-color image.

The application of satellite imagery for flood mapping using moderate resolution imaging spectroradiometer (MODIS) instruments aboard National Aeronautics and Space Administration's (NASA's) Terra and Aqua satellites offer a unique combination of near-global daily coverage with acceptable spatial resolution for major floods. These capabilities are being utilized for flood monitoring at regional and global scale. Brakenridge et al. (2003) demonstrated that MODIS data can be used to distinguish between flooded and non-flooded areas. Multispectral satellite sensors such as MODIS and Advanced Spaceborne Thermal Emission and Reflection Radiometer (ASTER) can be used to evaluate and validate hydrologic model predictions. Within the existing literature, satellites such as EO-1 (a 10 m panchromatic band plus 30 m multi-spectral bands), along with data from the ASTER sensor aboard Terra satellite or imagery from France's Systeme Pour l'Observation de la Terre (SPOT) satellite and other commercial satellites such as GeoEye, provide important land surface imagery which have been used for flood detection in different regions.

Similarly, other microwave (MW) sensors, for example the advanced synthetic aperture radar (ASAR) instrument aboard ENVISAT with a spatial resolution of 150–1000 m and a revisit time of a few days for higher resolution inundation maps (Di Baldassarre et al., 2009; Schumann et al., 2009, 2007) have been utilized for high-resolution flood mapping. Contrary to other MW instruments, visible (VIS)/infrared (IR) sensors aboard Terra satellite can detect floods globally with relatively high spatial (30 m ASTER, 250 m MODIS) and high temporal resolution (daily). The focus of the case studies discussed in this chapter is to integrate the best available satellite products to characterize the spatial extent of flooding over the sparsely gauged Indus River basin. A satellite remote sensing-based approach is proposed to simulate the spatial extent of flooding and evaluate the probability of detecting inundated areas.

This chapter provides an assessment on the utility of multi-sensor remote sensing data for flood monitoring over the Indus River basin. The datasets required in this study are readily accessible and, in fact, used for flood motoring applications around the world. As such, the applicability of the method and, thus, the capability to provide flood estimates is potentially global for sparsely observed regions and river basins. In this article, an integrated approach is developed by utilizing optical imagery from the MODIS and high resolution ASAR data to detect water inundation during the intense monsoon season in 2010. Moreover, the rainfall estimates from the Tropical Rainfall Measurement Mission (TRMM) along with passive MW-sensor-retrieved land surface brightness temperatures from Advanced Microwave Scanning Radiometer (AMSR-E) are used to monitor floods over the Indus River basin in South Asia.

9.2   SATELLITE SENSORS FOR FLOOD DETECTION

9.2.1   OPTICAL SENSORS FOR FLOOD EXTENT MAPPING

Advanced photogrammetric and pattern recognition methods for remote sensing data processing can provide objective information that may help to detect and monitor the progression of floods at high spatial resolution (Brakenridge et al.,

2003; Smith, 1997). For example, orbital sensors, such as MODIS, provide necessary data to help detecting floods with reasonable accuracy in regions where no other means are available for flood monitoring (Brakenridge and Anderson, 2006; Brakenridge et al., 2007). Such data, after certain processing are capable of providing timely information on flood extents with global coverage and frequent observations. The MODIS instruments on the NASA Aqua and Terra satellites together provide near-global observations of the Earth surface in daylight conditions twice each day with a spatial resolution between 250 and 1000 m. Optical sensors are a practicable choice for rapid response for large-scale flood events because of the high temporal resolution of Aqua and Terra. However, one limitation of optical sensors, such as MODIS, is that it cannot penetrate cloud cover and, thus, suffers from cloud contamination.

In this chapter, considering the advantages and limitation of orbital sensors, imagery from two different instruments with varying spatial resolution is used for flood mapping. First, a moderate resolution data from the MODIS sensor is used to map the extent of inundation throughout the Indus River basin. Second, higher resolution imagery from the ASAR is used for the lower Indus River basin. A MODIS imagery-based map for August 10 shows the inundation of the adjacent areas along the Indus River basin. Figure 9.1a shows the false-color composite (bands 7, 2, 1) from MODIS scenes. This image for August 10 is used to extract the flood extent using an unsupervised classification technique (Figure 9.1b). For that purpose, the iterative self-organizing data analysis technique (ISODATA) algorithm was applied to delineate the flood extent from the MODIS false-color composite image (Jensen, 2005). A detailed explanation and the step-by-step process for the image classification technique used are explained by Khan et al. (2011). A reference layer that includes water features such as lakes and rivers was used to identify and separate the evolution of the 2010 Indus flood (Figure 9.1a). To characterize the extreme inundation during overcast conditions over the lower Indus basin, the high-resolution ASAR image is used to estimate the possible maximum and minimum inundation during this 2010 flood over lower Indus River (Figure 9.1b; Khan et al., 2014).

9.2.2 MICROWAVE REMOTE SENSING-BASED FLOOD DETECTION

The motivation is to use unconventional data from advanced MW sensor such as brightness temperature to detect surface water changes at every pixel location from the satellite footprint. The AMSR-E-based proxy discharge is the ratio of brightness temperatures of a wet pixel (M) centered over the river and its floodplain and dry area near the site referred to as the calibration pixel (C) area. AMSR-E-detected water surface signal (M/C ratio) is an indirect means for river water surface change (Brakenridge et al., 2007). The Global Flood Detection System of the Joint Research Centre of the European Commission (http://old.gdacs.org/flood-detection/overview.aspx) and River Watch 2 (http://floodobservatory.colorado.edu/DischargeAccess.html) are using this technique for global flood detection (De Groeve, 2010; Kugler and De Groeve, 2007). The AMSR-E-based discharge measurements monitor water surface signals at more than 10,000 monitoring areas in different river basins throughout Africa, Asia, America, and Europe.

Passive MW sensors based river width change estimates are utilized for flood detection along the upper and lower Indus River. AMSR-E-detected water surface signal are compared with the river discharge data at multiple gauge stations throughout the Indus River basin. The gauged river discharge was obtained from the Federal Flood Commission that archived the daily river flows and water levels in the reservoirs in coordination with the Pakistan Meteorological Department (PMD). The river discharge data are selected based on the length of record, completeness, and reliability of data. The available discharge gauges maintained by PMD are listed in Table 9.1. The time period for this analysis was selected from 2008 to 2011, which includes the 2010 flood.

To assess the agreement between simulated and observed discharge, Pearson Correlation Coefficient (CC) is used. The AMSR-E-detected water surface signal matched closely with the observed runoff with a Pearson CC of 0.7, 0.72, and 0.7, at Tarbela, Kalabagh, and Chashma gauge discharge stations, respectively (Table 9.1). Similarly, for the lower Indus basin the flood signal corresponds well with observed flood hydrographs in monitoring stations where the river overflows the bank in the Sindh province. The CC was 0.82, 0.84, 0.88, and 0.83 for stations at Taunsa, Guddu, Sukkur, and Kotri, respectively (Khan et al., 2014).

In summary, the remotely sensed river discharge proxy shows high correlation with in situ data and detects the floods that produced havoc along the upper and lower reaches of the Indus River. The AMSR-E-detected water surface signal (M/C ratio) captures high flow peaks each year; however, the signal is insensitive to discharge fluctuations during low flows. The method was designed with the realization that the AMSR-E signal provides no direct information about discharge magnitude, but is highly correlated with observed discharge for significant flows. The results for all the tested sites show a good agreement with "ground truth" measured discharge at multiple locations along the upper and lower Indus River.

9.3 SATELLITE-BASED PRECIPITATION ESTIMATES FOR HYDROLOGIC MODELING

9.3.1 Space-Based Precipitation Estimates

Precipitation estimates at varying spatial and temporal scales are vital for climatic and hydrologic studies. With global coverage, high resolution, and near real-time availability, satellite-based precipitation data are critical for a wide range of sectoral applications studies such as terrestrial hydrology, climate change, agriculture, and natural hazards monitoring.

The underlying principle in global precipitation estimation from remote sensing is to observe the backscatter from different hydrometeor types (rain, hail, snow and ice crystals) in the atmosphere. Satellite-based precipitation estimation started with the use of VIS and IR instruments, by looking at the cloud top temperatures (Ba and Gruber, 2001; Hong et al., 2004; Krajewski and Smith, 2002; Petty, 1995). Since the late 1970s, IR satellite remote sensing techniques were first used for precipitation

TABLE 9.1

River Discharge from the Selected Gauge Stations Over the Indus River Basin for Monsoon and Pre-Monsoon Seasons-Based on the Daily River Flows

Gauge Station	Lat.	Long.	Elevation (m)	2008–2009 Discharge (m³/s)		2010 Discharge (m³/s)		Correlation AMSR-E and Gauge
				Monsoon	Pre-Monsoon	Monsoon	Pre-Monsoon	
Tarbela	34.12	72.75	430	5297	2410	7773	1595	0.7
Kalabagh	32.92	71.50	200	5909	3331	8989	2506	0.72
Chashma	32.43	71.39	180	6329	3354	10020	2377	0.7
Taunsa	30.50	70.86	120	5664	2861	9677	1995	0.82
Guddu	28.41	69.71	75	4838	1987	11476	1294	0.84
Sukkur	27.69	69.71	60	3222	1506	10488	1017	0.88
Kotri	25.47	68.31	22	1644	328	7606	156	0.83

estimation (Arkin et al., 1994; Meisner and Arkin, 1987). The majority of algorithms attempt to correlate the surface rain rate with IR cloud-top brightness temperatures (Tb) using the information obtained from IR imagery. The algorithms developed to date may be classified into three groups depending on the level of information extracted from the IR cloud images: cloud-pixel-based, cloud-window-based, and cloud-patch-based (Yang et al., 2011). Several examples of these algorithms may clarify this classification further.

In the past decade, a number of quasi-global scale estimates have been developed, including the TRMM-based multi-satellite precipitation analysis (TMPA; Huffman et al., 2007), the Naval Research Laboratory Global Blended Statistical Precipitation Analysis (Turk and Miller, 2005), Climate Prediction Center morphing algorithm (Joyce et al., 2004), precipitation estimation from remotely sensed information using artificial neural networks (PERSIANN; Hsu et al., 1997), PERSIANN Cloud Classification System (PERSIANN-CCS; Hong et al., 2004, 2005). To date, the most commonly available satellite global rain products are summarized in the Table 9.1. These quasi-global precipitation products are discussed in terms of accuracy and earth science applications throughout scientific literature; some of these products are mentioned in Table 9.1.

The standard TMPA blends information from several satellites and offers a high resolution precipitation product every 3 h at $0.25°$ spatial resolution with quasi-global coverage from $50°N$ and $50°S$ (Huffman et al., 2007). The algorithm derives its estimates primarily from MW information from several low-Earth orbiting satellites, and IR data of cloud-top temperatures from geostationary satellites are used to fill MW coverage gaps. The TMPA estimates are available in the form of two products, a real-time version (TMPA-RT) and a gauge-adjusted post-real-time research version (TMPA-V6). TMPA-RT estimates are processed with a latency time of 6 h, on the other hand, TMPA-V6 is a research and post-real-time product corrected with monthly ground gauge observation. Huffman et al. 2008 revealed that TMPA-V6 utilizes precipitation fields obtained by the TRMM Combined Instrument assessments, which uses data from both TRMM Precipitation Radar (PR) and the TRMM Microwave Imager (TMI) (Haddad et al., 1997a, 1997b), as well as the Global Precipitation Climatology Project based monthly rain gauge datasets generated through the Global Precipitation Climatology Center (Rudolf et al., 1994) are used to calibrate the TMPR-V6 product. The TMPA-RT and TMPA-V6 data records span over more than a decade with an uninterrupted availability of TMPA-RT since January 2002 and TMPA-V6 since January 1998 to present.

9.3.2 PRECIPITATION ESTIMATES FOR MONSOON MONITORING

In this analysis, remotely sensed satellite precipitation estimates used are from the latest version of the TMPA-gridded product (herein, 3B42RT) developed by the NASA Goddard Space Flight Center. The latest product, version 7 (i.e., 3B42RT-v7), over a 6-year period from 2005 to 2010 is used to analyze the diurnal cycle and spatio-temporal characteristics of the Asian monsoon over Pakistan. The algorithm for the 3B42 product is designed to produce gauge-adjusted, merged-IR rain rates. This product is produced by merging active and passive MW data from several low-orbit satellites with IR data collected by the international constellation of geo-synchronous Earth orbit satellites (Andermann et al., 2011; Huffman et al., 2007). TRMM monthly IR calibration parameters are used to make adjustments to the merged-IR rain rates, which are composited through the use of datasets from multiple satellites. The final results of the algorithm are adjusted, gridded, merged-IR rain rate estimates. The latest near-real-time version of TMPA (3B42RT-v7) is provided at a spatial resolution of $0.25° \times 0.25°$ within the global latitude belt 50°N–50°S. The detailed information on data processing and improvements can be found in the technical document available at ftp://precip.gsfc.nasa.gov/pub/trmmdocs/3B42_3B43_doc.pdf. This product became available from 1998 to present in early 2013, and replaces all previous versions.

During the monsoon season, the dominant peak of occurrences ranges from 55% to 65% and takes place in the early morning hours between 14 and 17 z, particularly during the 2010 monsoon season (Figure 9.2). This high-frequency peak is attributed to large scale, mesoscale convective systems that continue for hours through the evening and early morning hours. The rainfall frequency at 17 z is noticeably of a wider spatial coverage during the monsoon seasons.

The diurnal cycle of TRMM-retrieved summer convective rainfall during monsoon shows that most of the precipitation tends to occur in the afternoon and evening hours. The results show that spatial difference exhibited in 2010 was an increase in the intensity of rainfall at slightly higher elevations than normal, which can be attributed to the increase of moisture content in the atmosphere. Moreover, peaks of the precipitation occurred along the Himalayan foothills in the northeastern region of Pakistan. The temporal shift between pre-monsoon and monsoon seasons was enhanced in 2010, showing the shift from the deep convection associated with severe storms to strong, larger spatial extent convection associated with mesoscale convective systems (MCSs). Referring to the research completed by Romatschke et al. (2010), a maximum of events later in the day could be closely correlated with deep convective core systems which are powered by the heating of the land surface. Similar increase of atmospheric moisture content in 2010 can be observed in Figure 9.3, which exhibits an increase in the number of rainfall events.

The diurnal variability over the entire region of South Asia has been studied and reveals the temporal aspects of various classifications of convection. The three categories defined in this line of research are deep convective cores, wide convective cores, and broad stratiform convection. The categories have been discussed for this region by Barros et al. (2000) and Romatschke et al. (2010). Deep convective cores tend to occur between noon and midnight, due to a direct connection with land surface heating by solar radiation, which produces updrafts and deep, moist convection. This type of convection has occasionally been observed to occur during the evening hours, but it primarily occurs during monsoon seasons and takes place between midnight and the early morning hours. The third type, broad stratiform convection, is found with decaying MCSs and, as a result, has a shorter stature than the other two types. This type of convection tends to occur around midday and is primarily located over oceans, although its width is extensive and can encompass coastal land surfaces as well.

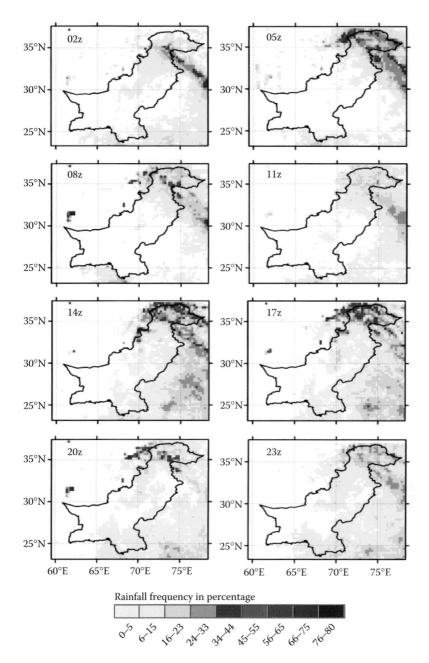

FIGURE 9.2 Diurnal variation of the percent frequency of rainfall occurrence during the 2010 monsoon season (JAS).

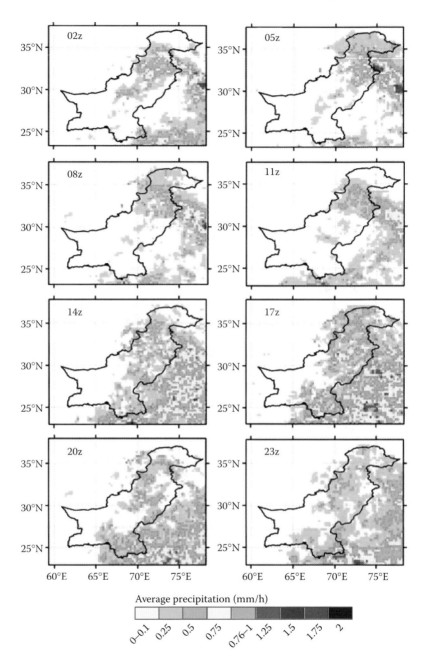

FIGURE 9.3 TRMM 3B42RT-based rain rates at every 3 h showing diurnal variability during the 2010 monsoon season (JAS).

9.4   SUMMARY

As hydrometeorological hazards and water security issues will not subside and may exacerbate due to changing climate. These challenges will demand advances in quantitative precipitation estimation and forecasting that will feed into a disaster modeling approach for the development of the next generation warning system. Therefore, satellite remote sensing techniques are of particular importance for those emerging regions where in situ gauge observations are sparse. This chapter has shown the variety of satellite remote sensing techniques, as a cost-effective but essential approach, that can play a critical role in observing the key hydrologic components such as precipitation. New remote sensing observations and models have triggered distributed hydrologic modeling at high resolution both in space and time. Moreover, recent technical advances in high-performance and parallel computing facilitates the combination of these individual modeling systems into an integrated forecasting chain.

In this chapter, river discharge proxy from the AMSR-E is used to detect and monitor flood over the Indus basin. Satellite-based surface water change signal is supplemented with the sparse gauge runoff observations to observe the evolution of the 2010 flood throughout the Indus River. Moreover, long-term, consistent, and sustained observation of discharge observation from selected gauge stations are used to cross-validate the AMSR-E-based flood signals. It is concluded that the passive MW sensor was able to detect flood (M/C signal) and corresponds very well with gauge discharge data (CC 0.7–0.8). There are ongoing efforts to assimilate these river discharge signals with other satellite data into hydrologic models for flood monitoring in ungauged basins (Khan et al., 2012; Robert Brakenridge et al., 2012). These studies revealed that MW sensors can be used to evaluate distributed hydrologic models for predicting floods in ungauged basins. The attractive feature of this approach is that it can reduce the dependency on gauged runoff and precipitation data to calibrate hydrologic models. Moreover, models are typically calibrated at point locations in the watershed; in contrast, the geo-spatio-temporal passive MW data allows monitoring of the watershed at every pixel throughout the river reach.

To delineate the extent of the 2010 flood along the Indus River, the MODIS sensor data is used, based on the advantage of frequent revisits and large areal coverage. Moreover, the all-weather and all-time capability of higher resolution imagery from the ASAR is utilized to detect floods over the lower Indus River basin. This binary approach to flood prediction will be very useful for providing simple flood versus no-flood estimates to any basin where in situ observations are scarce. Remote sensing precipitation estimates are uniquely suited to provide timely and uniform information during the monsoon season that are needed to evaluate flood hazards triggered by intense precipitation over upper Indus basin. The 2010 monsoon that occurred over the northeastern region of the Indus River basin is captured by the TMPA's latest version products. The overall precipitation pattern and intensity during the monsoon season was captured by the latest satellite precipitation estimates. Model performance was improved with a Nash–Sutcliffe efficiency of 0.74 and CC of 0.87, respectively. It is concluded that satellite data from MW sensors can be used to calibrated hydrologic model in a data-poor environment. It is anticipated that the planned successor mission to TRMM, the Global Precipitation Mission is designed to improve the measurement of light rainfall, and snowfall through improved radiometric capacities.

REFERENCES

Andermann, C., Bonnet, S., and Gloaguen, R. (2011). Evaluation of precipitation data sets along the Himalayan front. *Geochemistry, Geophysics, Geosystems,* 12(7). doi:10.1029/2011GC003513.

Arkin, P. A., Joyce, R., and Janowiak, J. E. (1994). The estimation of global monthly mean rainfall using infrared satellite data: The GOES Precipitation Index (GPI). *Remote Sensing Reviews,* 11(1–4), 107–124.

Ba, M. B. and Gruber, A. (2001). GOES multispectral rainfall algorithm (GMSRA). *Journal of Applied Meteorology,* 40(8), 1500–1514.

Barros, A., Joshi, M., Putkonen, J., and Burbank, D. (2000). A study of the 1999 monsoon rainfall in a mountainous region in central Nepal using TRMM products and rain gauge observations. *Geophysical Research Letters,* 27(22), 3683–3686.

Brakenridge, G., Anderson, E., Nghiem, S., Caquard, S., and Shabaneh, T. (2003). Flood warnings, flood disaster assessments, and flood hazard reduction: The roles of orbital remote sensing. In Paper presented at the *30th International Symposium on Remote Sensing of Environment,* Honolulu, HI.

Brakenridge, G. R., Nghiem, S. V., Anderson, E., and Mic, R. (2007). Orbital microwave measurement of river discharge and ice status. *Water Resources Research,* 43(4). doi: 10.1029/2006wr005238.

Brakenridge, R. and Anderson, E. (2006). MODIS-based flood detection, mapping and measurement: The potential for operational hydrological applications. In *Transboundary Floods: Reducing Risks Through Flood Management,* J. Marsalek, G. Stancalie, and G. Balint (eds.), Springer, Dordrecht, the Netherlands, pp. 1–12.

De Groeve, T. (2010). Flood monitoring and mapping using passive microwave remote sensing in Namibia. *Geomatics, Natural Hazards and Risk,* 1(1), 19–35.

Di Baldassarre, G., Schumann, G., and Bates, P. D. (2009). A technique for the calibration of hydraulic models using uncertain satellite observations of flood extent. *Journal of Hydrology,* 367(3–4), 276–282. doi: 10.1016/j.jhydrol.2009.01.020.

Haddad, Z. S., Short, D. A., Durden, S. L., Im, E., Hensley, S., Grable, M. B., and Black, R. A. (1997a). A new parametrization of the rain drop size distribution. *IEEE Transactions on Geoscience and Remote Sensing,* 35(3), 532–539.

Haddad, Z. S., Smith, E. A., Kummerow, C. D., Igushi, T., Farrar, M. R., Durden, S. L., Alves, M., Olson, W. S. (1997b). The TRMM "Day-1" radar/radiometer combined rain-profiling algorithm. *Journal of the Meteorological Society of Japan,* 75, 799–809.

Hong, Y., Hsu, K.-L., Sorooshian, S., and Gao, X. (2004). Precipitation estimation from remotely sensed imagery using an artificial neural network cloud classification system. *Journal of Applied Meteorology,* 43(12), 1834–1853. doi: 10.1175/jam2173.1.

Hong, Y., Hsu, K.-L., Sorooshian, S., and Gao, X. (2005). Self-organizing nonlinear output (SONO): A neural network suitable for cloud patch–based rainfall estimation at small scales. *Water Resources Research,* 41(3), W03008. doi: 10.1029/2004wr003142.

Hsu, K.-L., Gao, X., Sorooshian, S., and Gupta, H. V. (1997). Precipitation estimation from remotely sensed information using artificial neural networks. *Journal of Applied Meteorology,* 36(9), 1176–1190.

Huffman, G. J., Adler, R. F., Bolvin, D. T., and Nelkin, E. J. (2010). The TRMM multi-satellite precipitation analysis (TMPA). In *Satellite Rainfall Applications for Surface Hydrology,* Springer, Dordrecht, the Netherlands, pp. 3–22.

Huffman, G. J., Bolvin, D. T., Nelkin, E. J., Wolff, D. B., Adler, R. F., Gu, G., Bowman, K. P., Hong, Y., and Stocker, E. F. (2007). The TRMM multisatellite precipitation analysis (TMPA): Quasi-global, multiyear, combined-sensor precipitation estimates at fine scales. *Journal of Hydrometeorology,* 8(1), 38–55. doi: 10.1175/jhm560.1.

Jensen, J. (2005). *Introductory Digital Image Processing: A Remote Sensing Perspective,* Prentice Hall PTR Upper Saddle River, NJ, USA.

Jonkman, S. N. (2005). Global perspectives on loss of human life caused by floods. *Natural Hazards,* 34(2), 151–175.

Joyce, R. J., Janowiak, J. E., Arkin, P. A., and Xie, P. (2004). CMORPH: A method that produces global precipitation estimates from passive microwave and infrared data at high spatial and temporal resolution. *Journal of Hydrometeorology,* 5(3), 487–503.

Khan, S. I., Hong, Y., Gourley, J. J., Khattak, M. U., and De Groeve, T. (2014). Multi-sensor imaging and space-ground cross-validation for 2010 flood along Indus River, Pakistan. *Remote Sensing,* 6(3), 2393–2407.

Khan, S. I., Hong, Y., Vergara, H. J., Gourley, J. J., Brakenridge, G., De Groeve, T., Flamig, L., Policelli, F., and Yong, B. (2012). Microwave satellite data for hydrologic modeling in ungauged basins. *IEEE Geoscience and Remote Sensing Letters,* 9(4), 663–667.

Khan, S. I., Hong, Y., Wang, J., Yilmaz, K. K., Gourley, J. J., Adler, R. F., Policelli, F., Habib, S., Brakenridge, G. R., and Irwin, D. (2011). Satellite remote sensing and hydro-logic modeling for flood inundation mapping in lake victoria basin: implications for hydrologic prediction in ungauged basins. *IEEE Transactions on Geoscience and Remote Sensing,* 49(1), 85–95. doi: 10.1109/TGRS.2010.2057513.

Krajewski, W. F. and Smith, J. A. (2002). Radar hydrology: Rainfall estimation. *Advances in Water Resources,* 25(8–12), 1387–1394. doi: http://dx.doi.org/10.1016/S0309-1708(02)00062-3.

Kugler, Z. and De Groeve, T. (2007). *The Global Flood Detection System,* Office for Official Publications of the European Communities, Luxembourg.

McCarthy, J. (2001). *Climate Change 2001: Impacts, Adaptation, and Vulnerability: Contribution of Working Group II to the Third Assessment Report of the Intergovernmental Panel on Climate Change,* Cambridge University Press, Cambridge.

Meisner, B. N. and Arkin, P. A. (1987). Spatial and annual variations in the diurnal cycle of large-scale tropical convective cloudiness and precipitation. *Monthly Weather Review,* 115(9), 2009–2032. doi: 10.1175/1520-0493(1987)115<2009:saavit>2.0.co;2.

Petty, G. W. (1995). The status of satellite-based rainfall estimation over land. *Remote Sensing of Environment,* 51(1), 125–137.

Robert Brakenridge, G., Cohen, S., Kettner, A. J., De Groeve, T., Nghiem, S. V., Syvitski, J. P. M., and Fekete, B. M. (2012). Calibration of satellite measurements of river discharge using a global hydrology model. *Journal of Hydrology,* 475, 123–136. doi: http://dx.doi.org/10.1016/j.jhydrol.2012.09.035.

Romatschke, U., Medina, S., and Houze Jr, R. A. (2010). Regional, seasonal, and diurnal variations of extreme convection in the South Asian region. *Journal of Climate,* 23(2), 419–439.

Rudolf, B., Hauschild, H., Rueth, W., and Schneider, U. (1994). Terrestrial precipitation analysis: Operational method and required density of point measurements. *Global Precipitation and Climate Change,* 26, 173–186.

Schumann, G., Bates, P., Horritt, M., Matgen, P., and Pappenberger, F. (2009). Progress in integration of remote sensing–derived flood extent and stage data and hydraulic models. *Reviews of Geophysics,* 47(4). doi:10.1029/2008RG000274.

Schumann, G., Hostache, R., Puech, C., Hoffmann, L., Matgen, P., Pappenberger, F., and Pfister, L. (2007). High-resolution 3-D flood information from radar imagery for flood hazard management. *IEEE Transactions on Geoscience and Remote Sensing,* 45(6), 1715–1725. doi: 10.1109/tgrs.2006.888103.

Smith, L. C. (1997). Satellite remote sensing of river inundation area, stage, and discharge: A review. *Hydrological Processes,* 11(10), 1427–1439.

Turk, F. J. and Miller, S. D. (2005). Toward improved characterization of remotely sensed precipitation regimes with MODIS/AMSR-E blended data techniques. *IEEE Transactions on Geoscience and Remote Sensing,* 43(5), 1059–1069.

Yang, H., Sheng, C., Xianwu, X., and Gina, H. (2011). Global precipitation estimation and applications. In *Multiscale Hydrologic Remote Sensing*, Chang, N.-B. and Hong, Y. (eds.), CRC Press, London, pp. 371–386.

10 Evaluating the Diurnal Cycle of Precipitation Representation in West African Monsoon Region with Different Convection Schemes

Xiaogang He, Hyungjun Kim, Pierre-Emmanuel Kirstetter, Kei Yoshimura, Zhongwang Wei, Eun-Chul Chang, Yang Hong, and Taikan Oki

CONTENTS

10.1   INTRODUCTION

The intensification of the hydrologic cycle and extreme hydrologic events by natural factors and human activities call for great concern at the global scale (Oki and Kanae, 2006). The Sahel has been receiving attention because it suffered from persistent and severe drought conditions from the late 1960s to the mid-1980s, which dramatically impacted resident communities that depend on ecosystem services for their livelihoods. In the mid-1990s, the rainfall returned to near- or above-normal amounts (relative to the 1941–2012 period). However, the recent period (2005–2012) experienced three food crises triggered by severe drought linked to a recurrent lack of precipitation in this region (Boyd et al., 2013). More accurate prediction of West African monsoon (WAM) precipitation now, and under climate change, is required to better direct humanitarian aid responses as well as adaptive planning measures.

Current climate models are deficient in simulating the West African (WA) seasonal rainfall due to the complex monsoon system. Deficiencies at the seasonal scale may originate and accumulate from much smaller timescales, for example, at the diurnal cycle scale (Taylor and Clark, 2001; Parker et al., 2005; Shine et al., 2007). Precipitation frequency and intensity in WA, particularly in extreme events, are likely to change within the context of the changing climate (Trenberth et al., 2003). To assess the evolving characteristics of precipitation, it is essential to examine rain events and the diurnal cycle at a subdaily timescale (Roca et al., 2010). The diurnal cycle is a key test for model reliability and a tool for model development. For example, it can be used to validate physical parameterizations (Slingo et al., 1987; Garratt et al., 1993; Dai et al., 1999), to enhance the understanding of important mechanisms that drive the diurnal cycle (Randall et al., 1991), as well as to provide insights for improvements in the representation of subgrid-scale processes (Betts et al., 1996; Giorgi and Shields, 1999; Lin et al., 2000). Increased accuracy in the representation of these processes at the subdaily timescale can significantly improve short-term precipitation forecast, which is indispensable to mitigating the negative impacts of droughts in the WA region.

Precipitation in current climate models is generated by either grid-resolvable (large-scale) forcing or subgrid (convective parameterization) processes. The former can be simulated explicitly, whereas the latter must be parameterized because it is still generally computationally prohibitive to run global climate models (GCMs) and regional climate models at cloud-resolving resolutions (e.g., finer than 4 km). This consequently increases the uncertainties in reproducing observed precipitation. Most current GCMs are known to produce a diurnal cycle of continental precipitation that is in phase with insolation, with maximum rainfall occurring around midday instead of during the late afternoon (Yang and Slingo, 2001; Rio et al., 2009). The bias in the simulated diurnal cycle of precipitation (DCP) indicates that there is disparity in how convective processes are represented in models compared to the real world. The primary cause of the errors has been associated with convective parameterization schemes (CPSs; Fritsch and Carbone, 2004; Shine et al., 2007). A recent study conducted by Koo and Hong (2010) demonstrated that the CPS is the most important trigger in simulating DCP over land. Even more so, complex dynamic and thermodynamic features of the WAM climatological system, such as African easterly waves, African easterly jet, and mesoscale convective systems (MCSs), may be a challenge for CPSs' performance in terms of DCP.

The structure of this chapter is as follows. In Section 10.2, we introduce the precipitation verification data, model and experimental setup, and method of analysis. Section 10.3 includes a systematic description of the DCP amount, frequency, and intensity over land, and ocean and their sensitivities to CPSs. Section 10.4 provides a summary and our conclusions.

10.2 DATA AND METHODS

10.2.1 Precipitation Observations

Regional precipitation observations are obtained from the Tropical Rainfall Measurement Mission Multisatellite Precipitation Analysis (TMPA). TMPA combines precipitation estimates from multiple satellites and gauge observations to provide the best estimate of precipitation in each grid box (Huffman et al., 2007). In this study, the research product 3B42 V6 with a fine spatial (0.25° × 0.25°) and temporal (3 hourly; 0000, 0300,..., 2100 UTC) resolution within the global latitude belt from 50°S to 50°N is used. The 3-hourly average data are downscaled into hourly data by using a bilinear interpolation method for further analysis.

10.2.2 Model and Experimental Setup

10.2.2.1 Model Description
In this study, the NCEP/Environmental Modeling Center Regional Spectral Model (RSM) is used. A detailed model description of the RSM can be found in Juang and Kanamitsu (1994), Juang et al. (1997), and Hong and Leetmaa (1999). The RSM has been widely used over East Asia because of its ability to reproduce this region's summer monsoon. It has also been developed with the incorporation of stable water isotopes to study atmospheric river events over the United States (Yoshimura et al., 2010). In the RSM, perturbations of temperature, mixing ratio, pressure, and divergence are spectrally represented and nudged by a two-dimensional cosine series. A regional climate model with this spectral nudging technique has smaller large-scale biases and does not depend on the domain size. In addition, a global downscaling technique (Yoshimura and Kanamitsu, 2008) can also be physically well expressed to constrain the large-scale background in a realistic way, compared to the conventional lateral boundary nudging technique. Major physical processes adopted in the RSM include land surface effects (Noah land surface model; Ek et al., 2003), longwave and shortwave radiation (Chou scheme; Chou and Suarez, 1994), cloud–radiation interaction, planetary boundary scheme (Hong and Pan, 1996), large-scale condensation, gravity

wave drag (Alpert et al., 1988, 1996), and enhanced topography, as well as vertical and horizontal diffusion (Hong and Pan, 1996; Kim and Mahrt, 1992). Because deep and shallow convection are very important physical processes for triggering precipitation and are of special concern to this study, a more detailed description of the CPSs is introduced below.

Deep CPSs have proven to be one of the most challenging parts of numerical atmospheric modeling. Three CPSs are available in the current RSM: relaxed Arakawa–Schubert (RAS; Moorthi and Suarez, 1992), simplified Arakawa–Schubert (SAS; Pan and Wu, 1995), and the new Kain–Fritsch approach (KF2; Kain, 2004). In this study, these three CPSs are used to investigate the sensitivity of the diurnal variation of simulated precipitation. Both RAS and SAS make simplifications of the standard Arakawa and Schubert scheme (AS; Arakawa and Schubert, 1974), which postulate that the level of activity of convection is such that their stabilizing effect balances the destabilization by large-scale processes. RAS has been widely tested for seasonal forecasts at NCEP, whereas SAS has advantages for medium-range forecasts (Kalnay et al., 1996). The main difference between RAS and SAS lies in two aspects: (1) the cloud model and (2) the treatment of downdrafts (Kang and Hong, 2008). Specifically, the clouds in RAS have different tops, whereas SAS only has one type of cloud. RAS does not include any downdraft mechanisms, but SAS considers saturated downdrafts. Different from the AS scheme, the closure assumption in KF2 is based on the convective available potential energy (CAPE) for an entraining parcel, and this can provide more reasonable precipitation rates. KF2 makes modifications to the cloud model by permitting cloud development (variable cloud radius and cloud-depth threshold) for deep convection. KF2 has been widely tested for climate and mesoscale modeling studies (Wang and Seaman, 1997; Ridout et al., 2005).

10.2.2.2 Experimental Design

Figure 10.1 shows the model domain over West and North Africa based on a Mercator map projection with a 20-km horizontal resolution. The RSM uses 28 vertical sigma levels, consistent with the NCEP–US Department of Energy Atmospheric Model Intercomparison Project-II reanalysis (R2; Kanamitsu et al., 2002) data to avoid errors due to vertical interpolation of large-scale fields from the global model (Hong and Leetmaa, 1999).

The lateral boundary and initial conditions are taken from the reanalysis-nudged isotope-incorporated atmospheric GCM (AGCM) simulation conducted by Yoshimura and Kanamitsu (2008). In their method, surface boundary conditions, sea surface temperature, and sea ice distribution are taken from R2. The R2 datasets are used for the large-scale forcing to drive the AGCM to produce the large-scale analysis by using the global downscaling technique. The time steps for the model integration vary by season and are 40, 60, 40, and 30 s for spring, summer, autumn, and winter, respectively. Simulation results are generated at hourly intervals and interpolated onto 12 vertical pressure levels: 1000, 925, 850, 700, 600, 500, 400, 300, 250, 200, 150, and 100 hPa. Experiments start at 0000 UTC January 1, 2004 and terminate at 2300 UTC December 31, 2005. The first year of simulation (e.g.,

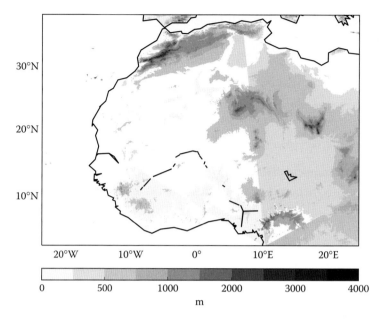

FIGURE 10.1 Topography for the regional model domain (excluding buffer zone).

January–December, 2004) is regarded as spin-up and discarded. The WAM season (April–September) of 2005 is chosen for the main analysis.

10.2.3 ANALYSIS METHOD

In this study, we applied harmonic analysis (Dai, 2001; Basu, 2007; Koo and Hong, 2010) to characterize the DCP, focusing on the relative amplitudes and phases of the diurnal (24-h period) S_1 and semidiurnal (12-h period) S_2 variation. The analysis procedures found in Haurwitz and Cowley (1973) were followed. The basic idea of harmonic analysis is to represent the fluctuations or variations in a time series by adding a series of sine and cosine functions (Wilks, 2005). A more accurate fit is obtained as more sinusoids are added. However, we will only focus on the first two harmonics, which have physical meaning and dominate the total precipitation. Details of the mathematical equations to fit the diurnal variation with the first and second harmonic function can be found in He et al. (2015).

10.2.4 PRECIPITATION CHARACTERISTICS

The diurnal variations of rainfall characteristics (amount, frequency, and intensity) can be mathematically defined according to Oki and Musiake (1994) and Jeong et al. (2011). Normalized amplitude is more appropriate for evaluating diurnal variation because the amplitude depends on the daily mean values. The values normalized by the daily average precipitation will be more useful for long-term datasets because they are less affected by interannual variability.

10.2.4.1 Amount—How Much Does It Rain?

The amount is the accumulated precipitation within a specific time period. The diurnal cycle of rainfall amount ($DP(m, h)$) can be defined as Equation 10.1:

$$DP(m, h) = \frac{P(m, h)}{(1/24)\sum_{h'=1}^{24} P(m, h')} \tag{10.1}$$

where $P(m, h)$ is the mean precipitation at hour h in the mth month.

10.2.4.2 Frequency—How Often Does It Rain?

The precipitation frequency for a selected hour (h) in the month m can be defined as follows:

$$N(m, h) = \frac{\sum_{d=1}^{N(m)} O(h, d)}{N(m)} \tag{10.2}$$

$$O(h, d) = \begin{cases} 1 & P(h, d) \geq 0.1 \text{ mm/h} \\ 00 & 0 \leq P(h, d) < 0.1 \text{ mm/h} \end{cases} \tag{10.3}$$

where:

h and d are the chosen hour (1–24) and day (days in the analysis period)
$N(m)$ is the total number of days for the mth month
$P(h, d)$ is the precipitation rate at hour h in a designated day (d)
$O(h, d)$ is a rainfall occurrence counter

The determination of the threshold (0.1 mm/h) is based on the study according to Shinoda et al. (1999), Dai et al. (1999), and Davis et al. (2003).

10.2.4.3 Intensity—How Strong Does It Rain?

The intensity is the mean rate over the precipitation periods and it can be defined as follows:

$$I(m, h) = \frac{P(m, h)}{N(m, h)} \tag{10.4}$$

10.3 SIMULATION RESULTS AND DISCUSSION

10.3.1 AMPLITUDE

Figure 10.2 describes the spatial distributions of the normalized amplitude explained by the DCP from TMPA and three simulations with different CPSs. From the observations, we can see a large part of the domain has amplitudes between ~0.4 and 0.8 in Central and West Africa. High values of amplitude (≥ 1.6) always occur in a vegetation gradient zone (transition from savanna to desert). Amplitudes are usually

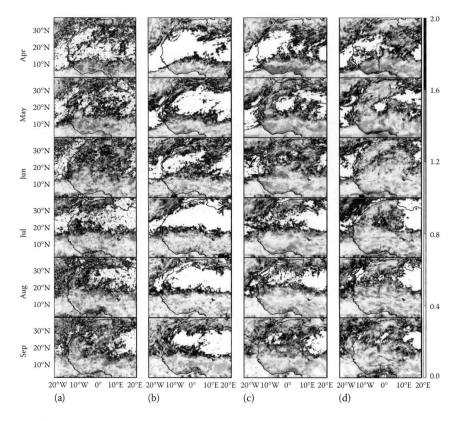

FIGURE 10.2 Normalized amplitudes of precipitation diurnal variations derived from (a) TMPA, (b) RAS, (c) SAS, and (d) KF2 during the monsoon season (April–September). (From He, X. et al., *Wea. Foresting*, 30, 424–445, 2015. ©American Meteorological Society. Used with permission.)

weakened over ocean compared to those over continental areas. In these regions, the mean precipitation is very low, but sudden rainfall is frequent (Koo and Hong, 2010). The amplitudes from the simulations (RAS, SAS, and KF2) are comparable to those from the observations, not only in terms of absolute value but also the spatial pattern. Obvious differences among the three CPSs can be found over the regions with higher amplitude. For example, KF2 produces the highest amplitudes among the three schemes over the North Atlantic Ocean and in particular during the pure monsoon season (July). RAS and SAS show very similar patterns after the monsoon onset, for example, during July and August. However, KF2 and SAS show similar patterns in premonsoon months, such as April and May. This may indicate that CPSs have different responses to the large-scale forcing and feature different characteristics of the monsoon dynamics for different months, although the large-scale monsoon circulations are similar in all the simulations.

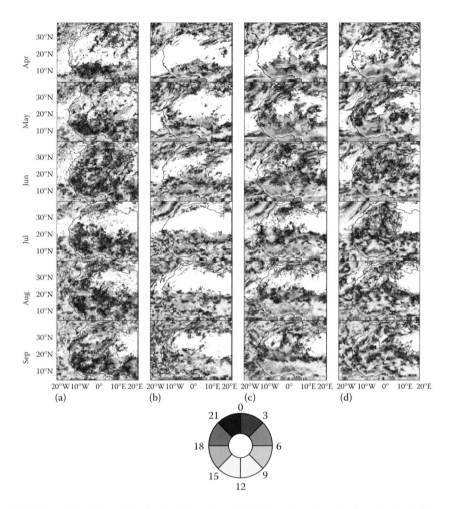

FIGURE 10.3 Phase (peak timing) of precipitation diurnal variations derived from (a) TMPA, (b) RAS, (c) SAS, and (d) KF2 during the monsoon season (April–September). LST is indicated by the color bar with a 3-h interval. (From He, X. et al., *Wea. Foresting*, 30, 424–445, 2015. ©American Meteorological Society. Used with permission.)

10.3.2 PHASE

The phase of DCP from observations and simulations is shown in Figure 10.3. The color bar indicates local solar time (LST) with a 3-h interval. Cyan represents early morning and orange indicates late afternoon. As shown in Figure 10.3a from satellite observations, the hour of maximum precipitation has embedded features appearing from the late afternoon to evening (~1500–2100 LST) and at nighttime (~2100–0300 LST) over the interior land including the intertropical convergence zone (ITCZ). As for the ocean, maximum precipitation mainly occurs during daybreak or (early) morning (~0300–0900 LST). The late afternoon peak over land is related to deep convection, which is driven by the increased sensible heat flux and will lead to significant upward motion of water vapor. The early morning peak over the ocean can be explained by the so-called static radiation–convection and dynamic radiation–convection mechanisms (Ramage, 1971; Gray and Jacobson, 1977). The nighttime peak may be related to nocturnal low-level jets (NLLJs), which bring moisture from the ocean to the land during the monsoon and impact rainfall variations (Sperber and Yasunari, 2006). Pu and Cook (2010) find that in the WA westerly jet region, winds exhibit a semidiurnal cycle, which peaks at 0500 and 1700 LST. Their findings help to explain why there is a mixed phase (~0300–0600 and ~1500–1800 LST) over the eastern Atlantic and the WA coast in TMPA observations as shown in Figure 10.3. The development of NLLJs would also help to sustain the strong diurnal cycle of organized deep convection, which initiates in the afternoon and further develops at night (Laing et al., 2008; Gounou, 2011). The nighttime peak can also be related to the effect of MCSs (Nikulin et al., 2012), which are usually initiated around 1700–1800 LST, but precipitate at maximum intensity several hours later during their mature phase (McGarry and Reed, 1978; Hodges and Thorncroft, 1997). Regarding model simulations, RAS, SAS, and KF2 all capture the distinct contrast between land (late afternoon) and ocean (early morning) well. However, the phases are ~3 h earlier and there is no obvious nighttime peak over coastal areas.

An interesting feature is that RAS and SAS have a distinct early morning peak over the transition area (~15°N–20°N) that is not shown in KF2 or the observations, both of which have the mixed timing. This phenomenon (mixed pattern) is related to the propagation of large-scale precipitation, whereas RAS and SAS cannot capture the propagation signal well. The propagation characteristics of the rainfall system can be depicted by using a Hovmöller diagram. To evaluate whether the model is able to replicate the propagating signal in the East–West direction, diurnally averaged Hovmöller diagrams (Figure 10.4) are constructed using the hourly normalized (by daily mean) rainfall between 2.5°N and 38°N from both observations and simulations. The Hovmöller diagram of TMPA (Figure 10.4a) clearly shows a coherent propagating rainfall axis along 20°W–5°W over ocean areas from July to September. From April to June, this signal appears to be fairly weak. Only simulations from KF2 capture this westward propagation in the corresponding region, although only for a few months (June, July, and September), and the signal is quite noisy. RAS, SAS, and KF2 even produce eastward propagation of rainfall systems in premonsoon months (April–May). These results indicate that the RSM with CPSs has difficulties simulating the propagating rainfall system, which can explain the missed mixed phase in the simulations. In addition, both observations and simulations exhibit a slow propagation signal in the eastern part of the analysis domain (from 5°W to 25°E). However, RAS and SAS simulate the afternoon rainfall ~2–3 h earlier, as seen in the shift of the propagating rainfall axis in simulations compared to that in the observations. This can further explain why the mixed phase is advanced in simulations (RAS and SAS) compared to that in the observations (TMPA), as shown in Figure 10.3. The other interesting observation is that the mixed early morning and late afternoon peaks shown in the three CPSs are close to the North Atlantic Ocean region, for example, around Morocco. This may be related to the high altitude, as shown by Dai and Trenberth (2004) and Koo and Hong (2010).

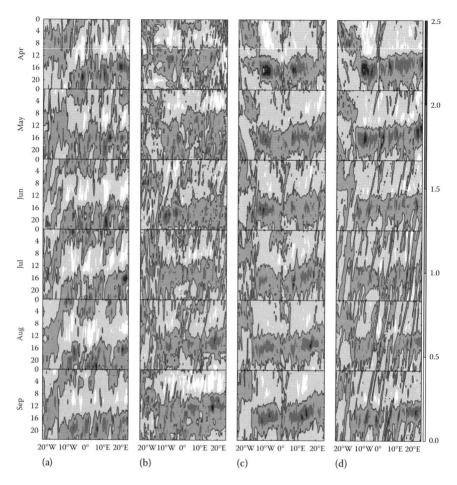

FIGURE 10.4 Hovmöller diagram (time–longitude distribution) for the normalized (by daily mean) diurnal variation of rainfall amount derived from (a) observations (TMPA) and (b)–(d) model simulations with different CPSs (RAS, SAS, and KF2). The y axis indicates LST. (From He, X. et al., *Wea. Foresting*, 30, 424–445, 2015. ©American Meteorological Society. With permission.)

10.3.3 LAND VERSUS OCEAN

10.3.3.1 Amount

Figures 10.5 and 10.6 compare the normalized diurnal variation of precipitation amount from TMPA, RAS, SAS, and KF2 over land and ocean as well as their corresponding first and second harmonics. According to TMPA observations, the time of maximum precipitation varies by region, with an afternoon peak (~1500–1800 LST) over land and a morning peak (~0800–1200 LST) over ocean. Peak time also varies among the 6 months and there is a ~1–2 h delay in the later monsoon months (July–September) compared to the premonsoon months (April–June). Large seasonal variability is not seen in the amplitudes, but differs by region. Land exhibits

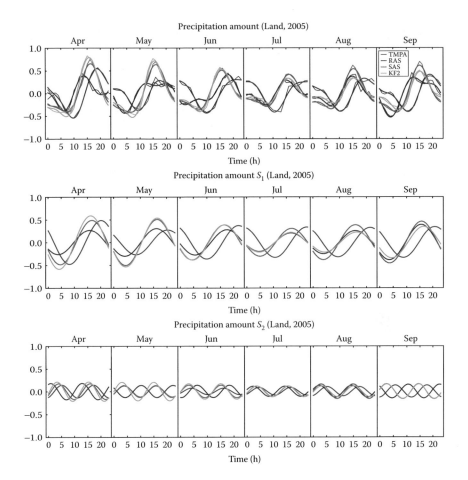

FIGURE 10.5 (Upper panel) Normalized diurnal variations derived from satellite and RSM output (dashed lines) and fitted models from the first two harmonics (continuous lines) and (middle panel) fitted diurnal (i.e., S_1) and (lower panel) fitted semidiurnal (i.e., S_2) cycles of precipitation amount from TMPA, RAS, SAS, and KF2 over land.

slightly higher (~0.4–0.5) amplitudes compared to those of the ocean (~0.25). For the simulations, the general features (i.e., periodicity and monthly variations of amplitude) are fairly well reproduced compared to the observations. In Figure 10.5, the simulated amplitude is higher than observed, and the phase is advanced ~2–3 h prematurely. Moreover, RAS shows the closest amplitude, but always advanced the phase too much compared to SAS and KF2. The only noticeable difference between SAS and KF2 is manifested by exaggerated amplitude; the peak time is almost the same. Sensitivity of diurnal variation to the CPSs is also different over land versus ocean. Large uncertainties can be found over land between the three CPSs; however, the disparity over ocean is quite small, as seen in Figure 10.6.

Diurnal and semidiurnal cycles can be described by the first and second harmonics (S_1 and S_2). From Figures 10.5 and 10.6, we can see the total variance is dominated

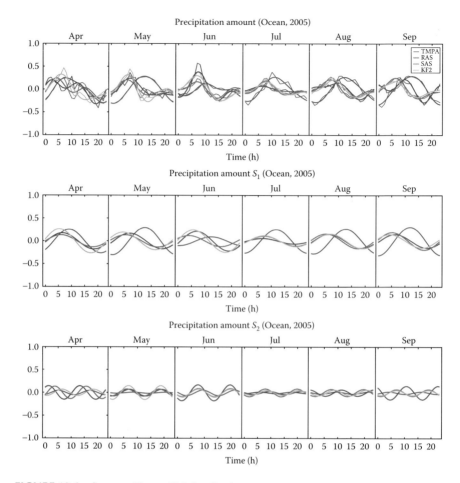

FIGURE 10.6 Same as Figure 10.5, but for the ocean.

by the first harmonic (diurnal cycle). The semidiurnal cycle is weaker than the diurnal cycle, and the amplitude is reduced to 1/3 over land and, in particular, almost 1/5 or less over ocean. The simulated variance explained by the diurnal and semidiurnal cycles is larger than the TMPA over land, but smaller (larger) for diurnal (semidiurnal) cycles over ocean. This differs from previous studies as shown by Koo and Hong (2010), whose simulation results reveal similar amplitudes of S_1 (over ocean) and S_2 (over land and ocean) compared to the observations. The afternoon peak shown in S_1 is shifted ~2 h ahead over land but shifted almost 5 h ahead over ocean. For the semidiurnal cycle, SAS and KF2 are similar compared to TMPA for almost all the months, not only in amplitude but also in phase. However, RAS sometimes has out-of-phase behavior, as we can see for May and September of Figures 10.5 (S_2) and 10.6 (S_2).

10.3.3.2 Frequency

The diurnal variation of precipitation frequency is shown in Figures 10.7 and 10.8. For TMPA over land, the variation of frequency closely resembles that of amount in

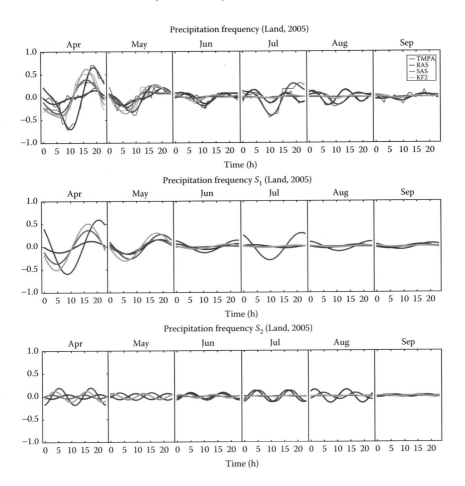

FIGURE 10.7   Same as Figure 10.5, but for precipitation frequency.

April, May, and July, but weakens in June, August, and September. For TMPA over ocean, the variation of frequency in April and May is more similar compared to that of amount, while in other months (June–September) the amplitude is largely reduced and even without obvious peak (July and August). The simulated diurnal variation of frequency has significant seasonal variability over both land and ocean, and the amplitude is weakened after the monsoon onset. As for the sensitivity to CPSs, large uncertainty exists over land compared to that over ocean. From Figure 10.7, RAS always produces comparable variability compared to TMPA; however, there is almost no variation from June to September. Similar results can be found in Figure 10.8 over ocean. The simulations are poor in representing the diurnal and semidiurnal variation over land (Figure 10.7 [S_1 and S_2]), especially in late monsoon seasons. Specifically, RAS performs a little better than SAS and KF2 for semidiurnal cycle. Although S_1 from simulations is weakened over ocean and the peak time is advanced ~5 h earlier, simulated S_2 shows a striking similarity although extremely flat (Figure 10.8 [S_2]).

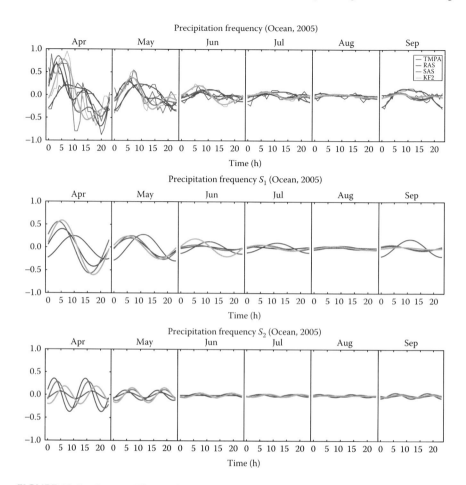

FIGURE 10.8 Same as Figure 10.7, but for the ocean.

The distinct fluctuations in April and May reveal that the occurrence of precipitation is sensitive to the parameterized physical mechanisms in CPSs.

10.3.3.3 Intensity

Figures 10.9 and 10.10 compare the diurnal variation of precipitation intensity. Different from frequency, the variation of intensity shows a similar trend with the total amount, not only over land but also over ocean. According to TMPA observations, the amplitude of intensity is weaker compared to the amount, but the phase displays very good consistency except for May and July over land and April, May, and June over ocean. In terms of simulation, all of the three CPSs generate stronger intensity over land from April to September (Figure 10.9), but slightly weaker intensity over ocean from July to September (Figure 10.10). To be specific, KF2 always produces the highest amplitude among the three CPSs, whereas amplitude in RAS is the smallest and earliest in terms of phase. It is intriguing that precipitation intensity in July has the maximum value around midnight, which is quite different from

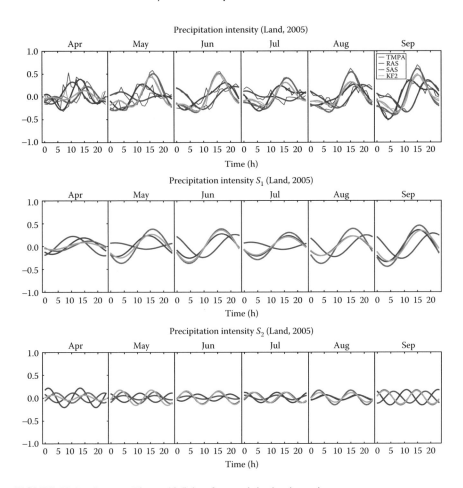

FIGURE 10.9 Same as Figure 10.5, but for precipitation intensity.

other months. The reason is not clear and needs further investigation. Precipitation intensity in terms of the first and second harmonic over land and ocean is compared in the middle and bottom panel in Figures 10.9 and 10.10. The simulated S_1 over land advanced the phase, and the amplitude is amplified slightly. For ocean area, similar conclusion can be drawn but the simulated amplitude is somewhat weakened compared to TMPA. As for the semidiurnal cycle (S_2), the simulated S_2 completely fails to reproduce the observed variation with opposite phase over land. However, simulations adequately reproduce the characteristics over ocean as can be seen in Figure 10.10.

In general, the above analysis tries to characterize the model's performance in terms of the diurnal variation of precipitation amount, frequency, and intensity over both land and ocean with adopting different CPSs. However, it should be noted that results are sensitive to the threshold for the examination of frequency and intensity. Besides, although both RAS and SAS adopt the quasi-equilibrium hypothesis from the AS scheme, they show very different behavior in terms of the amplitude and phase. Results are more

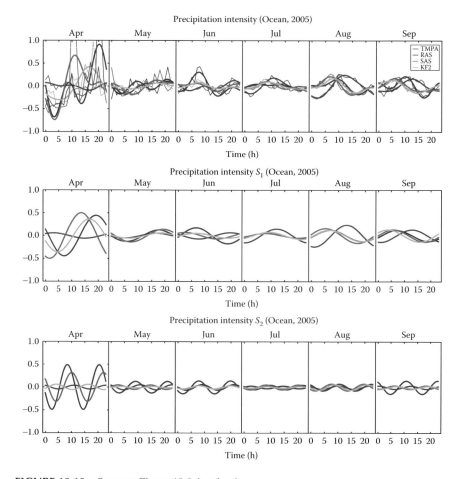

FIGURE 10.10   Same as Figure 10.9, but for the ocean.

similar between SAS and KF2 because the main physical mechanisms they represent are similar, which explains the similarities regarding the amplitude and phase.

10.3.4   CONVECTIVE AVAILABLE POTENTIAL ENERGY

To understand the performance differences in producing DCP among different CPSs, it is necessary to explore the key mechanisms that drive the DCP from a physical perspective. CAPE is one of the useful convective indices for investigating the trigger of deep convection. It is defined as the amount of potential energy of an air parcel lifted to a certain level vertically through the atmosphere. The higher the CAPE, the more energy is available for convection. In this section, an examination of CAPE will be discussed at the regional scale to evaluate the atmospheric instability and the relationship with the simulated diurnal variation of precipitation.

Figure 10.11 shows the spatial distribution of the timing of maximum CAPE at the regional scale. This can help us to get more insight compared to the spatial pattern of

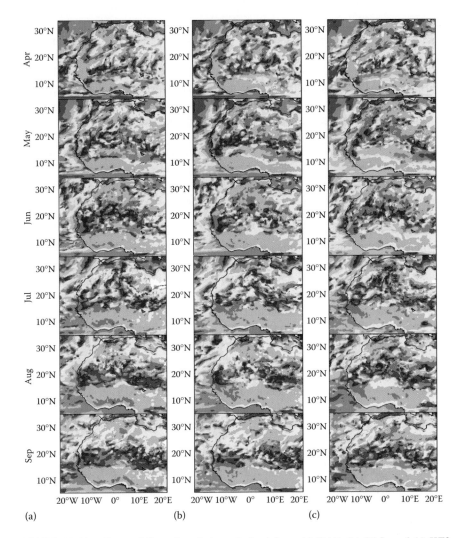

FIGURE 10.11 Phase of diurnal variations derived from (a) RAS, (b) SAS, and (c) KF2 in monsoon season (April–September) for CAPE. Timing is indicated by the color bar in Figure 10.3.

the phase of the precipitation amount (Figure 10.3). According to Figure 10.11, all of the three CPSs show distinct contrast of CAPE between land and ocean, with a late afternoon peak (~1500–1800 LST) over most continental areas and early morning peak (~0300–0600 LST) over most of the ocean area. An obvious nocturnal/early morning peak (~2100–0600 LST) can also be found within the ITCZ region. This transition area has a slight north–south movement associated with the oscillation of the ITCZ. Another special region to highlight is the North Atlantic Ocean, which has an early morning peak in April and May, but has a preference for mixed early morning/nocturnal/late afternoon peak after the monsoon onset (June–September). In contrast, the Guinea Sea has an opposite pattern with mixed early morning/

nocturnal/late afternoon peak in April and May, but only an early morning peak from June to September. By and large, the phase for CAPE is similar among the three CPSs and they are in agreement with the precipitation amount (Figure 10.3), although some disparity appears around the transition area (especially for KF2). The analysis of CAPE further demonstrates that the DCP in RSM is dominated by the convective precipitation and the discrepancy related to the simulation of convective precipitation therefore calls for more effort for improvement.

Figure 10.12 shows time series of the normalized CAPE (by daily mean) during the monsoon season over land and ocean. We can find that the diurnal cycle of CAPE is only dominated by a single peak, occurring at late afternoon (around 1800 UTC) over land and early morning (~0800 LST) over ocean. The diurnal cycle of CAPE among CPSs differs significantly over land with RAS having the largest amplitude and KF2 having the smallest amplitude. Nevertheless, this difference among the three CPSs is not significant over ocean, as we can see that RAS and SAS almost overlap each other, and KF2 shows a slightly larger difference compared to RAS and SAS. To summarize, the sensitivity of the diurnal cycle of CAPE to different CPSs only manifests with the amplitude. There is no significant shift of the peak time.

Furthermore, compared to the diurnal variation of precipitation amount over land (top panel of Figure 10.5), the diurnal cycle of CAPE in SAS and KF2 in most months shows a very good correspondence with precipitation in terms of the peak time. However, RAS advanced the maximum precipitation by about 5 h compared to CAPE in April, May, and September. This may be related to the large-scale precipitation, which occurs a little earlier and dominates the total precipitation. KF2, by opposition, lags the maximum precipitation compared to CAPE in April and May. This may be related to the cloud work function, which needs some time to trigger

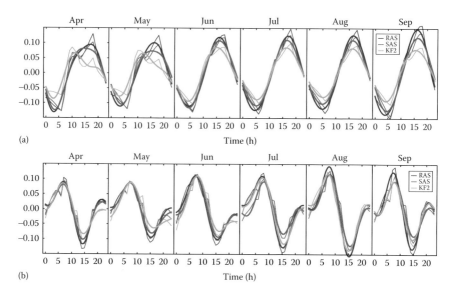

FIGURE 10.12 Normalized seasonal (April–September) evolution of diurnal cycle for CAPE averaged over (a) land and (b) ocean derived from RAS, SAS, and KF2.

the deep convection. As for the results over ocean area, although the amplitude of CAPE among three CPSs is extremely similar, this does not guarantee the similarity in precipitation as we can see from Figure 10.6 (top panel). Specifically, RAS has a lagged maximum of precipitation compared to CAPE in August and September. This is similar to the case for KF2 over land area and demonstrates that precipitation processes are not triggered immediately with the maximum of CAPE. In addition to all the analysis mentioned above, the amplitude manifested in CAPE increases in the order from KF2, SAS, and RAS. However, regarding precipitation amplitude, RAS has the smallest amplitude, which indicates that there is no direct proportional relationship between CAPE and precipitation.

In summary, the pattern of diurnal variation of precipitation follows that of CAPE, not only in terms of amplitude but also in terms of phase. Because models always advance CAPE over both land and ocean, the simulated DCP is advanced compared to observations. Besides, the diurnal cycle of CAPE is more sensitive to the cumulus parameterization over land area than ocean area.

10.4 CONCLUSIONS AND FUTURE WORK

In this study, the DCP during the 2005 WAM is simulated by the NCEP RSM and evaluated against satellite observations at the regional scale. The effects of CPSs on model simulations are discussed. Conclusions from the analysis can be summarized as follows.

10.4.1 GENERAL BEHAVIOR OF THE RSM

Comparison of model-simulated DCP with observations reveals that RSM could capture the diurnal variation of precipitation, with a late afternoon peak over land and an early morning peak over ocean. This spatial pattern is well captured compared to satellite observations in most areas. Regarding precipitation characteristics, RSM can well capture the amplitude of the amount, but the simulated frequency is too flat and the simulated intensity is too strong. In terms of the phase of the diurnal cycle, RSM advances the maximum precipitation too much, which is more apparent in diurnal variations (S_1) than semidiurnal variations (S_2), over land and ocean. This is the general deficiency of current climate models in consensus with previous research.

10.4.2 EFFECTS OF CPSs

Effects of CPSs on the simulated DCP are systematically examined at the regional scale with several CPSs adopted in RSM, which are RAS, SAS, and KF2. These three CPSs show very similar patterns in terms of the spatial distributions of the phase and amplitude at the regional scale, although a significant difference tends to occur within the ITCZ region. Regarding the seasonal evolution of the diurnal cycle, there is no obvious shift of the peak time at the regional scale, and the main difference is found for amplitude. It is noted that the degree of sensitivity of the diurnal cycle to CPS is different in the premonsoon season and pure monsoon season. Generally speaking, simulations among the three CPSs show smaller diversity in the premonsoon season than in the pure monsoon season. This may be related to the complex monsoon dynamics, which needs further investigation. These findings are of significance because they provide a foundation for future development of CPSs with the consideration of large-scale climate dynamics (monsoon systems).

10.4.3 FUTURE WORK

Some limitations still remain in this research and merit further consideration. Key feedback mechanisms involved in the diurnal cycle, including solar radiation heating, cloud formation (Trenberth et al., 2003), land–atmosphere interactions (Koster et al., 2004), as well as the model biases, planetary boundary layer scheme and superparameterization (Randall et al., 2003) deserve further progress.

REFERENCES

Alpert, J. C., M. Kanamitsu, P. M. Caplan, J. G. Sela, G. H. White, and E. Kalnay, 1988: Mountain induced gravity wave drag parameterization in the NMC Medium-Range Forecast Model. Preprints, *8th Conference on Numerical Weather Prediction*, Baltimore, MD, American Meteorological Society, pp. 726–733.

Alpert, J. C., S.-Y. Hong, and Y.-J. Kim, 1996: Sensitivity of cyclogenesis to lower tropospheric enhancement of gravity wave drag using the Environmental Modeling Center Medium Range Model. Preprints, *11th Conference on Numerical Weather Prediction*, Norfolk, VA, American Meteorological Society, pp. 322–323.

Arakawa, A. and W. H. Schubert, 1974: Interaction of cumulus cloud ensemble with the large-scale environment. Part I. *J. Atmos. Sci.*, 31, 671–701, doi:10.1175/1520-0469(1974)031,0674:IOA CCE.2.0.CO;2.

Basu, B. K., 2007: Diurnal variation in precipitation over India during the summer monsoon season: Observed and model predicted. *Mon. Wea. Rev.*, 135, 2155–2167, doi:10.1175/MWR3355.1.

Betts, A., J. H. Ball, A. C. M. Beljaars, M. J. Miller, and P. A. Viterbo, 1996: The land surface–atmosphere interaction: A review based on observational and global modeling perspectives. *J. Geophys. Res.*, 101, 7209–7225, doi:10.1029/95JD02135.

Boyd, E., R. J. Cornforth, P. J. Lamb, A. Tarhule, M. I. Lele, and A. Brouder, 2013: Building resilience to face recurring environmental crisis in African Sahel. *Nat. Climate Change*, 3, 631–637, doi:10.1038/nclimate1856.

Chou, M.-D. and M. J. Suarez, 1994: An efficient thermal infrared radiation parameterization for use in general circulation models, NASA Technical Memorandum TM-1994–104606, Series on Global Modeling and Data Assimilation, Vol. 3, 102 pp.

Dai, A., 2001: Global precipitation and thunderstorm frequencies. Part II: Diurnal variations. *J. Climate*, 14, 1112–1128, doi:10.1175/1520-0442(2001)014,1112:GPATFP.2.0.CO;2.

Dai, A., F. Giorgi, and K. E. Trenberth, 1999: Observed and model simulated diurnal cycles of precipitation over the contiguous United States. *J. Geophys. Res.*, 104, 6377–6402, doi:10.1029/98JD02720.

Dai, A. and K. E. Trenberth, 2004: The diurnal cycle and its depiction in the community climate system model. *J. Climate*, 17, 930–951, doi:10.1175/1520-0442(2004)017,0930:TDCAID. 2.0.CO;2.

Davis, C. A., K. W. Manning, R. E. Carbone, S. B. Trier, and J. D. Tuttle, 2003: Coherence of warm-season continental rainfall in numerical weather prediction models. *Mon. Wea. Rev.*, 131, 2667–2679, doi:10.1175/1520-0493(2003)131,2667:COWCRI.2.0.CO;2.

Ek, M. B., K. E. Mitchell, Y. Lin, E. Rogers, P. Grunmann, V. Koren, G. Gayno, and J. D. Tarpley, 2003: Implementation of noah land surface model advancements in the national centers for environmental prediction operational mesoscale eta model. *J. Geophys. Res.*, 108, 8851, doi:10.1029/2002JD003296.

Fritsch, J. M. and R. E. Carbone, 2004: Improving quantitative precipitation forecasts in the warm season: A USWRP research and development strategy. *Bull. Amer. Meteor. Soc.*, 85, 955–965, doi:10.1175/BAMS-85-7-955.

Garratt, J. R., P. Krummel, and E. A. Kowalczyk, 1993: The surface-energy balance at local and regional scales—A comparison of general circulation model results with observations. *J. Climate*, 6, 1090–1109, doi:10.1175/1520-0442(1993)006,1090:TSEBAL.2.0.CO;2.

Giorgi, F. and C. Shields, 1999: Tests of precipitation parameterizations available in latest version of NCAR regional climate model (RegCM) over continental United States. *J. Geophys. Res.*, 104, 6353–6376, doi:10.1029/98JD01164.

Gounou, A., 2011: The driving processes behind the diurnal cycles of the West African monsoon. Ph.D. thesis, 61–64, Universite Paul Sabatier, Toulouse III, Toulouse, France.

Gray, W. M. and R. W. Jacobson, 1977: Diurnal variation of deep cumulus convection. *Mon. Wea. Rev.*, 105, 1171–1188, doi:10.1175/1520-0493(1977)105,1171:DVODCC.2.0.CO;2.

Haurwitz, B. and A. D. Cowley, 1973: The diurnal and semidiurnal barometric oscillations, global distribution, and annual variation. *Pure Appl. Geophys.*, 102, 193–222, doi:10.1007/BF00876607.

He X., H. Kim, P.-E. Kirstetter, K. Yoshimura, E.-C. Chang, C. R. Ferguson, J. M. Erlingis, Y. Hong, and T. Oki, 2015: The diurnal cycle of precipitation in regional spectral model simulations over West Africa: sensitivities to resolution and cumulus schemes. *Wea. Foresting*, 30, 424–445, doi: http://dx.doi.org/10.1175/WAF-D-14-00013.1.

Hodges, K. I. and C. C. Thorncroft, 1997: Distribution and statistics of African mesoscale convective weather systems based on the ISCCP Meteosat imagery. *Mon. Wea. Rev.*, 125, 2821–2837, doi:10.1175/1520-0493(1997)125,2821:DASOAM.2.0.CO;2.

Hong, S.-Y. and A. Leetmaa, 1999: An evaluation of the NCEP RSM for regional climate modeling. *J. Climate*, 12, 592–609, doi:10.1175/1520-0442(1999)012,0592:AEOTNR.2.0.CO;2.

Hong, S.-Y. and H. L. Pan, 1996: Nonlocal boundary layer vertical diffusion in a medium-range forecast model. *Mon. Wea. Rev.*, 124, 2322–2339, doi:10.1175/1520-0493(1996)124, 2322:NBLVDI.2.0.CO;2.

Huffman, G. J., D. T. Bolvin, E. J. Nelkin, D. B. Wolff, R. F. Adler, and et al., 2007: The TRMM multisatellite precipitation analysis (TMPA): Quasi-global, multiyear, combined-sensor precipitation estimates at fine scales. *J. Hydrometeor.*, 8, 38–55, doi:10.1175/JHM560.1.

Jeong, J.-H., A. Walther, G. Nikulin, D. Chen, and C. Jones, 2011: Diurnal cycle of precipitation amount and frequency in Sweden: Observation versus model simulation. *Tellus* 63, 664–674.

Juang, H.-M. and M. Kanamitsu, 1994: The NMC nested regional spectral model. *Mon. Wea. Rev.*, 122, 3–26, doi:10.1175/1520-0493(1994)122,0003:TNNRSM.2.0.CO;2.

Juang, H.-M., S.-Y. Hong, and M. Kanamitsu, 1997: The NCEP regional spectral model: An update. *Bull. Amer. Meteor. Soc.*, 78, 2125–2143, doi:10.1175/1520-0477(1997)078,2125: TNRSMA.2.0.CO;2.

Kain, J. S., 2004: The Kain–Fritsch convective parameterization: An update. *J. Appl. Meteor. Climatol.*, 43, 170–181, doi:10.1175/1520-0450(2004)043,0170:TKCPAU.2.0.CO;2.

Kalnay, E., M. Kanamitsu, R. Kistler, W. Collins, D. Deaven, L. Gandin, and et al., 1996: The NCEP/NCAR 40-year reanalysis project. *Bull. Amer. Meteor. Soc.*, 77, 437–471, doi:10.1175/1520-0477(1996)077,0437:TNYRP.2.0.CO;2.

Kanamitsu, M., W. Ebisuzaki, J. Woollen, S.-K. Yang, J. J. Hnilo, M. Fiorino, and G. L. Potter, 2002: NCEP–DOE AMIP-II reanalysis (R-2). *Bull. Amer. Meteor. Soc.*, 83, 1631–1643, doi:10.1175/BAMS-83-11-1631.

Kang, H.-S. and S.-Y. Hong, 2008: Sensitivity of the simulated East Asian summer monsoon climatology to four convective parameterization schemes. *J. Geophys. Res.*, 113, D15119, doi:10.1029/2007JD009692.

Kim, J. and L. Mahrt, 1992: Simple formulation of turbulent mixing in the stable free atmosphere and nocturnal boundary layer. *Tellus* 44A, 381–394, doi: 10.1034/j.1600-0870.1992.t01-4-00003.x.

Koo, M.-S. and S.-Y. Hong, 2010: Diurnal variations of simulated precipitation over East Asia in two regional climate models. *J. Geophys. Res.*, 115, D05105, doi:10.1029/2009JD012574.

Koster, R. D., C. T. Gordon, Y. C. Sud, T. Oki, K. Oleson, and et al., 2004: Regions of strong coupling between soil moisture and precipitation. *Science*, 305, 1138–1140, doi:10.1126/science.1100217.

Laing, A. G., R. Carbone, V. Levizzani, and J. Tuttle, 2008: The propagation and diurnal cycles of deep convection in northern tropical Africa. *Quart. J. R. Meteor. Soc.*, 134, 93–109, doi:10.1002/qj.194.

Lin, X., D. A. Randall, and L. D. Fowler, 2000: Diurnal variability of the hydrological cycle and radiative fluxes: Comparisons between observations and a GCM. *J. Climate*, 13, 4159–4179, doi:10.1175/1520-0442(2000)013,4159:DVOTHC.2.0.CO;2.

McGarry, M. M. and R. J. Reed, 1978: Diurnal variations in convective activity and precipitation during phases II and III of GATE. *Mon. Wea. Rev.*, 106, 101–113, doi:10.1175/1520-0493(1978)106,0101:DVICAA.2.0.CO;2.

Moorthi, S. and M. J. Suarez, 1992: Relaxed Arakawa–Schubert: A parameterization of moist convection for general circulation models. *Mon. Wea. Rev.*, 120, 978–1002, doi:10.1175/1520-0493(1992)120,0978:RASAPO.2.0.CO;2.

Nikulin, G., C. Jones, F. Giorgi, G. Asrar, M. Büchner, and et al., 2012: Precipitation climatology in an ensemble of CORDEX-Africa regional climate simulations. *J. Climate*, 25, 6057–6078, doi:10.1175/JCLI-D-11-00375.1.

Oki, T. and K. Musiake, 1994: Seasonal change of the diurnal cycle of precipitation over Japan and Malaysia. *J. Appl. Meteor.* 33, 1445–1463.

Oki, T. and S. Kanae, 2006: Global hydrological cycles and world water resources. *Science*, 313, 1068–1072, doi:10.1126/science.1128845.

Pan, H.-L. and W.-S. Wu, 1995: Implementing a mass flux convective parameterization package for the NMC Medium-Range Forecast Model. NMC Office Note 409, Camp Springs, MD, 40 pp. Available online at http://www.lib.ncep. noaa.gov/ncepofficenotes/files/01408A42.pdf.

Parker, D. J., R. R. Burton, A. Diongue-Niang, R. J. Ellis, M. Felton, C. M. Taylor, and et al., 2005: The diurnal cycle of West African monsoon circulation. *Quart. J. R. Meteor. Soc.*, 131, 2839–2860, doi:10.1256/qj.04.52.

Pu, B. and K. H. Cook, 2010: Dynamics of the West African westerly jet. *J. Climate*, 23, 6263–6276, doi:10.1175/2010JCLI3648.1.

Ramage, C. S., 1971: *Monsoon Meteorology*. New York, Academic Press, pp. 295.

Randall, D., M. Khairoutdinov, and W. Grabowski, 2003: Breaking the cloud parameterization deadlock. *Bull. Amer. Meteor. Soc.*, 84, 1547–1564, doi:10.1175/BAMS-84-11-1547.

Randall, D. A., Harshvardhan, and D. A. Dazlich, 1991: Diurnal variability of the hydrological cycle in a general circulation model. *J. Atmos. Sci.*, 48, 40–62, doi:10.1175/15200469(1991)048,0040:DVOTHC.2.0.CO;2.

Ridout, J. A., Y. Jin, and C.-S. Liou, 2005: A cloud-base quasi-balance constraint for parameterized convection: Application to the Kain–Fritsch cumulus scheme. *Mon. Wea. Rev.*, 133, 3315–3334, doi:10.1175/MWR3034.1.

Rio, C., F. Hourdin, J.-Y. Grandpeix, and J.-P. Lafore, 2009: Shifting the diurnal cycle of parameterized deep convection over land. *Geophys. Res. Lett.*, 36, L07809, doi:10.1029/2008GL036779.

Roca, R., P. Chambon, I. Jobard, P. Kirstetter, M. Gosset, and J. C. Berges, 2010: Comparing satellite and surface rainfall products over West Africa at meteorologically relevant scales during the AMMA campaign using error estimates. *J. Appl. Meteor. Climatol.*, 49, 715–731, doi:10.1175/2009JAMC2318.1.

Shine, D. W., S. Cocke, and T. E. LaRow, 2007: Diurnal cycle of precipitation in a climate model. *J. Geophys. Res.*, 112, D13109, doi:10.1029/2006JD008333.

Shinoda, M., T. Okatani, and M. Saloum, 1999: Diurnal variations of rainfall over Niger in the West African Sahel: A comparison between wet and drought years. *Int. J. Climatol.* 19, 81–94.

Slingo, A., R. Wilderspin, and S. Brentnall, 1987: Simulation of the diurnal cycle of outgoing longwave radiation with an atmospheric GCM. *Mon. Wea. Rev.*, 115, 1451–1457, doi:10.1175/15200493(1987)115,1451:SOTDCO.2.0.CO;2.

Sperber, K. R. and T. Yasunari, 2006: Workshop on monsoon climate systems: Toward better prediction of the monsoon. *Bull. Amer. Meteor. Soc.*, 87, 1339–1403, doi:10.1175/BAMS-87-10-1399.

Taylor, C. M. and D. B. Clark, 2001: The diurnal cycle and African easterly waves: A land surface perspective. *Quart. J. R. Meteor. Soc.*, 127, 845–867, doi:10.1002/qj.49712757308.

Trenberth, K. E., A. Dai, R. M. Rasmussen, and D. B. P. Parsons, 2003: The changing character of precipitation. *Bull. Amer. Meteor. Soc.*, 84, 1205–1217, doi:10.1175/BAMS-84-9-1205.

Wang, W. and N. L. Seaman, 1997: A comparison study of convective parameterization schemes in a mesoscale model. *Mon. Wea. Rev.*, 125, 252–278, doi:10.1175/15200493(1997)125,0252:ACSOCP.2.0.CO;2.

Wilks, D. S., 2005: *Statistical Methods in the Atmospheric Sciences.* 2nd ed. Amsterdam: Boston, Academic Press, pp.627.

Yang, G. and J. Slingo, 2001: The diurnal cycle in the tropics. *Mon. Wea. Rev.*, 129, 784–801, doi:10.1175/15200493(2001)129,0784:TDCITT.2.0.CO;2.

Yoshimura, K. and M. Kanamitsu, 2008: Dynamical global downscaling of global reanalysis. *Mon. Wea. Rev.*, 136, 2983–2998, doi:10.1175/2008MWR2281.1.

Yoshimura, K., M. Kanamitsu, and M. Dettinger, 2010: Regional downscaling for stable water isotopes: A case study of an atmospheric river event. *J. Geophys. Res.*, 115, D18114, doi:10.1029/2010JD014032.

11 Multiscale Evaluation and Applications of Current Global Satellite-Based Precipitation Products over the Yangtze River Basin

Zhe Li, Dawen Yang, Yang Hong, Bing Gao, and Qinghua Miao

CONTENTS

11.1　INTRODUCTION

Global environmental problems are always characterized by multiscale patterns and processes, and this multiscale nature calls for the building of hierarchical observation facilities as well as modeling platforms with various support scales. As a key component of the global environment, the terrestrial hydrologic cycle also presents great heterogeneity across a variety of spatial and temporal scales. To address the needs for Earth's terrestrial water cycle monitoring, the satellite-based remote sensing technique has emerged as an essential method and provided unprecedented opportunities for global and regional hydrologic researches during the past two decades (Wood et al., 2011).

Precipitation, the primary input forcing of terrestrial hydrologic cycle, provides vital information to estimate land surface hydrologic fluxes and states (Nijssen and Lettenmaier, 2004). Therefore, accurate measurement of precipitation plays a critical role in determining hydrologic responses. However, due to the sampling uncertainty (in both time and space) associated with conventional rain gauges, it remains challenging for near-real-time precipitation monitoring in remote regions, ungauged basins, or complex terrains (Huffman et al., 2011). In recent 10 years, as an alternative approach, a growing number of satellite-based high-resolution (0.25 deg and 3 h, or finer) global precipitation products have been developed for large-scale precipitation quantitative monitoring in a near-real-time manner, and they include PERSIANN (Sorooshian et al., 2000), CMORPH (Joyce et al., 2004), PERSIANN-CCS (Hong et al., 2004), NRL-Blend (Turk and Miller, 2005), TMPA (Huffman et al., 2007), and GSMap (Kubota et al., 2007). With intensive investigations, now the hydrologic prediction community has recognized the opportunity these products offer to improve flood monitoring over medium to large river basins (Hossain and Lettenmaier, 2006), and it is also demonstrated the continuing improvements of multisensor blended algorithms have provided increasing potential to apply these products in hydrologic predictions (Su et al., 2008; Yong et al., 2012). Despite these long-lasting efforts to upgrade the input data sources as well as the retrieval algorithms, current satellite-based precipitation products are also subject to different error sources, arising from indirect measurements, retrieval algorithms, and sampling uncertainty (e.g., McCollum et al., 2002; Villarini et al., 2009). Therefore, before the products can be applied to hydrologic predictions, their error characteristics should be quantified so that these data can be used as effectively as possible (Turk et al., 2008).

The Yangtze River, the largest river in China, is known to be extremely susceptible to frequent floods ranging from local to regional scales due to the heavy rainstorm events in summer (Heike et al., 2012). The large-scale Asian monsoon activities combined with region-dependent terrains have shaped complicated precipitation regimes over the Yangtze River basin, thus a near-real-time precipitation monitoring technique is essential for developing an excellent basin-scale flood prediction and warning system (Sohn et al., 2012). The present work focuses on evaluation and application of several currently available global satellite-based precipitation products over the Yangtze River basin. First, with limited data from a dense gauge network, we implement a 4-year (2008–2011) statistical evaluation work to provide up-to-date insight into the skills and the error patterns of three most popular satellite products (TMPA, CMORPH, and PERSIANN), and select the best candidates for hydrologic applications over the Yangtze River; second, we extend our analysis into a decade-long (2003–2012) modeling-based evaluation of the precipitation products using a physically-based distributed model to discuss the potential and challenges for hydrologic applications of these satellite-based precipitation products, including basin-scale water budgeting and streamflow simulation (Li et al., 2015).

11.2 DATA AND MODEL

11.2.1 SATELLITE-BASED PRECIPITATION PRODUCTS

In the present study, we evaluate and intercompare four sets of most popular multisensor blended precipitation estimators: Version 7 of TMPA 3B42 research data (hereinafter referred to as 3B42V7) and of TMPA 3B42 near-real-time data (hereinafter referred to as 3B42RT), CMORPH and PERSIANN, all of which synthesize information from both passive microwave (PMW) data and infrared (IR) measurements. 3B42V7 utilizes ground gauge observations to remove the biases of satellite retrievals, while 3B42RT, CMORPH, and PERSIANN are derived solely from real-time satellite scanning images. All of these products offer precipitation estimates every 3 h on a $0.25° \times 0.25°$ grid with quasiglobal coverage (e.g., 50°S–50°N), and for our evaluation purpose, then we integrate the original 3-hourly precipitation retrievals into daily and larger timescale for each satellite dataset.

11.2.2 GROUND GAUGE DATA

Gauge observations play a critical role in the quantitative evaluation, and therefore, we adopted the China Daily Precipitation Grids (CDPG) produced routinely by the National Meteorological Information Center (NMIC) of the China Meteorological Administration (CMA) for statistical evaluation. This operational system takes in over 2400 gauge observations in China to generate nation-wide daily precipitation at $0.25° \times 0.25°$ grid scale since April 2008. The algorithm used to define the analyzed field of precipitation is adopted from the objective technique used by Xie et al. (2007). The NMIC operational system calculates the daily precipitation climatology first, then adjusts it using the PRISM approach (Daly et al., 1994), and yields the ratio of daily precipitation to climatological precipitation by interpolating the daily observations at gauges using the optimal interpolation method of Gandin (1965). More details about the NMIC processing algorithms can be found in Shen et al. (2010a, 2010b). In this study, as this CDPG data's resolution matches the spatial scale of satellite-based precipitation data, we implement a 4-year evaluation and directly compare CDPG with various satellite precipitation products during 2008–2011.

Although CDPG dataset is generated from a dense gauge network, it is only available since April 2008 and thus its data length remains limited. To extend our study into a decadal modeling-based evaluation, we also use another set of data, called China Daily Ground Climate Dataset (CDCD) from CMA. We select 141 CDCD gauges that are located in the Yangtze River basin (Figure 11.1a) and utilize their long-term (1961–2012) daily meteorological records, including precipitation, air temperature, relative humidity, wind speed, and sunshine hours, to calibrate model and evaluate the satellite precipitation products.

The daily discharge data (1961–2011) are collected from the Hydrologic Year Book published by the Hydrologic Bureau of the Ministry of Water Resources of China, and discharge data of 2012 is separately obtained from China Three Gorges Corporation (Li et al., 2015). We select 12 streamflow gauges in this study, most of which are the controlling gauges for major tributaries or mainstream of the Yangtze River (see Table 11.1).

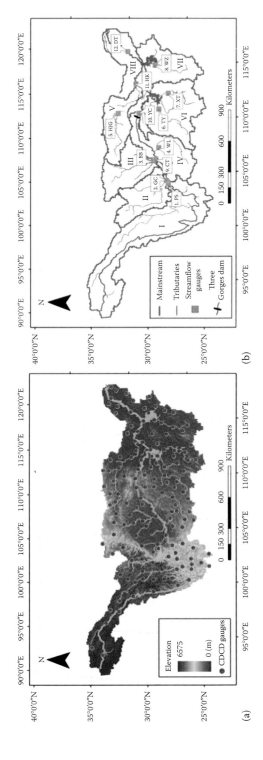

FIGURE 11.1 (a) Topography and CMA CDCD gauges and (b) streamflow gauges and subregions division in the Yangtze River basin. (Modified from Li et al., *J. Hydrometeorol.*, 16, 407–426, 2015. ©American Meteorological Society. With permission.)

TABLE 11.1

Streamflow Gauges and Hydrologic Regions in the Yangtze River Basin

Location	Gauge Name[a]	River Name	Drainage Area (10^4 km²)	Subregions[b]
Major tributaries	PS (1)	Jinsha	45.86	I
	GC (2)	Min	13.54	II
	BB (3)	Jialing	15.67	III
	WL (4)	Wu	8.30	IV
	HJG (5)	Han	9.52	V
	TY (6)	Yuan	8.52	VI
	XT (7)	Xiang	8.16	
	WZ (8)	Gan	8.09	VII
Mainstream	CT (9)	Yangtze	86.66	IV
	YC (10)	Yangtze	100.55	IV
	HK (11)	Yangtze	148.80	VIII
	DT (12)	Yangtze	170.54	

[a] Full names of the gauges: BB, Beibei; CT, Cuntan; DT, Datong; GC, Gaochang; HJG, Huangjiagang; HK, Hankou; PS, Pingshan; TY, Taoyuan; WL, Wulong; WZ, Waizhou; XT, Xiangtan; YC, Yichang.

[b] Full names of subregions: I: Jinsha River; II: Min and Tuo Rivers; III: Jialing River; IV: Wu River and Three Gorges Area; V: Han River; VI: Dongting Lake Region; VII: Poyang Lake Region; VIII: Middle and Lower Yangtze Mainstream.

To investigate the region-dependent performance of satellite products, we further divide the whole river basin into eight subregions (Figure 11.1b), according to the drainage system of the Yangtze (Li et al., 2015). From the upstream to the downstream, called Jinsha River, Min and Tuo Rivers, Jialing River, Wu River and Three Gorges Area, Han River, Dongting Lake region, Poyang Lake region, and the Middle and Lower Yangtze mainstream, respectively.

11.2.3 HYDROLOGIC MODEL

A geomorphology-based hydrologic model (GBHM; Yang and Musiake, 2003; Cong et al., 2009; Li et al., 2012) is established over the whole Yangtze River basin to serve as the critical platform for integrating different input forcing data. This distributed model takes advantage of the concept of geomorphologic similarity and characterizes the catchment by hillslope-stream formulation, and thus hillslope is the computational unit for hydrologic simulation in GBHM (Figure 11.2), and the hillslope-based simulation of hydrologic processes includes following parts: snowmelt, canopy interception, evapotranspiration, infiltration, surface and subsurface flow, and groundwater-river channel interaction (Yang and Musiake, 2003). A detailed description of GBHM physical representations can be found in several recent papers (Cong et al., 2009; Li et al., 2015).

The GBHM model of the Yangtze River basin is constructed with the grid size of 10 km, and the whole basin is divided into 137 sub-basins organized according to

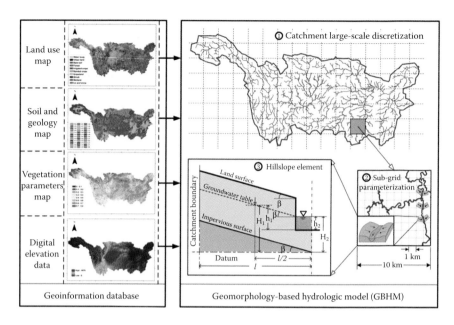

FIGURE 11.2 Schematic diagram of the model structure for Yangtze River GBHM. (From Li et al., *J. Hydrometeorol.*, 16, 407–426, 2015. ©American Meteorological Society. With permission.)

the Pfafstetter system (Yang and Musiake, 2003). All of the geographical information grid data (Figure 11.2, e.g., DEM, soil type, land use, and NDVI) are resampled into 10-km grids of the Yangtze, while 10-km meteorological forcing fields are also obtained by interpolation of 141 CDCD gauges records using an angular distance weighing or an elevation-corrected angular direction weighing method (Yang and Musiake, 2003).

We use daily discharge data and CDCD forcing data during the period of 1961–1965 to calibrate the model, and apply the same two datasets from 1966 to 2002 to perform a long-term validation. The results show GBHM model is able to simulate river flow regimes of the Yangtze River basin reasonably, with Nash-Sutcliffe coefficient of efficiency (*NSCE*) greater than 0.70, while relative bias (*RB*) ranging from −4.0% to 12.0% during the period of 1961–2002 (Li et al., 2015). Consequently, this distributed model can be employed as a rational modeling framework to benchmark different satellite precipitation products into hydrologic applications during the past decade (2003–2012).

11.3 STATISTICAL ASSESSMENT

11.3.1 ANNUAL SCALE COMPARISON

The performance of 3B42V7, 3B42RT, CMORPH, and PERSIANN over the Yangtze River are first evaluated and intercompared according to the CMA CDPG reference data during 2008–2011 at annual, seasonal, and daily scales, respectively.

Figure 11.3 shows the spatial distribution of estimation bias (mm/day) for annual mean precipitation retrieved by different satellite-based precipitation products. It is

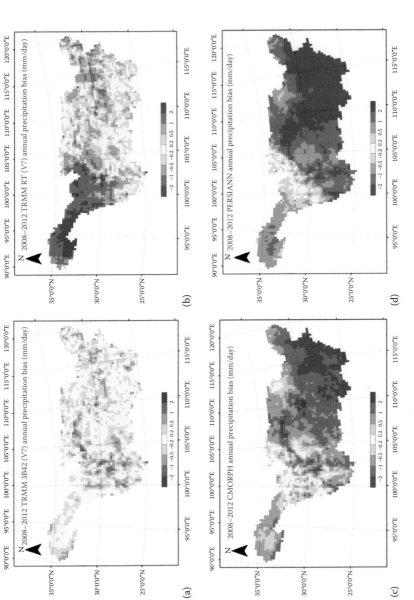

FIGURE 11.3 Annual mean precipitation estimation bias (mm/day) between the satellite precipitation products (a) 3B42 V7; (b) 3B42 RT; (c) CMORPH; and (d) PERSIANN, and the ground reference CDPG data.

clear that 3B42V7 presents the smallest bias among the four satellite-based estimators, indicating the critical role of gauge-based correction algorithm for bias removal. In contrast, without ground information, other three sets of near-real-time products show evident region-dependent biases. In the upper Yangtze River, 3B42RT and PERSIANN overestimate (by >1 mm/day and 0–2 mm/day, respectively) precipitation at annual scale. Considering both datasets are developed by PMW-calibrated IR technique, the local positive bias is attributed to the dominant usage of IR data without adequate calibration by PMW data because most PMW overpasses over the Tibetan Plateau have been screened out due to the snow covers (Yong et al., 2015). On the contrary, by using a different retrieval algorithm called Lagrangian interpolation approach, CMORPH estimation shows negative biases (−2 to 0 mm/day) in most area of the upper Yangtze River. Changing focus region to the middle and lower Yangtze River, CMORPH and PERSIANN tend to underestimate precipitation seriously (< −1 mm/day), while 3B42RT presents mixed error pattern (−1 to 1 mm/day) with strong local bias.

11.3.2 SEASONAL ESTIMATION AND ITS VARIATIONS

To study on the seasonal variations for the performance of satellite-based precipitation products, we divide the whole year into four seasons and analyze seasonal results separately. Here, due to the limited space, we take April–June and October–December to represent warm season and cold season, respectively. Figure 11.4 compares the spatial map of estimation relative bias in April–June and in October–December. We focus on near-real-time satellite-based products only because the monthly gauge-adjustment algorithm has reduced estimation bias substantially at larger temporal scales.

In the warm season (April–June, Figure 11.4a, c, and e), CMORPH and PERSIANN both underestimate precipitation (by −20% to −80% and −20% to −50%, respectively) in the middle and lower Yangtze River, while 3B42RT shows relative small bias with mixed error pattern (−20% to 50%) over this region. Compared with the results for the warm season, the performance of all the three datasets decline during the cold season (October–December, Figure 11.4b, d, and f): CMORPH and PERSIANN underestimate precipitation significantly (by −50% or even more) in most of the basin and 3B42RT's estimation bias increases to ranging from −20% to 200%.

In addition, we are able to separate the contribution from different seasons to annual total bias by comparing the seasonal error patterns. Obviously, the positive bias over the upper Yangtze River for 3B42RT is long lasting and presents through the whole year; however, the estimation biases over this region for CMORPH and PERSIANN are season dependent as the serious overestimation mainly comes from the cold season. It is also noted that all the products, especially 3B42RT and PERSIANN, indicate a strange "stratified" spatial pattern for estimation bias during the cold season, extending in the North–South direction. However, the actual accumulated precipitation in this season increases from the West to the East, which thoroughly differs from the errors pattern. This seasonal pattern perhaps is caused by the deficiencies of current satellite's detection ability to separate snow/ice from rainfall, or it is due to the overlapping of different satellite paths during this season, but we leave the indepth investigation of this issue for the future research.

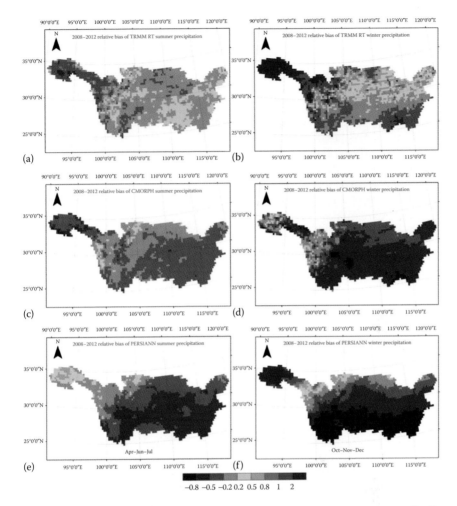

FIGURE 11.4 Relative bias of precipitation estimation during the warm season (April to June, left) and the cold season (October to December, right) by comparing satellite precipitation (a, b) 3B42 RT; (c, d) CMORPH; and (e, f) PERSIANN products, with CDPG data.

11.3.3 Daily-Scale Evaluation

Daily precipitation is strongly linked with watershed hydrologic response, so the performance of different satellite-based precipitation products at daily scale will directly determine the following hydrologic application results, especially for river flow and floods simulation. In this section, we discuss daily performance from three aspects, including time correlation, temporal variations of daily evaluation statistics, and regional variations of daily performance.

Figures 11.5 and 11.6 show the spatial maps of time correlation between satellites daily estimates and daily CDPG data in summer (April–June) and winter

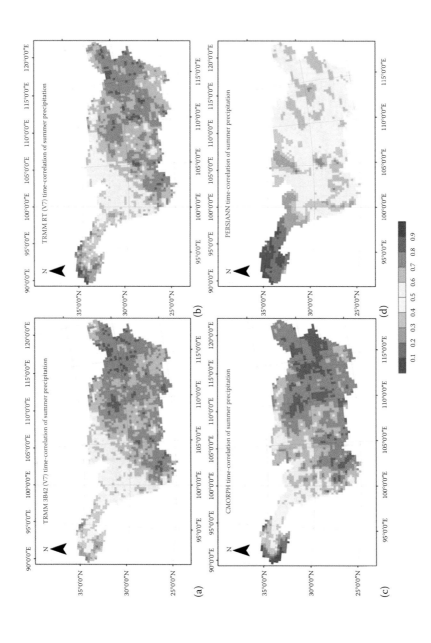

FIGURE 11.5 Time correlation of daily precipitation in summer over the Yangtze River between satellite estimates (a) 3B42 V7; (b) 3B42 RT; (c) CMORPH; and (d) PERSIANN, and CDPG observation.

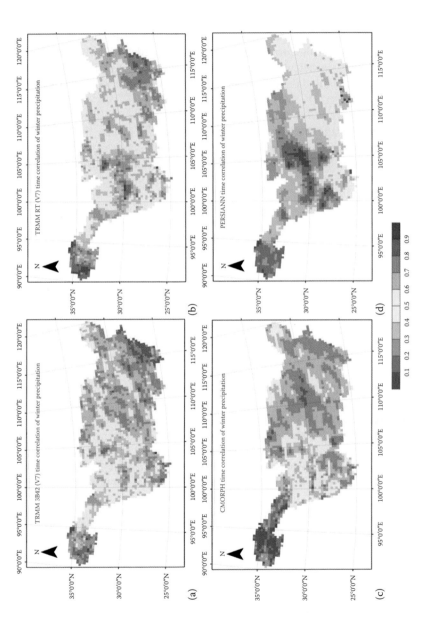

FIGURE 11.6 Time correlation of daily precipitation in winter over the Yangtze River between satellite estimates (a) 3B42 V7; (b) 3B42 RT; (c) CMORPH; and (d) PERSIANN, and CDPG observation.

(October–December), respectively. It is clear that PERSIANN shows the lowest correlation value in both summer and winter, while other three products can give reasonable result with temporal correlation larger than 0.5 over most area of the Yangtze River. CMORPH shows the largest correlation values (>0.6) during both summer and winter, and remarkably, gets even better results than 3B42V7 in summer. This suggests that gauge information works effectively to remove the bias, but its impact on improving correlation is perhaps limited. Similarly, comparing the time correlation coefficient maps, we also identify a decline in performance from summer to winter, and the extent of low-correlation region expands eastward during winter.

Furthermore, Figure 11.7 summarizes several basin-scale averaged evaluation statistics calculated in a time-varying manner, and the statistics include bias (mm/day), temporal correlation, root-mean-square error (*RMSE,* mm/day), and probability of detection (*POD*). For the high-frequency noises removal, we also apply a 30-day moving window averaging filter to obtain smoothed time series of the basin-scale statistics (Li et al., 2013).

With the basin-scale skill measures, we can clearly find that CMORPH and PERSIANN tend to underestimate precipitation in terms of basin-scale mean, while 3B42RT always overestimates it. In general, basin-scale evaluation shows the estimation bias of CMORPH is smallest among the three near-real-time satellite datasets, as 3B42RT presents extremely high overestimation during summer with pulse-like sharp peaks (up to 3–4 mm/day). Considering the averaged correlation at basin-scale, this skill metric indicates that the best dataset is CMORPH, which performs better than 3B42V7 in summer but works similar to 3B42V7 in winter. This time series analysis also reveals seasonal variations of the performance for satellite-based products because the correlation coefficient reaches its highest value in summer of each year and reduces to its lowest point in winter (Figure 11.7b). When discussing about the results of *RMSE,* it is found that 3B42RT and PERSIANN show the largest error level, and CMORPH and 3B42V7 work in a very similar manner (Figure 11.7c). Basically, *RMSE* is a good measure of accuracy, and thus this result implies that both CMORPH and 3B42V7 are better estimators compared with the ground reference data. Finally, Figure 11.7d shows that the detection ability of precipitation events for 3B42V7, 3B42RT, and CMORPH are similar, which suggests about 80% of the rainfall events can be detected from the space by current satellites products.

For a better understanding on the following results of modeling-based analysis, we also compare area-averaged precipitation at sub-basin scale. To directly link the precipitation estimates to the simulated results in next section, we analyze the scatter plots of CDCD gauge interpolated daily precipitation and the satellite estimates over the eight subregions (Table 11.1 and Figure 11.1) of the Yangtze River during the past decade (2003–2012). According to the results discussed above, it is suggested that TMPA series (3B42V7 and 3B42RT) and CMORPH are candidate datasets for hydrologic applications, and thus we only discuss the daily basin-scale performance for these three datasets (Figure 11.8).

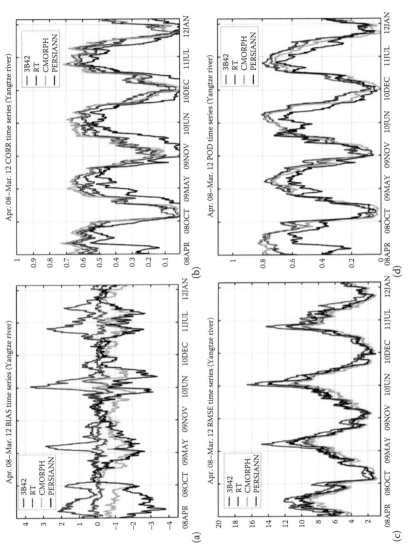

FIGURE 11.7 Smoothed daily time series of basin-scale averaged skill measures (a) bias; (b) correlation coefficient; (c) RMSE; and (d) POD for satellite products.

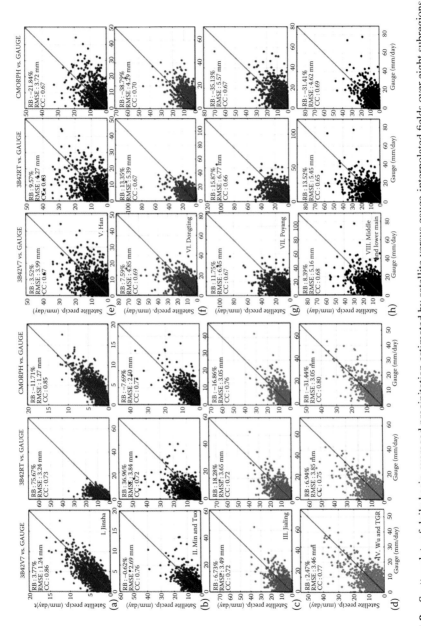

FIGURE 11.8 Scatter plots of daily basin-averaged precipitation estimated by satellite versus gauge-interpolated fields over eight subregions (a) Jinsha River; (b) Min and Tuo River; (c) Jialing River; (d) Wu River and Three Gorges region; (e) Han River; (f) Dongting Lake region; (g) Poyang Lake region; and (h) Middle and Lower mainstream. (From Li et al., *J. Hydrometeorol.*, 16, 407–426, 2015. ©American Meteorological Society. Used with permission.)

In general, 3B42V7 gives fairly good approximation to CDCD fields with slight overestimation over most subregions. Without gauge adjustment, 3B42RT is found to have overestimated daily basin-scale rainfall over the whole Yangtze River, especially in the upper Yangtze (with relative bias of 75.67%, 36.96%, and 18.28% for Jinsha River, Min, and Tuo River, and Jialing River, respectively). At the same time, CMORPH works as the complementary data for 3B42RT, as it slightly underestimates basin-scale daily precipitation over the upper Yangtze River but seriously underestimates over the lower Yangtze River (with relative bias of −31.44%, −38.79%, −35.13%, and −31.41% for Wu River and Three Gorges region, Dongting Lake region, Poyang Lake region and the Lower Yangtze mainstream, respectively). We also noted that 3B42V7 does not always show its superiority over other products at daily scale, in particular CMORPH, in terms of *RMSE* and correlation. This implies, although monthly gauge-based correction algorithm tends to make 3B42V7 statistically closer to the gauge-measured monthly precipitation, there is no guarantee for the improvement of daily precipitation estimates within a month.

11.4   MODELING-BASED EVALUATION

Taking the three sets of candidate satellite-based precipitation products (3B42V7, 3B42RT, and CMORPH) and CDCD precipitation records as the inputs to force the distributed GBHM model for the Yangtze River. With the same initial condition warmed up by a 10-year presimulation, we eventually use the modeling results during the period of 2003–2012 to assess the performance of various satellite-based precipitation products in regional hydrologic applications.

11.4.1   Annual Water Balance Simulation

Figure 11.9 compares annual water balance components, including precipitation, evapotranspiration, and runoff, at basin-scale over the eight subregions of the Yangtze. Employing the gauge-driven modeling results as the benchmark, we calculate the relative bias ($RB\_P, RB\_E,$ and RB_R) of the modeling results driven by satellite forcing data. As expected from the statistical analysis results, 3B42V7's simulation gives the closest agreement with benchmark modeling by gauge, and it is demonstrated that 3B42V7 should be the most appropriate satellite-retrieved precipitation data for annual scale water budgeting over the Yangtze River.

Because both 3B42RT and CMORPH present region-dependent bias of precipitation estimation, so this forcing error will be propagated into other simulated fluxes and states correspondingly. In summary, 3B42RT tends to substantially overestimate runoff in the upstream basins, while there is no obvious change for its modeled evapotranspiration (Figure 11.9a and b). Comparatively, CMORPH underestimates runoff seriously in the downstream regions of the Yangtze River, but its simulated evapotranspiration decreases correspondingly as well (Figure 11.9d–h). Comparing the relative bias of the estimated precipitation, simulated evapotranspiration, and simulated

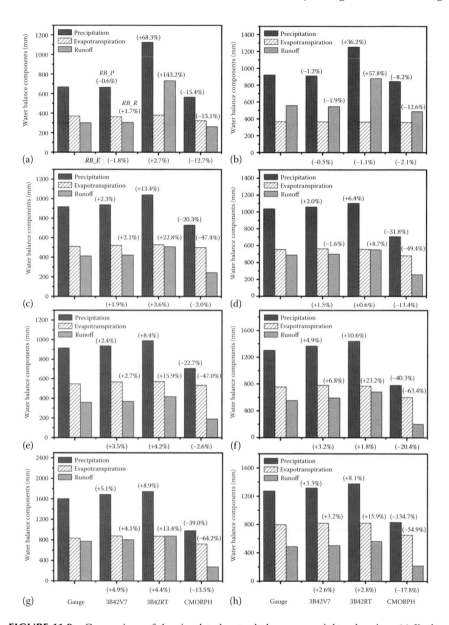

FIGURE 11.9 Comparison of the simulated water balance over eight subregions (a) Jinsha River; (b) Min and Tuo River; (c) Jialing River; (d) Wu River and Three Gorges region; (e) Han River; (f) Dongting Lake region; (g) Poyang Lake region; and (h) Middle and Lower mainstream, of the Yangtze River. (From Li et al., *J. Hydrometeorol.*, 16, 407–426, 2015. ©American Meteorological Society. Used with permission.)

runoff, it further provides implications on error propagation from the input forcing to model output.

Due to the energy-control nature for evapotranspiration processes over the Yangtze River during the warm season, the increased precipitation estimates by 3B42RT do not change the simulated evapotranspiration substantially, and we can find that the positive bias in 3B42RT estimates has been majorly propagated into the bias of simulated runoff. Looking into the relationship between the relative bias of precipitation (RB_P) and the relative bias of runoff (RB_R), it is concluded that the relative bias has been enhanced with a multiplier of 2 through hydrologic modeling. In contrast, the comparison results of CMORPH indicate that simulated evapotranspiration tends to be more sensitive to negative precipitation estimation bias over the Yangtze River, and similarly, this negative relative bias has also been enhanced by a factor of 1.5.

11.4.2 STREAMFLOW SIMULATION

Figure 11.10 compares the simulated monthly river discharge driven by different forcing datasets in major tributaries of the Yangtze River. First of all, it is found that 3B42V7 performs the best in capturing the streamflow dynamics of all tributaries of medium-size basin area ($8–46 \times 10^4 \, \text{km}^2$). As for near-real-time products, both 3B42RT and CMORPH show remarkable region-dependent errors, which have close correspondence to the local bias contained in precipitation estimates. In general, 3B42RT mainly overestimated streamflow in the Jinsha River and Min and Tuo Rivers during the warm season, which can be traced back to large positive bias in precipitation (Figure 11.8); also as a result of the substantially underestimated precipitation, the simulated streamflow by CMORPH cannot capture flood events during summer over the downstream sub-basins, such as Wu and Three Gorges area, Yuan River, and Gan River. In other words, this suggests 3B42RT is able to obtain better streamflow simulation results over the middle and lower Yangtze River, but CMORPH should be applied to the upper Yangtze River basin. Besides, considering the mixed error pattern of near-real-time satellite-based precipitation products as discussed above, special caution should be paid when 3B42RT or CMORPH is applied to streamflow modeling over the smaller-scale catchments compared to the scale of subregions discussed in this study because the local positive bias and negative bias perhaps cannot cancel each other out (Li et al., 2015).

Figure 11.11 compares the simulated monthly streamflow with observations along the mainstream of the Yangtze River. This result of mainstream stations with larger drainage area ($87–170 \times 10^4 \, \text{km}^2$) indicates that modeling performance will be better with larger watershed size. Generally, as the error of 3B42RT and CMORPH derived precipitation is dominated by positive and negative bias, 3B42RT tends to overestimate streamflow along the river mainstream, while CMORPH underestimates it overall. This error accumulation behavior indicates that 3B42RT and CMORPH need timely ground-based correction before their application to the mainstream streamflow modeling.

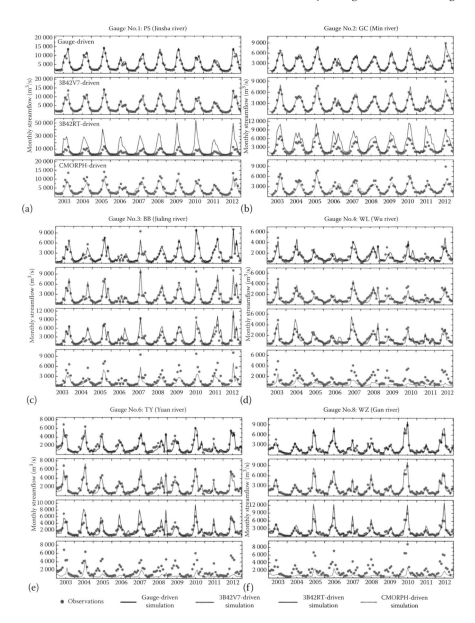

FIGURE 11.10 Comparison of simulated monthly discharge by different precipitation inputs over major tributaries of the Yangtze River (a) Jinsha River; (b) Min River; (c) Jialing River; (d) Wu River; (e) Yuan River; and (f) Gan River. (From Li et al., *J. Hydrometeorol.*, 16, 407–426, 2015. ©American Meteorological Society. With permission.)

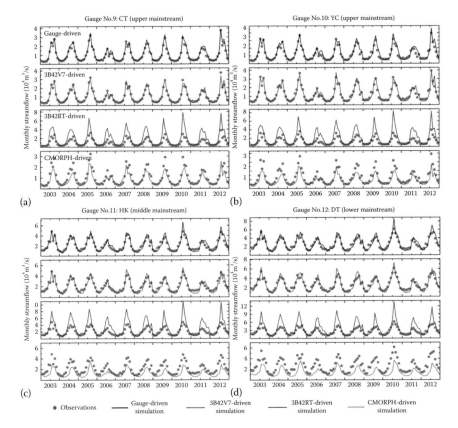

FIGURE 11.11 Comparison of simulated monthly discharge by different precipitation inputs over the mainstream of the Yangtze River (a, b) Upper mainstream; (c) Middle mainstream; and (d) Lower mainstream. (From Li et al., *J. Hydrometeorol.*, 16, 407–426, 2015. ©American Meteorological Society. With permission.)

11.5 CONCLUSIONS

High-resolution satellite-based precipitation products have provided unprecedented opportunities to improve our ability to understand large-scale precipitation dynamics in a near-real-time manner. In this study, we implement both statistics-based analysis and modeling-based evaluation of the performance of four most popular satellite precipitation products (3B42V7, 3B42RT, CMORPH, and PERSIANN) for their hydrologic applications in the Yangtze River basin. With a decade-long (2003–2012) observation data, this evaluation work aims to assess the multiscale hydrologic utilities of the latest satellite precipitation datasets. To accomplish this thorough evaluation work, we first statistically compare different satellite products with ground reference data at multiscales and then evaluate their utilities for hydrologic applications via a physically-based distributed hydrologic model over the Yangtze River basin. The major conclusions can be summarized as follows:

1. Generally, 3B42RT (CMORPH) overestimates (underestimates) precipitation, while 3B42V7 (PERSIANN) shows least (most) bias. Spatially, the upper Yangtze suffers more severe systematic errors for all datasets, showing relative large uncertainty for satellite-based precipitation estimates.

2. There are apparent season-dependent errors for all the satellite products, which presents better performance in the warm period. In the upper Yangtze River, there is a yearlong overestimation for 3B42RT, while the errors for CMORPH and PERSIANN at this region are season dependent, as the serious overestimation mainly comes from the cold season.

3. Considering the temporal correlations, RMSE, POD, and other occurrence statistics, CMORPH consistently shows good skill among the near-real-time products during the summer.

4. For regional water budgeting study, 3B42V7 works fairly well as the most appropriate dataset. It is also found that the bias in precipitation estimates has been mainly propagated into simulated runoff, and simulated evapotranspiration tends to be more sensitive to negative bias in the Yangtze River.

5. For monthly streamflow modeling, the assessment suggests that 3B42V7-driven simulations perform very well and show similar results as the gauge's. As for two near-real-time data, 3B42RT gets better performance in the midstream and downstream sub-basins, while CMORPH can be applied to the upstream watersheds. However, because the mixed error pattern exists, special caution should be taken when these products are applied to catchments with smaller spatial scale compared to sub-basins we discussed here. Along the mainstream, the modeling error will accumulate, which also indicates need to correct 3B42RT and CMORPH by ground observations before their application to the mainstream streamflow modeling.

With the contexts discussed above, we believe the present study will promote better utilization of satellite precipitation products in various hydrologic applications over the Yangtze River. Clearly, satellite precipitation products could provide useful information for regional hydrologic research, but there is an urgent need to develop multisource precipitation information (e.g., gauges, radar and satellites) merging techniques to substantially improve the performance of satellite-based precipitation estimates.

ACKNOWLEDGMENTS

This research was supported by the National Natural Science Funds for Distinguished Young Scholars (Project 51025931), the National Natural Science Foundation of China (Project 51190092), and the Science and Technology Promotion Project of Ministry of Water Resources of China (TG1528).

REFERENCES

Cong, Z. T., Yang, D. W., Gao, B., Yang, H. B., Hu, H. P., 2009. Hydrological trend analysis in the Yellow River basin using a distributed hydrological model. *Water Resour. Res.*, 45, W00A13.

Daly, C., Neilsen, R. P., Phillips, D. L., 1994. A statistical-topographic model for mapping climatological precipitation over mountainous terrain. *J. Appl. Meteorol.*, 33, 140–158.

Gandin, L. S., 1965. *Objective Analysis of Meteorological Fields.* Jerusalem: Israel Program for Scientific Translations, 242.

Heike, H., Becker, S., Jiang, T., 2012. Precipitation variability in the Yangtze River subbasins. *Water Int.*, 37, 16–31.

Hong, Y., Hsu, K. L., Gao, X., Sorooshian, S., 2004. Precipitation estimation from remotely sensed imagery using artificial neural network-cloud classification system (PERSIANN-CCS). *J. Appl. Meteorol. Climatol.*, 43, 1834–1853.

Hossain, F. and Lettenmaier, D. P., 2006. Flood prediction in the future: Recognizing hydrologic issues in anticipation of the global precipitation measurement mission. *Water Resour. Res.*, 42, W11301.

Huffman, G. J., Adler, R. F., Bolvin, D.T., Gu, G., Nelkin, E. J., Bowman, K. P., Hong, Y., Stocker, E. F., Wolff, D. B., 2007. The TRMM multi-satellite precipitation analysis (TMPA): Quasi-global, multi-year, combined-sensor precipitation estimates at fine scales. *J. Hydrometeorol.*, 8, 38–55.

Huffman, G. J., Bolvin, D. T., Nelkin, E. J., Adler, R. F., 2011. Highlights of version 7 TRMM multi-satellite precipitation analysis (TMPA). In: Klepp, C. and Huffman, G. J., (Eds.) *5th International Precipitation. Working Group Workshop, Workshop Program and Proceedings*, 11–15 October, 2010, Hamburg, Germany. (Reports on Earth System Science, 100/2011, Max-Planck-Institut für, Meteorologie, pp. 109–110.)

Joyce, R. J., Janowiak, J. E., Arkin, P. A., Xie, P., 2004. CMORPH: A method that produces global precipitation estimates from passive microwave and infrared data at high spatial and temporal resolution. *J. Hydrometeorol.*, 5, 487–503.

Kubota, T., Shige, S., Hashizume, H., Aonashi, K., Takahashi, N., Seto, S., Takayabu, Y. N., Ushio, T., Nakagawa, K., Iwanami, K., Kachi, M., Okamoto, K., 2007. Global precipitation map using satellite-borne microwave radiometers by the GSMaP Project: Production and validation. *IEEE Trans. Geosci. Remote Sens.*, 45, 2259–2275.

Li, M., Yang, D., Chen, J., Hubbard, S. S., 2012. Calibration of a distributed flood forecasting model with input uncertainty using a Bayesian framework. *Water Resour. Res.*, 48, W08510.

Li, Z., Yang, D., Gao, B., Jiao, Y., Hong, Y., Xu, T., 2015. Multiscale hydrologic applications of the latest satellite precipitation products in the Yangtze River basin using a distributed hydrologic model. *J. Hydrometeor.*, 16: 407–426.

Li, Z., Yang, D., Hong, Y., 2013. Multi-scale evaluation of high-resolution multi-sensor blended global precipitation products over the Yangtze River. *J. Hydrol.*, 500:157–169.

McCollum, J. R., Krajewski, W. F., Ferraro, R. R., Ba, M. B., 2002. Evaluation of biases of satellite rainfall estimation algorithms over the continental United States. *J. Appl. Meteor.*, 41, 1065–1080.

Nijssen, B. and Lettenmaier, D. P., 2004. Effect of precipitation sampling error on simulated hydrological fluxes and states: Anticipating the global precipitation measurement satellites. *J. Geophys. Res.*, 109, D02103, doi:10.1029/2003JD003497.

Shen, Y., Feng, M., Zhang, H., Gao, F., 2010b. Interpolation methods of China daily precipitation data (in Chinese). *J.Appl. Meteorol. Sci.*, 21, 279–286.

Shen, Y., Xiong, A., Wang, Y., Xie, P., 2010a. Performance of high-resolution satellite precipitation products over China. *J. Geophys. Res.*, 115, D02114.

Sohn, S.-J., Tam, C.-Y., Ashok, K., Ahn, J.-B., 2012. Quantifying the reliability of precipitation datasets for monitoring large-scale East Asian precipitation variations. *Int. J. Climatol.*, 32, 1520–1526.

Sorooshian, S., Hsu, K. L., Gao, X., Gupta, H., Imam, B., Braithwaite, D., 2000. Evaluation of PERSIANN system satellite-based estimates of tropical rainfall. *Bull. Am. Meteorol. Soc.*, 81, 2035–2046.

Su, F., Hong, Y., Lettenmaier, D. P., 2008. Evaluation of TRMM multisatellite precipitation analysis (TMPA) and its utility in hydrologic prediction in the la Plata Basin. *J. Hydrometeor.*, 9, 622–640.

Turk, F. J. and Miller, S. D., 2005. Toward improving estimates of remotely sensed precipitation with MODIS/AMSR-E blended data techniques. *IEEE Trans. Geosci. Remote Sens.*, 43, 1059–1069.

Turk, F. J., Arkin, P., Sapiano, M. R. P., Ebert, E. E., 2008. Evaluating high-resolution precipitation products. *Bull. Am. Meteor. Soc.*, 89, 1911–1916.

Villarini, G., Krajewski, W. F., Smith, J. A., 2009. New paradigm for statistical validation of satellite precipitation estimates: Application to a large sample of the TMPA 0.25° 3-hourly estimates over Oklahoma. *J. Geophys. Res.*, 114, D12106, doi:10.1029/2008JD011475.

Wood, E. F. et al. 2011. Hyperresolution global land surface modeling: Meeting a grand challenge for monitoring Earth's terrestrial water. *Water Resour. Res.*, **47**, W05301, doi:10.1029/2010WR010090.

Xie, P., Chen, M., Yatagai, A., Hayasaka, T., Fukushima, Y., Yang, S., 2007. A gauge-based analysis of daily precipitation over East Asia. *J. Hydrometeorol.*, 8, 607–626.

Yang, D. and Musiake, K., 2003. A continental scale hydrological model using the distributed approach and its application to Asia. *Hydrol. Processes*, 17, 2855–2869.

Yong, B., Hong, Y., Ren, L.-L., Gourley, J. J., Huffman, G. J., Chen, X., Wang, W., Khan, S. I., 2012. Assessment of evolving TRMM-based multisatellite real-time precipitation estimation methods and their impacts on hydrologic prediction in a high latitude basin. *J. Geophys. Res.*, 117, D09108.

Yong, B., Liu, D., Gourley, J. J., Tian, Y., Huffman, G. J., Ren, L.-L., Hong, Y., 2015. Global view of real-time TRMM multi-satellite precipitation analysis: Implication to its successor global precipitation measurement mission. *Bull. Am. Meteorol. Soc.*, 96, 283–296.

12 Uncertainty Analysis of Five Satellite-Based Precipitation Products and Evaluation of Three Optimally Merged Multialgorithm Products over the Tibetan Plateau

Yan Shen, Anyuan Xiong, Yang Hong, Jingjing Yu,
Yang Pan, Zhuoqi Chen, and Manabendra Saharia

CONTENTS

12.1 INTRODUCTION

The Tibetan Plateau (TP) known as the Earth's third pole is the world's highest plateau, averaging over 4000 m above sea level. It has a great influence on regional and even global climate change and disastrous weather arising from anomalous thermal and dynamic processes over the TP (Luo and Yanai, 1984). A high-quality precipitation observation is important for understanding the thermal and dynamic processes over the TP. Precipitation at each grid box is also required for driving land surface and hydrologic models. But, because of its complex terrain and harsh natural environment, very few gauges and meteorological radars are installed at the south-east of the TP, and there is no gauge or radar in the vast areas of the southern and western TP. Fortunately, with the development of remote sensing technology, several satellite-based precipitation retrieval algorithms have been developed and its related precipitation products are available now (Hsu et al., 1997, 1999; Sorooshian et al., 2000; Turk et al., 2003; Hong et al., 2004; Joyce et al., 2004; Huffman et al., 2007; Ali et al., 2009). However, satellite-based retrieval is an indirect way to obtain precipitation estimate, and different retrieval algorithms tend to have their own merits and demerits, quantified by uncertainty measures as a function of space, time, and rainfall intensity (Ebert et al., 2007; Tian et al., 2007, 2009; Shen et al., 2010a; Yong et al., 2012; Chen et al., 2013). Tian and Peters-Lidard (2010) proposed to estimate the global precipitation uncertainties as a variable of the rain rate at given locations among many other studies (Adler et al., 2001, 2009; Smith et al., 2006). To the best of our knowledge, uncertainty over the TP has not yet been fully studied with all available mesoscale satellite precipitation products. In this study, five of the mainstream satellite-based precipitation products are comprehensively examined for the first time over the TP for the period of 2005–2007. The uncertainty is examined and quantified with factors considering space, season, and rain rate as Tian and Peters-Lidard (2010) did. On the other hand, the influence of snow cover change and topographic features on the uncertainty is also investigated owing to the high elevation of the TP.

Building an ensemble average forecast field by the results from different methods, different models or the different forecast members of the same model has been widely studied and applied in meteorology (Sanders, 1963) and hydrology (McLeod et al., 1987). The advantage of the ensemble average is that it is able to effectively synthesize forecast information of multiple members to obtain higher forecast skill than a single member. However, Li (2011) studied that the ensemble prediction does not always provide a more accurate forecast field than a single forecast. In this study, ensemble prediction is employed to provide the best satellite precipitation data for numerous research studies and applications over the sparsely gauged TP. Here, three ensemble methods: arithmetic mean, inverse-error-square weight and one-outlier-removed arithmetic mean, are introduced and results are compared to the five individual satellite data sources both at the seasonal and annual timescales.

The following sections will first describe the study area, data, and method, followed by uncertainty analysis, data ensemble investigation, and conclusions.

12.2   STUDY AREA, DATA, AND METHOD

The study area is 25°N–40°N and 75°E–105°E confining the TP region over the three-year period of 2005–2007. A dense national gauge network of ~330 gauges has been established within the research region as shown in Figure 12.1. The average

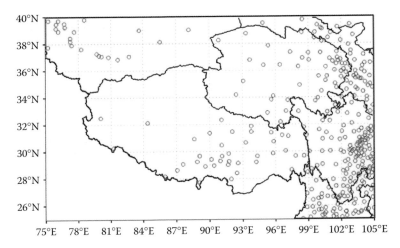

FIGURE 12.1 Gauge distribution over the research area (one red circle represents one gauge, black line at the position of 100°E is used to divide the research area into east–west two parts.)

distance between the gauges is 78.9 km. If the vertical line of 100°E is used to divide the research area into east–west two parts, the gauge number (average gauge-to-gauge distance) for the western and eastern part is ~110 (133.1 km) and 220 (55.5 km), respectively, indicating a much higher gauge network in the east and relatively sparse in the west. The gauge observations have gone through three levels of quality control and then accumulated into daily gauge-site precipitation data. These quality-controlled point data are used to generate the gauge-based precipitation analysis (GPA) at 0.25° gird box (Shen et al., 2010b) using the optimal interpolation method first proposed by Xie et al. (2007), with additional topographic corrections. In this process, the topographic effect is adjusted by following the method used in PRISM data analysis (Daly et al., 1994). Figure 12.2a–e shows the spatial distribution of the 3-year mean annual and seasonal precipitation from GPA. A mean precipitation distribution over the TP is characterized by an east-to-west decreasing trend. A large amount of precipitation is observed over the eastern and southeastern TP, and the temporal distribution has seasonal dependence with high (low) amount in warm (cold) seasons.

Five level-3 satellite-based precipitation estimates by blending passive microwave (PMW) and infrared (IR) sensors are used in this paper. They are (1) global precipitation fields generated by the NOAA CPC morphing technique (CMORPH; Joyce et al., 2004); (2) Precipitation Estimation From Remotely Sensed Information Using Artificial Neural Network (PERSIANN; Hsu et al., 1997, 1999); (3) the Naval Research Laboratory (NRL) blended satellite precipitation estimates (Turk et al., 2003); (4) Tropical Rainfall Measurement Mission (TRMM) precipitation products 3B42 version 7, and (5) its real-time version 3B42RT (Huffman et al., 2007). The main differences in the five products arise from two factors: One is the input satellite

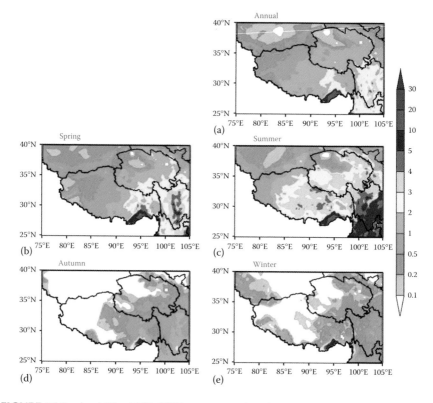

FIGURE 12.2   (a–e) The 2005–2007 mean annual and seasonal precipitation distribution from GPA over the TP (unit: mm day^{-1}).

sources and the other is the IR–PMW merging algorithms. The related information about the five satellite products is included in Table 12.1.

The method used in this paper is the three merging approaches: Arithmetic mean, inverse-error-square weight, and one-outlier-removed arithmetic mean, respectively, as follows:

$$R_1 = \frac{1}{n}\sum_{k=1}^{n} S_k \tag{12.1}$$

$$R_2 = \frac{1}{n}\sum_{k=1}^{n} W_k \times S_k, \quad W_k = \frac{1}{\sigma_K^2} \tag{12.2}$$

$$R_3 = \frac{1}{n-1}\sum_{k=1}^{n-1} S_k \tag{12.3}$$

TABLE 12.1

Information of Five Satellite-Based Precipitation Products Used in This Research

Product Name	Provider	Input Data	Retrieval Algorithm
Precipitation Estimation from Remotely Sensed Information using Artificial Neural Networks (PERSIANN)	UC Irvine (Hsu K.-L.)	IR: GOES-IR PMW: TRMM 2A12	Adaptive artificial neural network
Naval Research Laboratory blended algorithm (NRL)	NRL (Turk J.)	IR: Geo-IR PMW: SSM/I, TRMM, AMSU, AMSR	Histogram matching method
TRMM Multisatellite precipitation analysis (3B42RT for Real Time or 3B42 Version 7)	GSFC (Huffman G.)	IR: Geo-IR PMW: TMI, SSMI, SSMIS, AMSR-E, AMSU-B, MHS	3B42RT: Histogram matching method 3B42: GPCC monthly gauge observations to correct the bias of 3B42RT
CPC Morphing Technique (CMORPH)	NOAA CPC (Joyce B.)	IR: Geo-IR PMW: SSMI, AMSU-B, TMI, AMSR-E	CPC Morphing technique: First, the vector of the cloud motion is calculated by the IR data. Then the rainfall from the PMW exclusively is transported based on the motion vector.

where:

R_1, R_2, and R_3 are the precipitation obtained from the arithmetic mean, inverse-error-square weight, and one-outlier-removed arithmetic mean, respectively

S_k is a satellite-based precipitation product

n is the number of satellite products examined and $n = 5$ in this paper

W_k is the weighting factor that is a function of the inverse proportion to the error square (σ_K^2) for each satellite product

All values are calculated both at the spatial and temporal scales. The spatial scale is a 0.25° grid box with at least one gauge available over the TP and the temporal scale is daily for 2005–2007.

Several common statistical indices are used to quantitatively evaluate individual algorithms and the merged ensemble estimates with GPA including Bias, relative bias (R_{Bias}), root-mean-square error (RMSE), relative RMSE (R_{RMSE}), and correlation coefficient (CC). Additionally, a set of contingency table statistics is used in this study. They are probability of detection (POD), false-alarm ratio (FAR), critical success index (CSI), and equitable threat score (ETS) (Ebert et al., 2007). All analyses in this paper are done just for grid boxes with at least one gauge available to ensure statistical significance and a less interpolated error.

12.3 UNCERTAINTY ANALYSIS

12.3.1 SPATIAL DISTRIBUTION

GPA is conventionally used to quantify the uncertainty of satellite products, but its limited distribution makes it less representative. Satellite products have full spatial coverage and so they can be used to examine the uncertainty of the satellite-based precipitation estimates over the whole region. The standard deviation of five satellite products is calculated first and the uncertainty is defined as the ratio of standard deviation to mean daily precipitation from five satellite products. Because precipitation with a rain rate less than 0.5 mm day^{-1} accounts for almost 80% of the total amount of rainfall, especially in winter, which will lead to unreliable uncertainty, standard deviation is used when plotting the spatial distribution of uncertainty. The spatial distribution of annual and seasonal uncertainties over the TP for 2005–2007 is shown in Figure 12.3b–f. The elevation of the research domain is also shown in Figure 12.3a. The spatial uncertainty is dependent on seasons. The maximum uncertainty is more than 5.0 mm day^{-1} and is always distributed in the south-eastern TP where the

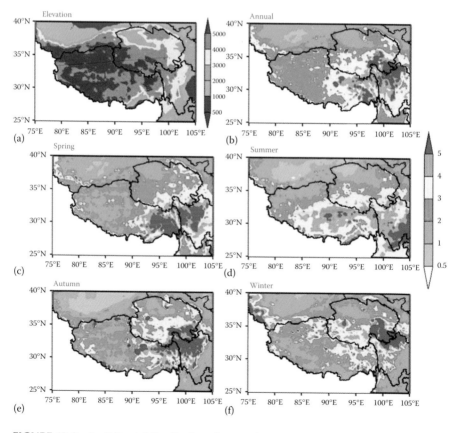

FIGURE 12.3 (a–f) Spatial distribution of uncertainty (unit: mm day^{-1}) for 2005–2007 and the elevation (unit: meter) over the TP.

elevation changes sharply from 3000 to more than 5000 m. In the western and northern TP, uncertainty is relatively small with values less than 3.0 mm day^{-1}. It indicates that precipitation estimates from different satellite retrieval algorithms generally has better agreement in the western and northern TP than in the south-eastern TP.

12.3.2 SCF-Dependent Seasonal Uncertainty

Figure 12.4 shows the seasonality time series of mean relative uncertainty of satellite precipitation estimates for 2005–2007. The uncertainty decreases from January to April, then remains relatively low, and increases after October. The uncertainty of satellite precipitation estimates shows a fluctuation from 136.02% to 319.63%, with a mean value of 211.99%. Given the current limitation of satellite precipitation sensors in the detection of solid precipitation, we suspect that the uncertainty is probably associated with snowfall seasonality over the TP. Thus, we further investigated the uncertainty connection with snow cover fraction (SCF) derived from MODIS daily snow data (MODIS user guide 2003) in Figure 12.4, where the SCF is defined as the ratio between snow-covered grid boxes and total grid boxes over the research domain. As shown in Figure 12.4, the SCF temporally fluctuates between 5.27% and 32.15%, with a mean value of 14.40%. As anticipated, the time series of SCF shows very good agreement with the seasonal uncertainty of satellite precipitation given their relatively high correlation coefficient (between two time series is 0.75). The warm season from April to September is a period with relatively low uncertainty and SCF with the mean of 179.23% and 10.04%, respectively. However, the related value for the cold season from October to following March is 244.93% and 18.67%, respectively. The higher/lower uncertainty is related to the relatively high/low SCF. This clearly indicates that the current space-borne quantitative precipitation estimation (QPE) is incapable of adequately resolving winter precipitation. Meanwhile, the recently launched Global Precipitation Measurement (GPM) mission, with a dual-frequency precipitation radar and multifrequency PMW channels, holds promising potential in this regard.

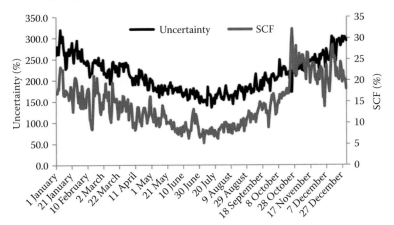

FIGURE 12.4 Seasonality of mean relative uncertainty (unit: %) of satellite precipitation estimates and mean of SCF (unit: %) from MODIS data averaged for the 2005–2007 period over the TP.

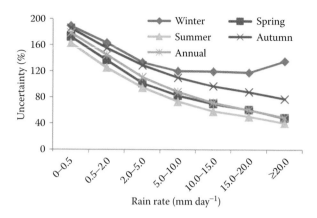

FIGURE 12.5 The relationship between uncertainties and rain rate for annual mean and four seasons over the TP for 2005–2007.

12.3.3 RAIN RATE-DEPENDENT UNCERTAINTY

The influence of rainfall rate on the uncertainty is previously investigated by Tian and Peters-Lidard (2010) among others in different parts of the world except in the TP. In this study, different rain rate categories are classified based on the mean data of five satellite products. According to the rain rate, the uncertainty is calculated at each grid box for four seasons and annual mean (Figure 12.5). It shows that the uncertainty decreases with the rain rate. In summer, the uncertainty is 160% with the rain rate category in 0.0–0.5 mm day^{-1} while the uncertainty is reduced to 40% at the rain rate greater than 20.0 mm day^{-1}. Moreover, the uncertainty is seasonally dependent with the smallest in summer and the largest in winter. For example, when the rain rate has fallen to 5.0–10.0 mm day^{-1}, uncertainty is 73.3% and 110.2% for summer and winter, respectively.

12.3.4 TOPOGRAPHY-DEPENDENT UNCERTAINTY

Topography has a great and complex effect on precipitation. Four elevation categories are classified as follows: (1) elevation < 2000 m; (2) 2000 m ≤ elevation < 3000 m; (3) 3000 m ≤ elevation < 4000 m; and (4) elevation ≥ 4000 m. The uncertainty is calculated at each grid box. Then, the seasonal and annual uncertainty is averaged according to the elevation categories (Figure 12.6). In general, the effect of topography on the uncertainty tends to gradually decrease when the elevation is less than 4000 m, and uncertainty increases fast with the elevation larger than 4000 m. The trend is relatively small in summer but large in winter. Moreover, the seasonal and annual uncertainty of satellite-based precipitation products is investigated as a function of the different elevation points. A region encompassed by 28°–35°N and 92°–104°E is selected because of its dramatic change of elevation and relatively dense network. Based on the gauge analysis over the grid box at least one gauge available, the meridional distribution from 92°E to 104°E of seasonal and annual uncertainty for five satellite-based precipitation products is shown in Figure 12.7 together with the average

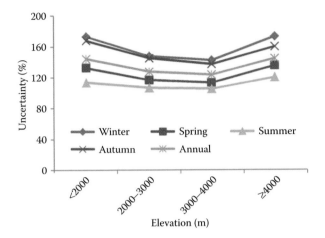

FIGURE 12.6 The relationship between uncertainty (unit: %) and elevation (unit: meter) for annual mean and four seasons over the TP for 2005–2007.

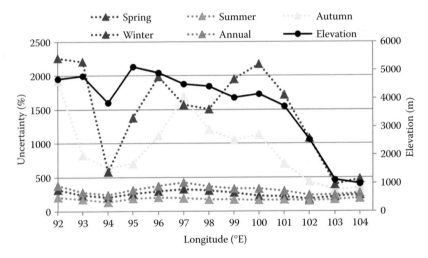

FIGURE 12.7 The meridional distribution from 92°E to 104°E of seasonal and annual uncertainty for five satellite-based precipitation products together with the average elevation taking 28°N–35°N as the cross section for 2005–2007.

elevation taking 28°–35°N as the cross section. The mean elevation is decreased from 5118.56 to 987.54 m (Figure 12.7), and the range of elevation is more than 4000 m when the longitude is changed from 92°E to 104°E. The uncertainty of five satellite-based precipitation products is dependent on the elevation points. Taking the 94°E and 95°E as the dividing points, the elevation is decreased at first, then increased, and then decreased gradually. The same feature is exhibited for the seasonal and annual uncertainty of satellite products, especially for the winter and autumn seasons. The correlation coefficient between uncertainty for winter and autumn seasons and the elevation reaches at 0.77 and 0.67, respectively. It is 0.51 for the spring season. The influence of the elevation on the uncertainty is minimum in summer and the correlation coefficient is only 0.17. The correlation coefficient between the annual uncertainty and the elevation is still 0.63 indicating the great effect of the elevation on the annual uncertainty. In addition to the relatively small effect of the elevation on the uncertainty in summer, the effect of the elevation on the other seasonal uncertainty is significant, particularly in the winter and autumn.

12.4 UNCERTAINTY ANALYSIS OF THE MERGED MULTIALGORITHM DATA ENSEMBLES

Owing to relatively large uncertainty and limited observations over the TP, it is beneficial to further improve the precipitation quality by capitalizing on the strengths of all available satellite precipitation products. Thus, we evaluate the performance of three ensemble methods for generating the best possible merged satellite precipitation data. They are arithmetic mean, inverse-error-square weight, and one-outlier-removed arithmetic mean, respectively, as shown in Section 12.2.

12.4.1 STATISTICAL ANALYSIS

Tables 12.2 and 12.3 summarize the comparison statistics for the five satellite products and the three ensembles at daily and 0.25° resolution over the TP for the summer and winter of 2005–2007, respectively. The Bias of each individual satellite precipitation estimation ranges from −0.736 to 1.830 mm day^{-1} and the RMSE ranges from 4.492 mm day^{-1} for CMORPH to 6.793 mm day^{-1} for PERSIANN. However, Bias (RMSE) between the ensembles and GPA are much reduced from −0.475 to 0.208 mm day^{-1} (4.184–4.247 mm day^{-1}), which is better than that between any of five individual satellite estimates and gauge observations. The ensemble data produced by the inverse-error-square weight has the best performance with Bias (relative bias) of −0.056 mm day^{-1} (−1.9%) among all three ensemble products while the ensemble data produced by the one-outlier-removed method provides the smallest

TABLE 12.2
Evaluation Results of GPA versus Five Satellite Estimates and Three Satellite Ensembles over the TP for the Summer Period of 2005–2007

	Bias	R_{Bias}	RMSE	R_{RMSE}	C.C.
CMORPH	−0.736	−0.253	4.492	1.546	0.568
PERSIANN	1.830	0.630	6.793	2.338	0.502
NRL	0.713	0.246	6.188	2.130	0.478
TRMM/3B42	−0.252	−0.087	5.337	1.837	0.507
3B42RT	0.324	0.112	5.391	1.855	0.506
Arithmetic mean	0.208	0.072	4.247	1.462	**0.634**
Inverse-error-square	**−0.056**	**0.019**	4.228	1.455	0.633
One-outlier-removed	−0.475	−0.164	**4.184**	**1.440**	0.625

TABLE 12.3

Same as Table 12.2 Except for the Winter

	Bias	R_{Bias}	RMSE	R_{RMSE}	C.C.
CMORPH	0.035	0.147	1.840	7.625	0.051
PERSIANN	1.686	6.990	4.228	17.527	0.153
NRL	1.684	6.981	5.721	23.713	0.056
TRMM/3B42	**−0.003**	**−0.014**	**1.728**	**7.161**	**0.265**
3B42RT	0.873	3.618	3.414	14.152	0.163
Arithmetic mean	0.767	3.179	2.151	8.916	0.184
Inverse-error-square	0.419	1.738	1.930	7.999	0.196
One-outlier-removed	0.185	0.766	1.733	7.185	0.197

RMSE (4.184 mm day^{-1}). Correlation coefficients from the three ensembles rang-ing from 0.625 to 0.634 are much larger than the best value from the individual satellite product (0.568 in CMORPH). These results suggest that the quality of assembling different satellite products is further improved by capitalizing on each individual product with proper ensemble methods, resulting in a relatively lower bias and RMSE, and a higher correlation. For the winter period, the Bias from the one-outlier-removed method presents the smallest value of 0.185 mm day^{-1} in the three merging algorithms but it is still larger than the best value of −0.003 mm day^{-1} obtained from the TRMM 3B42. The similar characteristics is for the correlation coefficient changing from 0.184 to 0.197 from the ensembles but it is lower than the best value from TRMM 3B42 (0.265). The RMSE obtained from the one-outlier-removed method is 1.733 mm day^{-1}, which is at the same level as the best value from TRMM 3B42 (1.728 mm day^{-1}). The difference of three merging algorithms is increased and the result from the one-outlier-removed method is better than other two methods but it is still worse than that from the TRMM 3B42. For each single satellite precipitation estimate, the bias shows a relatively large range from only −0.003 mm day^{-1} for TRMM 3B42 to 1.686 mm day^{-1} for PERSIANN with the dif-ference of 1.689 mm day^{-1}. The same characteristic is for the RMSE changing from 1.728 mm day^{-1} for TRMM 3B42 to 5.721 mm day^{-1} for NRL with the difference of 3.993 mm day^{-1}. Bias and RMSE among five satellite products are a large varying amplitude and TRMM 3B42 is significantly better than other single satellite prod-ucts in winter, which leads to the failure of the ensemble data. This result has been proved in the field of the ensemble prediction for numerical weather models (Yoo and Kang, 2005; Jeong and Kim, 2009; Winter and Nychka, 2010; Li, 2011) and is first confirmed and extended in the field of satellite precipitation estimates.

When the rainfall threshold is selected 0.1 mm day^{-1}, values of ETS, CSI, POD, and FAR for five satellite precipitation estimates and three merged ensembles over the TP for the summer and winter period during 2005–2007 are shown in Tables 12.4 and 12.5, respectively. These indices are calculated against GPA grids that contain at least one gauge. In summer, ETS, CSI, and FAR for five satellite precipitation estimates are very close, except that the POD for five satellite estimates has a large amplitude from 0.73 for TRMM/3B42 to 0.82 for PERSIANN. The CSI and POD for the ensembles

TABLE 12.4

Values of ETS, CSI, POD, and FAR for Five Satellite Estimates and Three Satellite Ensembles over the TP for the Summer Period of 2005–2007. The Rainfall Threshold Is 0.1 mm day⁻¹

	ETS	CSI	POD	FAR
CMORPH	0.24	0.64	0.79	0.23
PERSIANN	**0.28**	0.67	0.82	0.22
NRL	0.24	0.65	0.81	0.24
TRMM/3B42	0.24	0.61	0.73	**0.21**
3B42RT	0.21	0.61	0.75	0.24
Arithmetic mean	0.24	**0.69**	**0.93**	0.27
Inverse-error-square	0.27	**0.69**	0.89	0.25
One-outlier-removed	**0.28**	**0.69**	0.87	0.23

TABLE 12.5

Same as Table 12.4 Except for the Winter Period

	ETS	CSI	POD	FAR
CMORPH	0.03	0.12	0.33	0.84
PERSIANN	0.07	0.18	0.85	0.81
NRL	0.05	0.16	0.62	0.83
TRMM/3B42	**0.14**	**0.21**	0.39	**0.69**
3B42RT	0.09	0.19	0.58	0.78
Arithmetic mean	0.05	0.17	**0.94**	0.83
Inverse-error-square	0.09	0.20	0.82	0.79
One-outlier-removed	0.10	0.20	0.74	0.78

are increased compared with those for five satellite estimates. The highest CSI in five satellite estimates is 0.67 while that for ensembles is 0.69. It indicates that the ensembles can detect more rainfall events than individual satellite estimates. The FAR for the ensembles in the rainfall threshold of 0.1 mm day⁻¹ is little enhanced compared to that from individual products; however, further investigation shows when higher rainfall threshold such as 5.0 mm day⁻¹ is selected that the FAR for the ensembles will be smaller. The merged ensemble is an effective way to correctly detect more rainfall events and reduce missed and false events. In winter, however, ETS, CSI, POD, and FAR from the five satellite products present a relatively large changing amplitude. The ETS (TS) changes from 0.03 (0.12) for CMORPH to 0.14 (0.21) for TRMM 3B42 while the POD (FAR) varies from 0.33 (0.84) for CMORPH to 0.85 (0.69) for PERSIANN (TRMM 3B42). The best value for each of the four parameters is not consistently obtained from the three merging algorithms, especially for the ETS and FAR. The ETS (FAR) changes from 0.05 (0.78) to 0.1 (0.83) from the three merging algorithms and it is still lower (higher) than the best value of 0.14 (0.69) from the TRMM 3B42. The CSI changing from 0.17 to 0.2 by the three merging algorithms is almost at the same level as the best value in five individual satellite products (0.21 from TRMM 3B42). The improvement of the POD is obvious for the highest value of 0.85 from PERSIANN to 0.94 from the ensemble of the arithmetic mean.

12.4.2 Spatial Distribution

The spatial distribution of the annual mean uncertainty for the ensemble is shown in Figure 12.8 together with the five satellite-based precipitation products to examine which of the satellite products contribute more to the uncertainty. Compared with the uncertainty of PERSIANN, NRL, and TRMM 3B42RT, uncertainty of TRMM 3B42 and CMORPH is relatively small but it is still larger than that of the ensemble over the entire research. There is no existing gauge in the western part of the TP where uncertainty from each of the five satellite products is more than 3 mm day^{-1}. The ensemble data show a very small uncertainty with the value less than 3 mm day^{-1} after merging the five products. It indicates that the ensemble has the smallest uncertainty among the five products. The same feature is depicted at the four

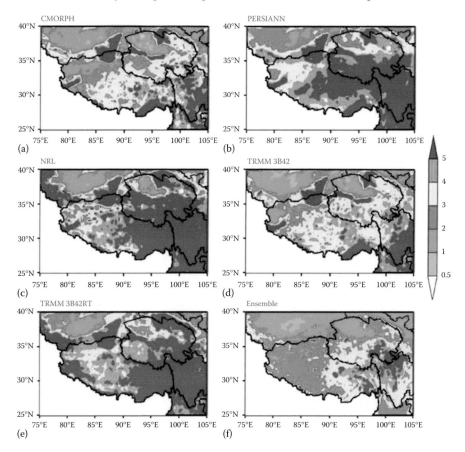

FIGURE 12.8 (a–f) The 2005–2007 annual mean spatial distribution of uncertainty for five satellite products and the ensemble from the inverse-error-square weight method over the TP.

TABLE 12.6

Number of Each Satellite Products Being Taken as the Outlier over the TP for 2005–2007

	CMORPH	PERSIANN	NRL	3B42	3B42RT
Number	489,789	2,326,270	1,295,385	455,129	1,305,663

seasonal timescales, which are omitted to avoid redundancy. To further investigate which satellite products contribute more to the uncertainty in the qualitative sense, the number of each satellite product being taken as the outlier is calculated for 2005–2007 and is shown in Table 12.6. The outlier satellite product has the largest deviation from the arithmetic mean at each grid box and time. Among the five satellite products, only TRMM 3B42 uses the gauge data to correct the bias and the minimum outlier number is obtained from the TRMM 3B42 with the number of 455129 in the research domain and period. The procedure of cloud classification and snow screening is not considered in the PERSIANN algorithm, which leads to the maximum outlier number of 2,326,270. The result of outlier number for each satellite product can demonstrate that the PERSIANN product contributes the most to the uncertainty qualitatively, followed by the TRMM 3B42RT, NRL, CMORPH, and TRMM 3B42.

CMORPH and TRMM 3B42 products show smaller RMSE and lower number being taken as outlier over the spatial scale, so another question arises whether combining CMORPH with TRMM 3B42 can provide better estimates than the three ensembles used in this paper. Table 12.7 shows the evaluation results of three ensembles and the combination of CMORPH and TRMM 3B42 versus the GPA over the grid box with at least one gauge available at the seasonal and annual timescales. Bold values are the best results obtained for a particular season and each statistical parameter. Although the evaluation results are dependent on the merging methods for a particular season, the result from the one-outlier-removed arithmetic mean is more stable and reliable. For example, the statistics index of Bias, RMSE, and C.C. from the one-outlier-removed arithmetic mean is consistently the best among the four methods at the annual timescale. Bias is 0.044 mm day^{-1}, RMSE is 3.096 mm day^{-1}, and C.C. is 0.597. In summer, C.C. from the arithmetic mean (0.634) and Bias from the inverse-error-square weight (−0.056 mm day^{-1}) are the best but RMSE from the one-outlier-removed arithmetic mean (4.184 mm day^{-1}) is the smallest. However, the result from combining CMORPH with TRMM 3B42 has advantage in cold seasons of autumn and winter, which shows the smallest Bias (0.016 mm day^{-1}), RMSE (1.652 mm day^{-1}), and the highest correlation coefficient (0.226) among the four ensemble data but it is still worse than the result from TRMM 3B42.

12.5 CONCLUSIONS

In this study, uncertainty of five state-of-the-art satellite-based precipitation estimates have been comprehensively evaluated for the first time with respect to region, season, elevation, rain intensity, snow cover, and topography over the TP spanning the period of 2005–2007 due to the satellite products availability. Also, three

TABLE 12.7

Evaluation Results of GPA versus three Satellite Ensembles and Combining CMORPH with TRMM 3B42 (CMORPH+3B42) over the TP at the Seasonal and Annual Timescales for 2005–2007. The Rainfall Threshold Is 0.1 mm day^{-1}

	Arithmetic Mean			Inverse-Error-Square			One-Outlier-Removed			CMORPH+3B42		
	Bias	RMSE	C.C.	Bias	RMSE	C.C.	Bias	RMSE	C.C.	Bias	RMSE	C.C.
Spring	0.892	3.671	0.511	0.567	3.589	0.520	0.110	3.408	0.533	−0.301	3.374	0.518
Summer	0.208	4.247	0.634	−0.056	4.228	0.633	−0.475	4.184	0.625	−0.494	4.338	0.601
Autumn	1.059	2.579	0.362	0.738	2.256	0.398	0.448	1.963	0.433	0.263	1.943	0.362
Winter	0.767	2.151	0.184	0.419	1.930	0.196	0.185	1.733	0.197	0.016	1.652	0.226
Annual	0.712	3.351	0.573	0.400	3.231	0.589	0.044	3.096	0.597	−0.148	3.127	0.583

merging methods are further investigated to provide one best possible space-borne precipitation product for climate and hydrology researches. Major conclusions are summarized below:

1. The uncertainty map over the TP is produced for five satellite precipitation estimates, which shows strong regional and seasonal dependencies. Larger uncertainty is distributed in the east-southern TP and relatively small uncertainty is in the western and northern TP. Uncertainty has high seasonality, temporally changing with a decreasing trend from January to April, then remaining a relatively low value and increasing after October, with an obvious winter peak and summer valley.

2. Overall, the uncertainty also shows an exponentially decreasing trend with higher rainfall rates. Additionally, the effect of topography on the uncertainty tends to rapidly increase when the elevation is higher than 4000 m while the impact slowly decreases in areas lower than that topography. The effect of the elevation on the uncertainty is significant for all seasons except the summer.

3. Further cross-investigations indicate that the uncertain seasonality has a very strong correlation with time series of MODIS-based SCF over the TP, correlation coefficient as high as 0.75. This clearly indicates the limitation of current satellite-based QPE incapable of adequately resolving winter precipitation.

4. To reduce the still relatively large and complex uncertainty over the TP, three data merging methods are examined to provide the one best possible satellite precipitation data by optimally combining the five state-of-art products. The three merging methods: arithmetic mean, inverse-error-square weight, and one-outlier-removed arithmetic mean, show insignificant yet subtle differences. Bias and RMSE of the three merging methods show seasonal dependence but the one-outlier-removed method is more robust and its result notably outperforms the five individual products at the four seasons except the winter. The correlation coefficient obtained by the three merging methods is consistently higher than any of five individual satellite estimates, indicating an effective and great improvement. However, because of the large difference among satellite products in winter, the result from the ensemble is not always better than the best one among the five satellite products.

5. Finally, the spatial distribution of the ensemble data is present against the five individual satellite estimates, which indicates that the ensemble can provide a general improvement over the entire studied region both at the seasonal and annual timescales. Comparing the number of each satellite product being taken as the outlier demonstrates that the PERSIANN, TRMM 3B42RT, and NRL contribute more uncertainty while TRMM 3B42 and CMORPH contribute less uncertainty. In warm seasons and annual timescale, combining CMORPH with TRMM 3B42 may not provide the overall better results than that from the one-outlier-removed method but in winter it is still inferior to the result obtained from TRMM 3B42. TRMM 3B42 shows the best performance in winter over the TP. We recommend the result from the one-outlier-removed method as the best over the TP for

the seasonal and annual timescales except for the winter, although these optimally merging multialgorithm data appear to be a cost-effective way to provide better quality satellite precipitation data presently. The recently launched GPM mission, with dual-frequency precipitation radar and multifrequency PMW channels, holds promising potential in this complex and high-altitude TP region, and the data from the GPM can further evaluate and verify the results in this paper.

ACKNOWLEDGMENTS

This work was funded by National Science Foundation of China Major research program (91437214). This work was also partially supported by grants from Chinese Ministry of Science and Technology (2012BAC22B04), by China Meteorological Administration (GYHY201406001), and by National Natural Science Foundation of China (51379056, 41101375).

REFERENCES

Adler, R. F., C. Kidd, G. Petty, M. Morissey, and H. M. Goodman. 2001. Intercomparison of Global Precipitation Products: The Third Precipitation Intercomparison Project (PIP-3). *Bulletin of the American Meteorological Society* 82:1377–1396.

Adler, R. F., J. J. Wang, G. Gu, and G. J. Huffman. 2009. A Ten-Year Tropical Rainfall Climatology Based on a Composite of TRMM Products. *Journal of Meteorological Society of Japan* 87A:281–293.

Ali, B., K.-L. Hsu, B. Imam, S. Sorooshian, G. J. Huffman, and R. J. Kuligowski. 2009. PERSIANN-MSA: A Precipitation Estimation Method from Satellite-Based Multispectral Analysis. *Journal of Hydrometeorology* 10:1414–1429.

Chen, S., P. E. Kirstetter, Y. Hong, J. J. Gourley, Y. D. Tian, Y. C. Qi, Q. Cao, J. Zhang, K. Howard, J. J. Hu, and X. W. Xue. 2013. Evaluation of Spatial Errors of Precipitation Rates and Types from TRMM Space-borne Radar over the Southern CONUS. *Journal of Hydrometeorology* 14:1884–1896.

Daly, C., R. P. Neilsen, and D. L. Phillips. 1994. A Statistical Topographic Model for Mapping Climatological Precipitation over Mountainous Terrain. *Journal of Applied Meteorology* 33:140–158.

Ebert, E. E., J. E. Janowiak, and C. Kidd. 2007. Comparison of Near-Real-Time Precipitation Estimates from Satellite Observations and Numerical Models. *Bulletin of the American Meteorological Society* 88:47–64.

Hong Y, K. L. Hsu, and S. Sorooshian. 2004. Precipitation Estimation from Remotely Sensed Information Using an Artificial Neural Network-Cloud Classification Systems. *Journal of Applied Meteorology* 43:1834–1852.

Hsu, K. L., H. V. Gupta, X. Gao, and S. Sorooshian. 1999. Estimation of Physical Variables from Multichannel Remotely Sensed Imagery Using Neural Networks: Application to Rainfall Estimation. *Water Resource Research* 35:1605–1618.

Hsu, K. L., X. Gao, S. Sorooshian, and H. V. Gupta. 1997. Precipitation Estimation from Remotely Sensed Information using Artificial Neural Networks. *Journal of Applied Meteorology* 36:1176–1190.

Huffman, G. J., R. F. Adler, and D. T. Bolvin. 2007. The TRMM Multisatellite Precipitation Analysis (TMPA): Quasi-Global, Multiyear, Combined-Sensor Precipitation Estimates at Fine Scales. *Journal of Hydrometeorology* 8:38–55.

Jeong, D. and Y. O. Kim. 2009. Combining Single-Value Streamflow Forecasts–A Review and Guidelines for Selecting Techniques. *Journal of Hydrology* 377:284–299.

Joyce, R. J., J. E. Janowiak, P. A. Arkin, and P. P. Xie. 2004. CMORPH: A Method That Produces Global Precipitation Estimates from Passive Microwave and Infrared Data at High Spatial and Temporal Resolution. *Journal of Hydrometeorology* 5:487–503.

Li, S. J. 2011. Forecasting Skill of Ensemble Mean: Theoretical Study and the Applications in the T213 and TIGGE EPSs. Department of Atmospheric Sciences, PhD degree, Nanjing University, Nanjing, China, (in Chinese).

Luo, H. B. and M. Yanai. 1984. The Large-Scale Circulation and Heat Sources over Tibetan Plateau and Surrounding Areas during the Early Summer of 1979. Part II: Heat and Moisture Budgets. *Monthly Weather Review* 112:966–989.

McLeod, A. I., D. J. Noakes, K. W. Hipel, and R. M. Thompstone. 1987. Combining Hydrologic Forecasts. *Journal of Water Resource Planning and Management*, 113:29–41.

Sanders, F. 1963. On Subjective Probability Forecasting. *Journal of Applied Meteorology* 2:191–201.

Shen, Y., A. Y. Xiong, Y. Wang, and P. P. Xie. 2010a. Performance of High-Resolution Satellite Precipitation Products over China. *Journal of Geophysical Research* 115, D02114, doi:10.1029/2009JD012097.

Shen, Y., M. N. Feng, H. Z. Zhang, and X. Gao. 2010b. Interpolation Methods of China Daily Precipitation Data. *Journal of Applied Meteorological Science* 21:279–286.

Smith, T. M., P. A. Arkin, J. J. Bates, and G. J. Huffman. 2006. Estimating Bias of Satellite-Based Precipitation Estimates. *Journal of Hydrometeorology* 7:841–856.

Sorooshian, S., K. L. Hsu, X. Gao, H. Gupta, B. Imam, and D. Brainthwaite. 2000. Evaluation of PERSIANN System Satellite-Based Estimates of Tropical Rainfall. *Bulletin of the American Meteorological Society* 81:2035–2046.

Tian, Y. D. and C. D. Peters-Lidard. 2010. A Global Map of Uncertainties in Satellite-Based Precipitation Measurements. *Geophysical Research Letters* 37:L24407, doi:10.1029/2010GL046008.

Tian, Y., C. D. Peters-Lidard, B. J. Choudhury, and M. Garcia. 2007. Multitemporal Analysis of TRMM-Based Satellite Precipitation Products for Land Data Assimilation Applications. *Journal of Hydrometeorology* 8:1165–1183.

Tian, Y. D., C. D. Peters-Lidard, J. B. Eylander, R. J. Joyce, G. J. Huffman, R. F. Adler, K. Hsu, F. J. Turk, M. Garcia, and J. Zeng. 2009. Component Analysis of Errors in Satellite-Based Precipitation Estimates. *Journal of Geophysical Research* 114, D24101, doi: 10.1029/2009JD011949.

Turk, F. J., E. E. Ebert, B. J. Sohn, H. J. Oh, V. Levizzani, E. A. Smith, and R. Ferraro. 2003. Validation of an Operational Global Precipitation Analysis at Short Time Scales. *Proceedings of 12th Conference on Satellite Meteorology and Oceanography*, American Meteorological Society, Long Beach, CA.

Winter, C. L. and D. Nychka. 2010. Forecasting Skill of Model Averages. *Stochastic Environmental Research and Risk Assessment* 24:633–638.

Xie, P. P., A. Yatagai, M. Chen, T. Hayasaka, Y. Fukushima, C. Liu, and S. Yang. 2007. A Gauge-Based Analysis of Daily Precipitation over East Asia. *Journal of Hydrometeorology* 8:607–626.

Yong, B., Y. Hong, L.-L. Ren, J. J. Gourley, G. J. Huffman, X. Chen, W. Wang, and S. I. Khan. 2012. Assessment of Evolving TRMM-Based Multisatellite Real-Time Precipitation Estimation Methods and Their Impacts on Hydrologic Prediction in a High Latitude Basin. *Journal of Geophysical Research* 117:D09108, 1–21, doi:10.1029/2011JD017069.

Yoo, J. H. and I. S. Kang. 2005. Theoretical Examination of a Multi-Model Composite for Seasonal Prediction. *Geophysical Research Letter* 32: doi:10.1029/2005GL023513.

13 Use of Remote Sensing-Based Precipitation Data for Flood Frequency Analysis in Data-Poor Regions
Case of Blue Nile River Basin, Ethiopia

Abebe Sine Gebregiorgis, Semu Ayalew Moges, and Seleshi Bekele Awulachew

CONTENTS

13.1 INTRODUCTION

In water resource development, the availability of sufficient historical time series data is an important aspect of flood frequency analysis (FFA). From the statistical point of view, estimation from small samples may give unreasonable or physically unrealistic parameter estimates, especially for distributions with a large number of parameters (three or more) (Rao and Hamed, 2000). Practically, however, in most part of the globe, observation data are limited, especially in developing countries, such as Ethiopia, or in some cases unavailable, such as in remote regions. In such cases, the estimation of design parameters for hydraulic structures, roads, and similar projects may suffer from under- or overestimation of the parameter values. Therefore, for the estimation of design floods in regions where little or no data are available, developing regional flood frequency curves is the best solution to overcome the problem (Gebeyehu, 1989; Rosbjerg and Madsen, 1995). This analysis is often supported through a regionalization procedure to convey information from gauged to ungauged location (Gebregiorgis et al., 2012).

The accuracy of the estimated flow quantile using flood frequency models, in general, depends on the length of continuous historical flow data we have (Chow, 1964). The gauging network and the quality of streamflow data also have a significant impact on FFA. To generate synthetic streamflow from observed rainfall information using a hydrologic model has also been a challenge for the past decades. This is because the conventional way of rainfall observation (ground based measurement) has limitations for hydrologic modeling with respect to global coverage, and temporal availability. First, surface-based (gauged) measurements represent the magnitude of rainfall on the ground at a point location rather than across a catchment area. Second, ground-based measurements give the quantity accumulated over a long duration rather than an instant magnitude or accumulated in short span duration. Third, a large part of the globe is ungauged, and there is no rainfall information over data-scarce regions and water bodies for the purpose of hydrologic and other applications. Finally, in situ rainfall observation networks continue to decline worldwide (Stokstad, 1999; Shiklomanov et al., 2002).

In contrast, satellite precipitation estimates provide excellent coverage both on the ground and in water bodies with high spatial and temporal resolutions. Beyond this, satellite rainfall estimate has numerous advantages as compared to in situ rainfall measurement: it resolves the challenge of data integration across multiple national agencies for transboundary flow modeling in large international river basins, it also provides continuous and consistent measurements, and it avoids the operational cost of in situ observation networks. Therefore, in this era, rainfall estimates from space are the most valuable source of rainfall data for hydrologic application, especially in parts of the world where surface observation networks are sparse.

Currently, several global high-resolution satellite precipitation products are publicly available. For example, during the era of Tropical Rainfall Measurement Mission (TRMM) Multisatellite Precipitation Analysis, 3B42V7 and 3B42RT (Huffman et al., 2010) are the most commonly and widely used satellite rainfall products for numerous hydrometeorological applications such as quantitative precipitation forecasting and numerical weather prediction models (Turk et al., 2010); flood forecasting and water resources monitoring (Hong et al., 2007, Gebregiorgis and Hossain, 2011, 2013); land data assimilation (Gottschalck et al., 2005); and landslide prediction (Hong et al., 2007). Now the Integrated Multi-satellite Retrievals for Global Precipitation Measurement/GPM is the future promising precipitation data source with a wide array of societal applications. Through an improved precipitation measurement technique, using an advanced radar and a radiometer and other constellation operational satellites, the GPM mission has already started providing global precipitation measurement at $0.1°$ spatial and 30 min temporal resolutions. Such high-resolution precipitation measurement will help to advance our knowledge on atmosphere–land surface interactions and improve our ability to forecast extreme events (flood and drought).

Despite the applications discussed above, most of the studies have also indicated that the uncertainty present in satellite rainfall estimates greatly impacted the hydrologic prediction. In general, satellite rainfall data have large uncertainties depending on the type of sensors and resolution (Kidd et al., 2003). As satellite rainfall estimation is not a direct observation, the uncertainty that is inherent in the satellite estimates originate from sampling and retrieval algorithm errors. For example, Nijssen and Lettenmaier (2004) demonstrated the effect of the sampling error on simulated hydrologic fluxes and states by imposing synthetic error fields on a gridded precipitation data set. The result of the study showed that the streamflow error were significantly large for a small watershed area and rapidly reduced for large watersheds (above 50,000 km²). Hong et al. (2006) examined the influence of the precipitation error on the uncertainty of hydrologic response using the Monte Carlo simulation. In line with the concerns of many studies on satellite rainfall uncertainty, so far many efforts have been made to improve the accuracy of satellite rainfall products by combining PMW and IR sensors rainfall products using different algorithms (Sorooshian et al., 2000; Joyce et al., 2004; Huffman et al., 2007). More importantly, the launching of the GPM satellite is expected to improve global rainfall measurement significantly through the deployment of a "Core" satellite carrying an advanced radar/radiometer system to measure precipitation from space and serve as a reference standard to unify precipitation measurements from a constellation of research and operational satellites. Hence, the progresses that have been made so far indicate a complete paradigm shift from the conventional to remote sensing measurement system.

Therefore, this study is primarily intended to examine the possibility of using satellite rainfall information to analyze the flood frequency of streamflow by using hydrologic and statistical models. The study incorporates two major tasks. The first task involves the use of satellite rainfall estimates and other meteorological forcings to simulate the streamflow using a hydrologic model at selected locations that have ground observations for the validation purposes. Second, fitting the simulated and ground measurement annual peak flow to a regional flood frequency model developed for the region (Gebregiorgis et al., 2012).

The remainder of this chapter is organized as follows: Descriptions of the study area and research methodology are presented in the next subsequent sections. Following these sections, the major work of this research, satellite-based streamflow simulation and application of FFA of the basin are presented. The paper finally discusses the summary and conclusions of the work.

13.2 DESCRIPTION THE STUDY REGION AND METHODOLOGY

13.2.1 THE BLUE NILE RIVER BASIN

The Blue Nile River Basin (BNRB) lies in the western part of Ethiopia, between 7°45′ and 12°45′N, and 34°05′ and 39°45′E as shown in Figure 13.1 (left). The study area covers about 192,953 km² with total perimeter of 2440 km. It accounts for almost 17.1% of Ethiopia's land area and about 50% of the total average annual runoff in Ethiopia. The climate of the basin is dominated by the effects of topography

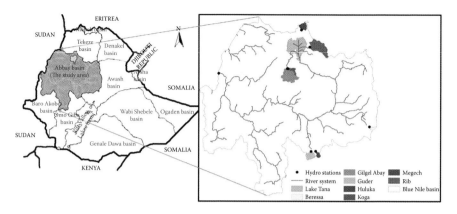

FIGURE 13.1 Location of Blue Nile River Basin, Ethiopia, and the selected study watersheds.

ranging from 590 m to more than 4000 m. This determines a rich variety of local climates ranging from hot to desert-like climate along the Sudan border, to temperate on the high plateau, and cold on the mountain peaks. The annual rainfall varies from about 800–2220 mm with a mean of about 1420 mm.

The Blue Nile River is a major tributary of the Nile River, contributing up to 60% of the total flow at Aswan in Egypt (Shahin, 1985). It is characterized by a unimodal streamflow pattern, with the majority of flow occurring between mid-June–October. Extreme events of an unknown recurrence interval have been known to affect both upstream and downstream areas of the Blue Nile, including Lake Tana in Ethiopia and the entire stretch of the river up to Khartoum (Sutcliffe and Parks, 1999). Floods have a considerable impact on the livelihoods of people and on the infrastructure along this reach. Therefore, it is important to know the flood magnitudes in the region for various recurrence intervals to provide mitigation measures and reduce flood damage.

Extreme high flow series data have been collected from the Ministry of Water Resource, Ethiopia. The available stations in the basin are unevenly distributed with high concentration of stations in the central, northern, and northeastern part of the basin. The northwestern part of the basin has scattered and irregularly distributed stations. Eight gauging stations, including the station at Ethio-Sudan boarder, are considered for this study (Figure 13.1, right). Based on the availability of satellite rainfall estimates and observation streamflow data, the study period is limited from 2001 to 2013.

13.2.2 RESEARCH METHODOLOGY

An extensive study was conducted on BNRB related to FFA. To mention, hydrologic similar regions were identified; for each identified region, the frequency curves were developed (Gebregiorgis et al., 2012). To summarize the previous work, the method used for regionalization and frequency analysis was the index flood method (Burn and Goel, 2000; Rao and Hamed, 2000), which comprises the standardized

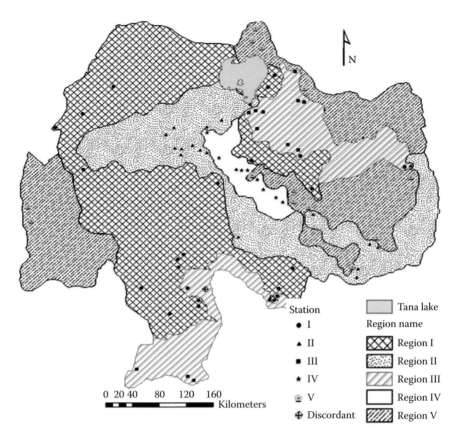

FIGURE 13.2 The five delineated homogenous regions of Blue Nile River basin (BNRB). (From Gebregiorgis et al., *J. Hydrol. Eng.*, 18, 1349–1359, 2012.)

annual maximum flow series of each station divided by site-averaged annual maximum flow values. The L-moment method was applied to derive statistical parameters of each station. The L-Moment Ratio Diagram (LMRD) was used as a tool to identify a priori homogeneous regions and distributions based on the premise of the study. The main hypothesis used in the study was that if the annual maximum flows of different stations come from a single parent distribution model, then these stations belong to the same group and they form a hydrologically homogeneous region (Figure 13.2). The clusters of stations were categorized into different regions based on the proximity of stations in the LMRD. The derived regions and stations included in the group are tested by different homogeneity tests.

The purpose of regionalization based on statistical parameters was principally important to identify a single common best-fitting distribution for a single region. In general, Gebregiorgis et al. (2012) focused on FFA of BNRB based on ground observation data (rainfall, streamflow) by involving the following steps: Computation of statistical parameters of stations within the basin, regionalization of the basin, selection of the best-fitting distribution models, and development of regional flood frequency curves (Figure 13.3).

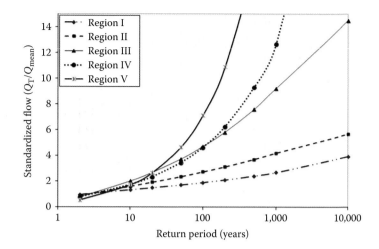

FIGURE 13.3 Regional flood frequency curves for five homogeneous regions of BNRB after Gebregiorgis et al., 2012. Q_T: Maximum flow quantile of T years return period; Q_{mean}: Mean of annual maximum flow.

In this study, we focused on the application of remote sensing-based rainfall estimates (3B42RT) to simulate streamflow at selected gauging stations (Figure 13.1) using a Variable Infiltration Capacity (VIC) hydrologic model (Liang and Xie, 2001). VIC was used to simulate the hydrologic fluxes of the basin (soil moisture, runoff, evapotranspiration, etc.,) at 0.125° spatial resolution and 24 h (daily) temporal scales by using 3B42RT rainfall estimate, other meteorological forcing data (maximum temperature, minimum temperature, and wind speed), soil data, and vegetation cover as the major inputs. The land surface fluxes produced by the VIC model were then utilized by the Horizontal Routing Model (Lohmann et al., 1998) in the offline mode to rout the runoff from grid box as streamflow at selected stations. By ejecting the aforementioned input data, both models were calibrated against the observed streamflow data at Blue Nile El-diem station (Ethio-Sudan boarder). This provided us the opportunity to adjust/calibrate the soil parameters over the entire basin. Once the model was calibrated, the streamflow was generated at eight stations at daily timescale for 2001–2013. Then the annual peak flow series were extracted from the simulated streamflow data. Finally, the peak flow data were used to fit to the proposed flood frequency models by Gebregiorgis et al. (2012).

13.3   SIMULATED AND OBSERVED FLOW QUANTILES

For streamflow simulation, 3B42RT is used as input for the hydrologic model (VIC). Figure 13.4 shows that the mean and total annual precipitation distribution over the region. In fact, the Western part of the basin receives the maximum precipitation. The primary driving force for the formation of precipitation within the

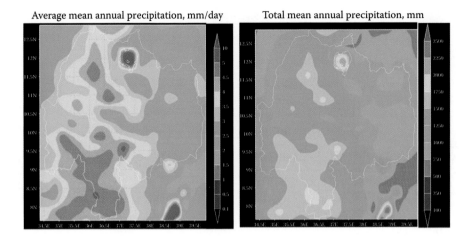

FIGURE 13.4 The spatial distribution of mean average and total annual rainfall for Blue Nile region from TRMM-3B42RT product.

region is due to the flow of moist air masses from different directions (Indian Ocean, Mediterranean, and Red Seas, Western Africa region) that concentrated in the southwest part of the region (Conway, 2000). The spatial variability in precipitation within the region is highly related to the general pattern of humid conditions in the southern highlands of the basin grading to drier conditions in the north and west.

As seen on Figures 13.5 and 13.6 notably, the precipitation has mono modal distribution, which generally peaks during the month of June, July, and August (rainy season). However, for some of the basin, the rainfall season is extended 1 or

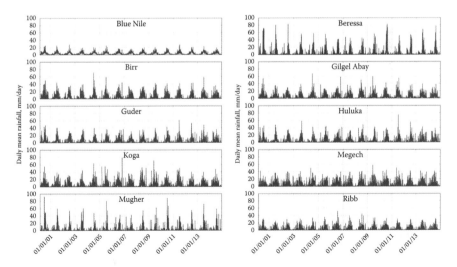

FIGURE 13.5 Time series of rainfall spatially averaged over different watersheds and the entire Blue Nile Basin.

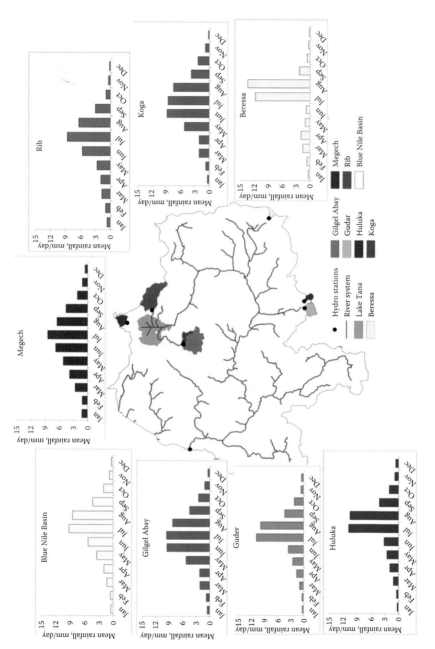

FIGURE 13.6 The distribution of mean monthly rainfall for selected seven watersheds and the entire Blue Nile Basin.

2 months before and after the major wet season, such as Megech, Koga watersheds (Figure 13.5). In general, the monthly variability of rainfall for all basins is high which indicates sensitivity of the available water resource for transboundary and regional consumptions. In addition, it is an indication for the occurrence of flash and seasonal river floods.

The quantiles estimated using in situ observed discharge and simulated discharge using remotely sensed precipitation products give encouraging results for five out of eight stations (Figure 13.7). In most of the stations, the lower return period quantiles demonstrate good agreement between the observed and simulated results. The discrepancy between the two becomes significant for the higher quantiles (e.g., Figure 13.7e and g). In general, the correlation coefficient between the observed and simulated peak flow quantiles is above 0.9 for El-diem, Beressa, Gilgel Abay, Guder,

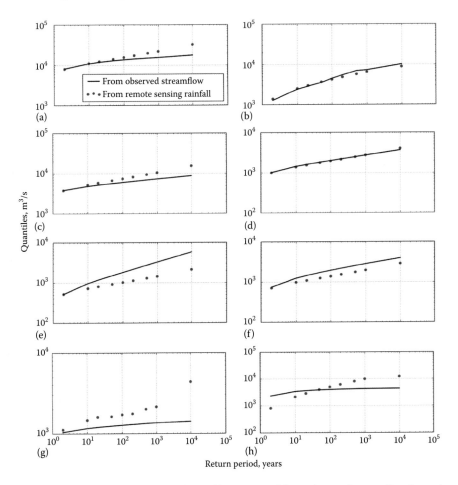

FIGURE 13.7 Comparison of flood quantiles generated from observed streamflow (smooth black line) and using TRMM satellite rainfall product based on the method proposed on Gebregiorgis et al. 2012. (a) Blue Nile at El-diem station; (b) at Beressa; (c) at Gilgel Abay; (d) at Guder; (e) at huluka; (f) at Koga; (g) at Megech; and (h) at Rib station.

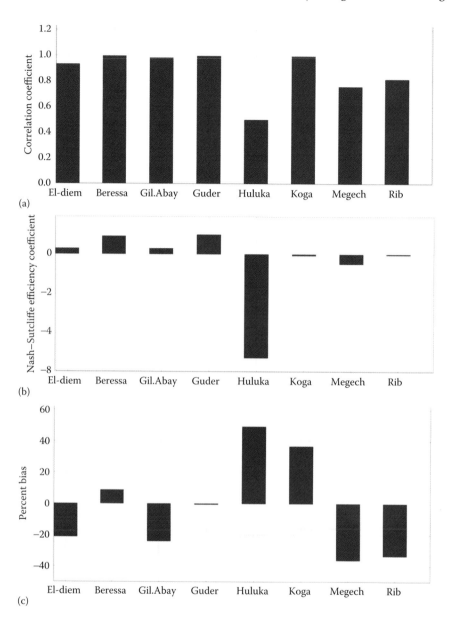

FIGURE 13.8 Performance measure between simulated from satellite rainfall and observed peak flow quantiles: (a) Correlation coefficient; (b) Nash–Sutcliffe efficiency; and (c) Percent bias.

and Koga stations (Figure 13.8a). Similarly, strong positive Nash–Sutcliffe coefficients have been obtained for these stations (Figure 13.8b). In contrast, the lowest correlation coefficient is 0.5 for the Huluka station with a high negative Nash–Sutcliffe result. The bias result (Figure 13.8c) shows that the remotely sensed simulated peak flow quantiles are overestimated at Huluka and Koga.

In regions such as BNRB, one of the bottlenecks for regional water resource project development and flood hazard/risk mitigation programs is the presence of little or no observation data, which is inadequate to predict the design flood information. To deal with such a problem, the approach involves data transfer from gauged to ungauged, regionalization, extrapolation of very high return period quantiles from short data records, and so on, producing quantile estimates with a high level of uncertainty. Given that most of the Upper Blue Nile is not gauged, the promise of remotely sensed precipitation products is demonstrated. However, further investigation of the quality of the in situ observed discharge data are also required. For instance, our discussion with Ministry of Water experts in Ethiopia indicates that Rib and Megech river stations have a known case of underestimation of extreme flows due to measurement uncertainty. Overall, the study shows the advancement of the remotely sensed precipitation in reproducing extreme events.

13.4 CONCLUSION

This study is to evaluate the application of remote sensing precipitation data for FFA. In the view of this, the flood quantiles are estimated for different return periods using observed streamflow and simulated streamflow from remote sensing precipitation product. The use of satellite rainfall data for FFA is advantageous from many perspectives. For instance, if a dimensionless frequency curve has already been developed for the region (which represents the ratio of the flood of any frequency to the mean annual flood), then the remaining simple task is to find out the mean annual simulated flow from satellite data at any station of interest. There is no need to develop a relationship between topographic characteristics of the watershed areas and the mean annual flood to be predicted at any ungauged locations. The existence of continuous and consistent length of record of data at any point provide better estimate of the frequency relationship and ultimately reduces the sampling error. As has been indicated, the quantiles estimated using in situ observed discharge and simulated discharge using remotely sensed precipitation products give encouraging results for many of the stations. The result proves that the advancement of the remotely sensed precipitation could be used to reproduce extreme events in data-scarce regions including in physically inaccessible areas. However, a perfect method to evaluate flood quantiles is still lacking. Further investigations are required related to the quantity and quality of observed data used for comparison. It is important also to understand the reliability of the model in predicting the streamflow from satellite precipitation. Because the progress seen in the global satellite rainfall estimation method is promising, our hydrologic prediction capability is more pragmatic as we enter the GPM era.

REFERENCES

Burn, D. H. and N. K. Goel. (2000). The formation of groups for regional flood frequency analysis. *Hydrol. Sci. J.,* 45(1), 97–112.
Chow, V. T. (1964). *Handbook of Applied Hydrology.* McGraw-Hill Book Company, USA.
Conway, D. (2000). The climate and hydrology of the Upper Blue Nile, Ethiopia. *Geogr. J.,* 166, 49–62.

Gebeyehu, A. (1989). Regional flood frequency analysis. PhD thesis, Royal Institute of Technology, Sweden.

Gebregiorgis, A. S. and F. Hossain. (2011). How much can a priori hydrologic model predictability help in optimal merging of satellite precipitation products? *J. Hydrometeor.*, doi:10.1175/JHM-D-10-05023.1.

Gebregiorgis, A. S. and F. Hossain. (2013). Performance evaluation of merged satellite rainfall products based on spatial and seasonal signatures of hydrologic predictability, *Atmos. Res.*, doi.org/10.1016/j.atmosres.2013.05.003.

Gebregiorgis, A. S. Moges, and S. Awulachew. (2012). Basin regionalization for the purpose of water resource development in a limited data situation: Case of Blue Nile River Basin, Ethiopia. *J. Hydrol. Eng.*, 18(10), 1349–1359.

Gottschalck, J., J. Meng, M. Rodell, and P. Houser. (2005). Analysis of multiple precipitation products and preliminary assessment of their impact on global land data assimilation system land surface states. *J. Hydrometeor.*, 6, 573–598, doi:10.1175/JHM437.1.

Hong, Y., K.-L. Hsu, H. Moradkhani, and S. Sorooshian. (2006). Uncertainty quantification of satellite precipitation estimation and Monte Carlo assessment of the error propagation into hydrologic response. *Water Resour. Res.*, 42, W08421, doi:10.1029/2005WR004398.

Hong, Y., R. F. Adler, F. Hossain, S. Curtis, and G. J. Huffman. (2007). A first approach to global runoff simulation using satellite rainfall estimation. *Water Resour. Res.*, 43, W08502, doi:10.1029/2006WR005739.

Huffman, G. J., R. F. Adler, D. T. Bolvin, and E. J. Nelkin. (2010). The TRMM multisatellite precipitation analysis (TMPA). In *Satellite Rainfall Applications for Surface Hydrology*, Gebremichael, M. and Hossain, F. (eds.), Springer, New York (ISBN: 978-90-481-2914-0), pp. 3–22.

Huffman, G. J., R. F. Adler, D. T. Bolvin, G. J. Gu, E. J. Nelkin, K. P. Bowman, Y. Hong, E. F. Stocker, and D. B. Wolff. (2007). The TRMM multisatellite precipitation analysis (TMPA): Quasi-global, multiyear, combined-sensor precipitation estimates at fine scales. *J. Hydrometeorol.*, 8(1), 38–55.

Joyce, R., J. E. Janowiak, P. A. Arkin, and P. Xie. (2004). CMORPH: A method that produces global precipitation estimates from passive microwave and infrared data at high spatial and temporal resolution, *J. Hydrometeorol.*, 5, 487–503.

Kidd, C., D. R. Kniveton, M. C. Todd, and T. J. Bellerby. (2003). Satellite rainfall estimation using combined passive microwave and infrared algorithms. *J. Hydrometeorol.*, 4, 1088–1104.

Liang, X. and Z. Xie. (2001). A new surface runoff parameterization with subgrid-scale soil heterogeneity for land surface models. *Adv. Water Resour.*, 24(9–10), 1173–1193.

Lohmann, D., E. Raschke, B. Nijssen, and D. P. Lettenmaier. (1998). Regional scale hydrology: I. Formulation of the VIC-2L model coupled to a routing model. *Hydrol. Sci. J.*, 43(1), 131–141.

Nijssen, B. and D. P. Lettenmaier. (2004). Effect of precipitation sampling error on simulated hydrological fluxes and states: Anticipating the global precipitation measurement satellites. *J. Geophys. Res.*, 109(D2), D02103, doi:10.1029/2003JD003497.

Rao, A. R. and Hamed, K. H. (2000). *Flood Frequency Analysis*. CRC press LLC, FL, Boca Raton.

Rosbjerg, D. and Madsen, H. (1995). Uncertainty measures of regional flood frequency estimators. *J. Hydrol.*, 167, 209–224.

Shahin, M. (1985). *Hydrology of the Nile Basin*. Elsevier, Amsterdam, the Netherlands.

Shiklomanov, A. I., R. B. Lammers, and C. J. Vorosmarty. (2002). Widespread decline in hydrological monitoring threatens pan-arctic research. *EOS Trans.*, 83, 16–17.

Sorooshian, S., K. L. Hsu, X. Gao, H. V. Gupta, B. Imam, and D. Braithwaite. (2000). Evaluation of PERSIANN system satellite based estimates of tropical rainfall. *Bull. Amer. Meteor. Soc.*, 81(9), 2035–2046.

Stokstad, E. (1999). Scarcity of rain, stream gages threatens forecasts. *Science,* 285, 1199–1200.

Sutcliffe, J. V. and Parks, Y. P. (1999). *The Hydrology of the Nile.* IAHS Special Publication, IAHS Press, Wallingford, UK.

Turk, J. T., G. V. Mostovoy, and V. Anantharaj. (2010). The NRL-blend high resolution precipitation product and its application to land surface hydrology. In *Satellite Rainfall Applications for Surface Hydrology,* Gebremichael, M. and Hossain, F. (eds.), Springer, New York (ISBN: 978-90-481-2914-0).

Section III

Hydrologic Capacity Building for Improved Societal Resilience

14 Real-Time Hydrologic Prediction System in East Africa through SERVIR

Manabendra Saharia, Li Li, Yang Hong, Jiahu Wang, Robert F. Adler, Frederick S. Policelli, Shahid Habib, Daniel Irwin, Tesfaye Korme, and Lawrence Okello

CONTENTS

14.1 INTRODUCTION

Rainstorm-induced floods and landslides have caused staggering global losses of life and property (Negri et al., 2005). Floods, often coupled with landslides, lead to more than 20,000 fatalities and adversely affect more than 140 million people yearly for the past 10 years (Adhikari et al., 2010). As such, there is a great need globally for accurate and timely hydrologic simulations to mitigate loss of life and property. But often the countries most affected by floods are also least prepared for it. Modest national weather services and lack of technology hinders informing the public about impending severe events in advance. Most developing countries lack real-time information about hydrologic conditions of rivers. It is well known that the ability to predict floods in advance and disaster preparedness can substantially mitigate its adverse impacts. To this end, the Regional Monitoring and Visualization System (SERVIR), as the primary enterprise of NASA Applied Sciences Program, has taken a leading role in building skill and capabilities in disaster warning systems for developing countries. SERVIR, which is a joint venture between NASA and US Agency for International Development, partners with regional organizations around the world to facilitate the transfer of technology regarding the use of satellite-based Earth observation data in environmental decision making. SERVIR is a regional visualization and monitoring system using Earth observations to support environmental management, climate adaptation, and disaster response in developing countries.

14.1.1 SERVIR GLOBAL

SERVIR (www.servirglobal.net) started in 2004 as a collaboration between NASA, World Bank and the Central American Commission for Environment and Development. It started with Panama as its first hub to serve Central America, and since then has expanded into multiple continents with other hubs being established in Nairobi (Kenya) to serve East Africa and Kathmandu (Nepal) for the Hindu Kush-Himalaya (HKH) region. These hubs were established in partnerships with the Regional Center for Mapping of Resources for Development in Nairobi, and the International Centre for Integrated Mountain Development in Kathmandu, respectively. The SERVIR Coordination Office, located at the NASA MSFC, serves in a coordinating role and maintains a test bed for innovative applications in developing countries. SERVIR is currently active in three regions: Mesoamerica, East Africa, and the HKH regions.

For example, SERVIR-East Africa has identified flood and drought modeling with early warning as one of its major focuses. In response to these needs, over the past several years, as the investigator of NASA-SERVIR-East Africa project and NASA Applied Science "Global Hurricane-Flood-Landslide Disaster Prediction" project, the proposal team at OU HyDROS Lab (http://hydro.ou.edu), together with collaborators at NASA Marshall and Goddard centers, have jointly developed the Coupled Routing and Excess Storage (CREST, Wang et al., 2011) distributed hydrologic model and the SLope-Infiltration-Distributed Equilibrium slope-stability model (Liao et al., 2011, 2010). These two models have become the basis to prototype Flood and Landslide Disaster Early Warning Systems.

14.1.2 SERVIR AFRICA

The CREST model has been evaluated and implemented globally and regionally and is currently operationally running at University of Oklahoma (http://eos.ou.edu), NASA GSFC (http://trmm.gsfc.nasa.gov/), and SERVIR-East Africa (http://www.servirglobal.net). As shown in Figure 14.1 lower-right panel, we have also implemented a higher resolution (1-km) CREST water balance model forced by satellite 3B42RT near real-time rainfall estimates and 72-h Weather Research Forecast rainfall forecasts to improve flood forecasting covering eight eastern African nations. Apart from this, currently, SERVIR is also running several versions of the tool including historical model runs, near real-time, short-term forecast, and long-term forecast to assess the impacts of climate scenarios on regional water resources. Ultimately, the modeling tools and platforms developed by SERVIR provide valuable information about streamflow that can enable regional water managers in Africa make better decisions regarding water resources, floods, and agriculture.

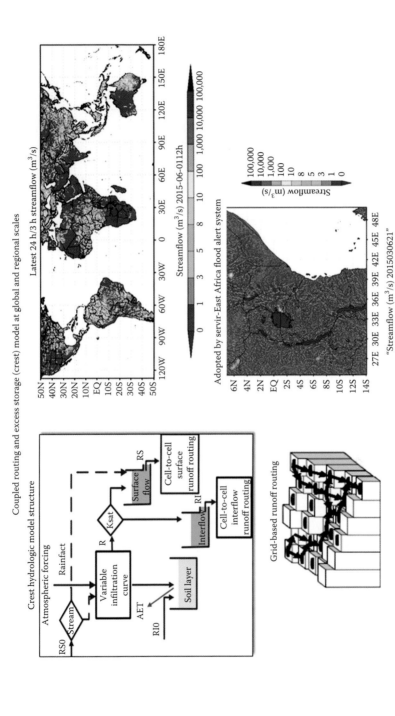

FIGURE 14.1 OU-NASA CREST hydrologic model has been applied at global and regional scale via NASA TRMM/GPM and SERVIR program.

14.2 CASE STUDY OF OPERATIONAL FLOOD PREDICTION SYSTEM IN NZOIA BASIN, AFRICA

To serve the mission of SERVIR, Li et al. (2008) evaluated the applicability of integrating NASA's standard precipitation product, TRMM-based Multi-satellite Precipitation Analysis 3B42 Real-Time (TMPA 3B42RT) rainfall data, with a flood prediction system for disaster management in Nzoia, sub-basin of Lake Victoria, Africa. The key datasets enabling the development of a distributed hydrologic model in Africa include TMPA 3B42RT Real-time rainfall estimates, the digital elevation data from the NASA Shuttle Radar and Topography Mission (SRTM), hydrologic parameter files from Hydrologic data and maps based on SHuttle Elevation Derivatives at multiple Scales (HydroSHEDS), the Moderate Resolution Imaging Spectroradiometer (MODIS) land cover, and soil parameters provided by Food and Agricultural Organization (FAO).

14.2.1 STUDY AREA

In general, repeated flooding is a serious problem in East Africa, particularly in the Lake Victoria basin (Figure 14.2), which affects the lives of 30 million people of that region (Osano et al., 2003). The region around Lake Victoria is prone to flooding because of heavy rains and overflowing of the tributary rivers and streams. People in the heavily populated regions of Kenya, Uganda, and Tanzania live their life under a constant threat of flooding every year. In late May of 2002 alone, widespread flooding throughout Kenya displaced up to 60,000 people. In the year 2006, it was reportedly claimed the lives of 1000 people and displaced another 150,000 people.

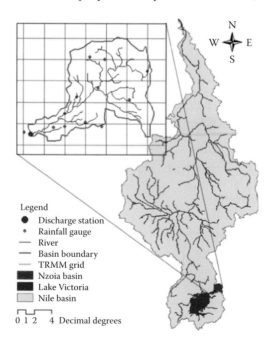

FIGURE 14.2   Study region.

The initial focus area of this project was to implement flood early warning information for Nzoia, a sub-basin of the Lake Victoria region because of its territorial, geographic, and epidemiological importance for the region. The Nzoia River sub-basin, lying in East Africa, covers approximately 12,696 km^2, and lies in the upstream of the Lake Victoria basin and Nile River basin (Figure 14.1). The basin elevation ranges from 1134 to 2700 m. It encompasses three geographical regions: the highlands around Mount Elgon and the Cherangany Hills, the upper plateau, which includes Eldoret, and the lowlands. The region receives an average of 1350 mm of rain annually and is an important cereal and sugarcane-farming region of Kenya, producing at least 30% of the national output of both maize and sugar. The total length of the river is 252 km with an average gradient of 4 m/km.

14.2.2 DATA USED

The key datasets enabling the development of a distributed hydrologic model in Africa include TMPA 3B42RT (Huffman et al., 2007; http://trmm.gsfc.nasa.gov), the digital elevation data from SRTM (Rabus et al., 2003; http://www2.jpl.nasa.gov/srtm/). SRTM-derived hydrologic parameter files of HydroSHEDS (Lehner et al., 2008), soil parameters provided by the FAO 2003 (http://www.fao.org/AG/agl/agll/dsmw.htm), the MODIS land classification map is used as a surrogate for land use/cover, with 17 classes of land cover according to the International Geosphere–Biosphere Programme classification (Friedl et al., 2002).

14.2.3 A CONCEPTUAL PHYSICAL DISTRIBUTED HYDROLOGIC MODEL

Module-Structured Gridded Xinanjiang Model (Zhao and Liu, 1995) is a conceptual, physically based, distributed hydrologic model. It was set up in the basin according to the resolution of GTOPO30 DEM and coupled with a flow-routing scheme with direction file from HydroSHEDS. The model has been successfully and widely applied in humid and semihumid regions in China since its development in the 1970s. It has been used for analyzing the impacts of climate change (Jiang et al., 2007) and other purposes mainly for flood forecasting (Jayawardena and Zhou, 2000). Its runoff generation method has been used widely in distributed hydrologic simulations recently, especially; its function to describe the soil moisture variation is employed in the Variable Infiltration Capacity model (Wood et al., 1992).

The structure and flow chart of the distributed model is shown in Figure 14.3. First, runoff of each grid was calculated with runoff-generation module of the Xinanjiang model. Then they were concentrated to the outlet along the flow path with a runoff-routing scheme according to the concentration time and multilinear reservoirs (Li et al., 2005).

14.2.4 CALIBRATION AND SIMULATION RESULTS

Initially, the Xinanjiang model parameters were estimated using digital datasets of soil, MODIS land cover types, SRTM elevation, and HydroSHEDS. Afterward, we went through two phases of model set-up: (1) 1985–2001 hydrologic model warming up period and (2) 2002–2006 hydrologic model calibration and verification.

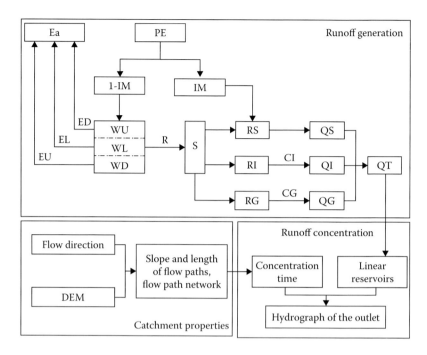

FIGURE 14.3 The structure and flow chart of the distributed hydrologic model.

14.3 MODEL CALIBRATION AND SIMULATION

Because TMPA 3B42RT rainfall data were only available after 2001, we implemented the hydrologic model for the period 2002–2006 during which both satellite and gauge data are available. After recalibration against gauge rainfall for this period, we observed that the bias ratio and the NSCE score became 0.94% and 0.84%, respectively (see Figure 14.4a). Afterward, the calibrated hydrologic model was forced by 3B42RT rainfall data, and the results shown in Figure 14.4b overpredicted daily streamflows 20.4% but with promising Nash–Sutcliffe coefficient of efficiency (NSCE) score (0.67). We further simulated the streamflows forced by the bias-corrected 3B42RT, that is, 3B42V6 (Huffman et al., 2007). This reduced the overall bias of discharge prediction to 3.6% at the same time improving the NSCE score to 0.71 (Table 14.1). Showing the accumulation of daily discharges from 2002 to 2006, Figure 14.4c also demonstrates the improvement of NSCE and the reduction of bias with forcing data from bias-corrected 3B42RT over the original 3B42RT data.

14.4 CONCLUSIONS

Satellite-based rainfall and geospatial datasets are potentially useful for cost-effective detection and early warning of natural hazards, such as floods, specifically for regions of the world where local data are sparse or nonexistent. Many researchers seek to take advantage of the recently available and virtually uninterrupted supply of satellite-based rainfall information as an alternative and supplement to the ground-based observation

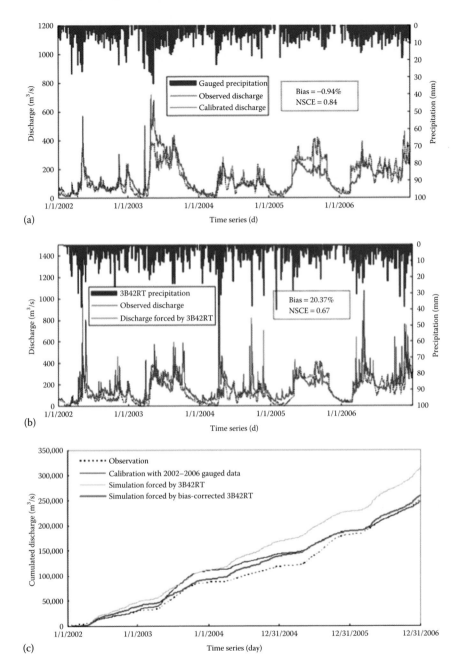

FIGURE 14.4 Hydrograph of daily streamflow for (a) warming up period (1985–2001) forced by gauge rainfall. (b) Prediction period (2002–2006) forced by 3B42RT. (c) Accumulation of discharge data over the period of 2002–2006.

TABLE 14.1

Summary of the Calibration and Overall Results

Phases	Calibration and Results				Prediction and Results			
	Time Span	Forcing Rainfall	Bias (%)	NSCE	Time Span	Forcing Data	Bias (%)	NSCE
Warming up	1985–2001	Gauge	−5.20	0.73	2002–2006	3B42RT	26.6	0.53
Implementation	2002–2006	Gauge	−0.94	0.84	2002–2006	3B42RT	20.37	0.67
					2002–2006	Bias-corrected 3B42RT	3.60	0.71

to implement a cost-effective flood prediction in many undergauged regions around the world. The goal of this study is to build disaster management capacity in East Africa by providing local governmental officials and international aid organizations a practical decision-support tool so as to better assess emerging flood impacts and to quantify spatial extent of water hazard risk, as well as to respond to such flood emergencies more expediently.

In this study, the applicability of TMPA 3B42RT rainfall estimates for distributed hydrologic modeling over a flood-prone region, Nzoia, a sub-basin of Lake Victoria and Nile River (Figure 14.1) was evaluated. This research first validated the TMPA 3B42RT rainfall products with gauged daily rainfall observations from the year 2002 to 2006 (Table 14.1; Figures 14.3 and 14.4). Afterward, a spatially distributed hydrologic model was set up and calibrated to the Nzoia basin. The key datasets enabling the development of a distributed hydrologic model there include TMPA 3B42 Real-time rainfall estimates, the digital elevation data from the NASA SRTM mission, hydrologic parameter files of HydroSHEDS, the MODIS land cover, and soil information provided by FAO. Then benchmark streamflow simulation performance was produced by the calibrated hydrologic model using the rain gauge and observed streamflow data provided by African Regional Centre For Mapping Resource For Development (RCMRD), local partner of this project. Finally, continuous discharges forced by TMPA Real-time data from 2002 to 2006 were simulated and reasonable results were obtained in comparison with the benchmark performance according to the designated statistical indices such as bias ratio and NSCE.

Although, the TMPA 3B42RT has performed satisfactorily for flood prediction purpose in the Nzoia basin, further research will be necessary to investigate the optimal calibration strategy for integrating remote sensing data into a hydrologic model. On the other hand, results of this study also demand continuous progress in space-borne rainfall estimation technology in terms of both the accuracy and spatio-temporal resolutions of rainfall estimates. In this regard, the recently launched Global Precipitation Measuring mission (http://gpm.gsfc.nasa.gov) and Soil Moisture Active Passive mission (http://smap.jpl.nasa.gov/) would largely facilitate the improvement of hydrologic modeling based on space-based remote sensing data. In many countries around the world, satellite-based remote sensing estimations may be the best readily available data source due to insufficient surface networks, long delays in data transmission, and the absence of data sharing in many transboundary river basins. The NASA-SERVIR system aims to facilitate practical applications of NASA remote sensing products and Earth system models to advance disaster decision-making system.

REFERENCES

Adhikari, P., Hong, Y., Douglas, K.R., Kirschbaum, D.B., Gourley, J., Adler, R., Brakenridge, G.R., 2010. A digitized global flood inventory (1998–2008): Compilation and preliminary results. *Nat. Hazards* 55, 405–422. doi:10.1007/s11069-010-9537-2.

Friedl, M.A. et al., 2002. Global land cover mapping from MODIS: Algorithms and early results. *Remote Sens. Environ.* 83, 287–302. doi:10.1016/S0034-4257(02)00078-0.

Huffman, G.J., Bolvin, D.T., Nelkin, E.J., Wolff, D.B., Adler, R.F., Gu, G., Hong, Y., Bowman, K.P., Stocker, E.F., 2007. The TRMM multisatellite precipitation analysis (TMPA): Quasi-global, multiyear, combined-sensor precipitation estimates at fine scales. *J. Hydrometeorol.* 8, 38–55. doi:10.1175/JHM560.1.

Jayawardena, A.W., Zhou, M.C., 2000. A modified spatial soil moisture storage capacity distribution curve for the Xinanjiang model. *J. Hydrol.* 227, 93–113. doi:10.1016/S0022-1694(99)00173-0.

Jiang, T., Chen, Y.D., Xu, C., Chen, X., Chen, X., Singh, V.P., 2007. Comparison of hydrological impacts of climate change simulated by six hydrological models in the Dongjiang Basin, South China. *J. Hydrol.* 336, 316–333. doi:10.1016/j.jhydrol.2007.01.010.

Lehner, B., Verdin, K., Jarvis, A., 2008. New global hydrography derived from spaceborne elevation data. *Eos Trans. Am. Geophys. Union* 89, 2. doi:10.1029/2008EO100001.

Liao, Z., Hong, Y., Kirschbaum, D., Liu, C., 2011. Assessment of shallow landslides from Hurricane Mitch in central America using a physically based model. *Environ. Earth Sci.* 66, 1697–1705. doi:10.1007/s12665-011-0997-9.

Liao, Z., Hong, Y., Wang, J., Fukuoka, H., Sassa, K., Karnawati, D., Fathani, F., 2010. Prototyping an experimental early warning system for rainfall-induced landslides in Indonesia using satellite remote sensing and geospatial datasets. *Landslides* 7, 317–324. doi:10.1007/s10346-010-0219-7.

Li, K.Y., Coe, M.T., Ramankutty, N., 2005. Investigation of hydrological variability in West Africa using land surface models. *J. Clim.* 18, 3173–3188. doi:10.1175/JCLI3452.1.

Li, L., Hong, Y., Wang, J., Adler, R.F., Policelli, F.S., Habib, S., Irwn, D., Korme, T., Okello, L., 2008. Evaluation of the real-time TRMM-based multi-satellite precipitation analysis for an operational flood prediction system in Nzoia Basin, Lake Victoria, Africa. *Nat. Hazards* 50, 109–123. doi:10.1007/s11069-008-9324-5.

Negri, A.J., Burkardt, N., Golden, J.H., Halverson, J.B., Huffman, G.J., Larsen, M.C., McGinley, J.A., Updike, R.G., Verdin, J.P., Wieczorek, G.F., 2005. The hurricane–flood–landslide continuum. *Bull. Am. Meteorol. Soc.* 86, 1241–1247. doi:10.1175/BAMS-86-9-1241.

Osano, O., Nzyuko, D., Tole, M., Admiraal, W., 2003. The fate of chloroacetanilide herbicides and their degradation products in the Nzoia Basin, Kenya. *AMBIO J. Hum. Environ.* 32, 424–427. doi:10.1579/0044-7447-32.6.424.

Rabus, B., Eineder, M., Roth, A., Bamler, R., 2003. The shuttle radar topography mission—a new class of digital elevation models acquired by spaceborne radar. *ISPRS J. Photogramm. Remote Sens.* 57, 241–262. doi:10.1016/S0924-2716(02)00124-7.

Wang, J. et al., 2011. The coupled routing and excess storage (CREST) distributed hydrological model. *Hydrol. Sci. J.* 56, 84–98. doi:10.1080/02626667.2010.543087.

Wood, E.F., Lettenmaier, D.P., Zartarian, V.G., 1992. A land-surface hydrology parameterization with subgrid variability for general circulation models. *J. Geophys. Res. Atmos.* 97, 2717–2728. doi:10.1029/91JD01786.

Zhao, R.J., Liu, X.R., 1995. The Xinanjiang model. In: *Computer Models of Watershed Hydrology*, V.P. Singh (ed.), pp. 215–232. Water Resources Publications, Littleton, CO.

15 Satellite Remote Sensing Drought Monitoring and Predictions over the Globe

Zengchao Hao, Yang Hong, Qiuhong Tang, Youlong Xia, Vijay P. Singh, Fanghua Hao, Hongguang Cheng, Wei Ouyang, and Xinyi Shen

CONTENTS

15.1 INTRODUCTION

Drought may impose serious challenges to natural systems and many sectors of society and is among the most costly disasters leading to tremendous losses around the globe. With the climate change accompanied by climate extremes, it is expected that drought events are likely to become more intensified and frequent in the future (Dai, 2011; Trenberth et al., 2003, 2004). Accurate and reliable drought monitoring and prediction are of critical importance to the drought mitigation efforts and preparedness plans to address drought-related issues, such as food security and water scarcity across the local and global scales.

Drought is a complicated nature phenomenon due to its wide impact on many sectors, and there is not a unique definition of drought. It can be classified into different types, including meteorological drought, agricultural drought, hydrologic drought, and socioeconomical drought (American Meteorological Society, 1997; Heim, 2002). Meteorological drought is related to the precipitation deficit over an extended period of time, whereas agricultural drought corresponds to the insufficient water to meet the needs of plant growth or crop production. Agricultural drought is generally caused by low precipitation, insufficient water-holding capacity in the root zone of the soil, and/or high atmospheric water demand resulting in the crop yield reduction (Marshall et al., 2012). Hydrologic drought is related to the deficiency of surface and subsurface water to meet the need of water supply. Substantial efforts have been devoted to the drought characterization of each type, for which the development of drought indices to characterize different drought properties, including duration, severity, and spatial extent, is among the most active areas (Hao and AghaKouchak, 2014; Hao and Singh, 2013, 2015a; Hao et al., 2016; Heim, 2002; Kao and Govindaraju, 2010; Mishra and Singh, 2010; Quiring, 2009; Zargar et al., 2011).

Due to complicated drought conditions, a variety of drought-related variables, such as precipitation, temperature, soil moisture, and runoff, from various sources are needed for an accurate assessment of drought. Currently, *in situ* observations generally fall short in drought monitoring due to the limited density of existing observational networks and the lack of resources to enhance, maintain, and expand these networks on a global scale (Hayes et al., 2012). Site-based data do not meet the need for drought assessment on a regional scale and inherent drawbacks exist in drought tracking with site-based drought indices due to the lack of continuous coverage. To monitor the propagation of drought developments through the hydrologic cycle, it is required to continuously track multiple hydrologic variables, such as precipitation, soil moisture, and runoff, which can be achieved through hydrologic model simulations that are of particular importance for agricultural and hydrologic droughts. The main limitation of land-surface simulations in drought characterization is that errors in forcing variables may be transferred to simulated variables and, even with identical forcing datasets, substantially discrepant simulations may be obtained by different models due to different parameterizations of the model (Mitchell et al., 2004; Wang et al., 2009; Xia et al., 2012), resulting in biases in drought characterizations.

Recent advances in Earth observation (EO) satellites in low Earth orbit (LEO) and geostationary orbit (GEO) have improved global observations of key components of the climate system and the hydrologic cycle. Most of the key land-surface and atmospheric variables, including precipitation, evapotranspiration, soil moisture, water vapor, sea-surface temperature, wind speed, snow/ice, and terrestrial water-storage variations, have been observed or estimated with various temporal and spatial resolutions via remote sensing (Tang et al., 2009), which provide invaluable information to track drought conditions across the globe (Wardlow et al., 2012b). The advantages of remote sensing products are that they provide continuous temporal and spatial information for drought assessment and monitoring. Satellite-based sensors and missions, such as Advanced Very High Resolution Radiometer (AVHRR), Tropical Rainfall Measurement Mission (TRMM; Huffman et al., 2007), Moderate Resolution Imaging Spectroradiometer (MODIS; Justice et al., 1998), Advanced Microwave Scanning Radiometer–Earth Observation System (AMSR-E) (Jackson et al., 2010; Kawanishi et al., 2003; Njoku et al., 2003), and Gravity Recovery and Climate Experiment (GRACE; Tapley et al., 2004), have been collecting global data products at various spatial resolutions, leading to tremendous efforts in drought monitoring and assessment across different regions. The forefront effort in drought monitoring is the use of the Normalized Difference Vegetation Index (NDVI) from the AVHRR instrument launched in 1979 to track the effect of drought on vegetation health and growth conditions. The remote-sensed precipitation from TRMM products has been used to compute Standardized Precipitation Index (SPI) for meteorological drought characterization, and related studies have grown recently (Ezzine et al., 2014; Naumann et al., 2012; Sahoo et al., 2015). The estimation of evaporation from the satellite remote sensing has led to the development of remote sensing-based drought indices, such as Evaporative Stress Index (ESI) and Drought Severity Index (DSI), for drought characterization at large scales (Anderson et al., 2011; Mu et al., 2013). The drought condition related to groundwater has also been assessed based on water-storage variations from NASA's GRACE Mission (Houborg et al., 2012; Thomas et al., 2014; Zaitchik et al., 2008). Apart from drought severity (or occurrence), the remote sensing data have also been shown to play an important role in identifying trends and emergence of new drought-prone areas exposed to a greater risk of multiyear drought events (Funk et al., 2012).

Accurate and reliable prediction of drought conditions would be of critical importance to provide essential information to initiate the drought early warning. Along with tremendous efforts in drought monitoring, the drought prediction with remote sensing products has been emerging (Funk, 2009; Liu and Juárez, 2001). The advantage of remote sensing products in characterizing drought conditions is the enhanced drought early warning in large regions to take proactive measures. Statistical methods have been used to predict drought conditions, based solely on remote sensing products with drought indices, such as NDVI (Liu and Juárez, 2001; Tadesse et al., 2014) or SPI (Hao et al., 2015a). Meanwhile, remote sensing products have been integrated with *in situ* observations to drive hydrologic models to simulate streamflow and soil moisture for agricultural and hydrologic drought prediction (Nijssen et al., 2014; Sheffield et al., 2014). Overall, remote sensing products provide great opportunities for drought prediction to aid drought early warning.

A challenge of the drought monitoring and prediction at the regional and global scales is the insufficiency of data, especially for regions with sparse or no observation networks. The application of remote sensing data products integrated with other data types has been expanding to aid drought monitoring and the prediction and development of drought information systems (DIS) at the regional and global scales. Based on the integration of remote sensing products, *in situ* information, and land-surface modeling, several drought monitoring and/or prediction systems have been developed at the global scale, including Global Drought Monitor Portal (Heim and Brewer, 2012), University College London Global Drought Monitor (Lloyd-Hughes and Saunders, 2007), Standardized Precipitation Evapotranspiration Index (SPEI) Global Drought Monitor (Vicente-Serrano et al., 2010), Global Drought Information System (GDIS; http://www.drought.gov/), Global Drought Monitoring System (http://eos.ou.edu/Global_Drought.html), Global Agricultural Drought Monitoring and Forecasting System (GADMFS; Deng et al., 2013), Princeton's Experimental Global Water Cycle and Drought Monitor (http://hydrology.princeton.edu), Global Integrated Drought Monitoring and Prediction System (GIDMAPS; Hao et al., 2014), and multimodel GDIS (Nijssen et al., 2014). Meanwhile, various drought monitoring and/or prediction systems at the regional scale have also been established (Heim and Brewer, 2012; Nijssen et al., 2014), including the US Drought Monitor (Svoboda et al., 2002), NLDAS drought monitor (Sheffield et al., 2012; Xia et al., 2014a), NOAA's Seasonal Drought Outlook (SDO) for the United States, European Drought Observatory (EDO; Acácio et al., 2013; Vogt et al., 2011), the Famine Early Warning Systems Network (FEWS NET; Funk, 2009), Princeton's US Drought Monitoring and Prediction System (DMAPS; Luo and Wood, 2007), The University of Washington Surface Water Monitor (Shukla and Wood, 2008; Wood, 2008; Wood and Lettenmaier, 2006), US—Mexico Drought Prediction Tool (Lyon et al., 2012), Sub-Sahara African Drought Monitoring and Forecasting System (Sheffield et al., 2008; 2014), and Experimental Drought Monitor for India (Shah and Mishra, 2015). The remote sensing products have their own limitations, which necessitate the combination with the *in situ* observations and model simulations for accurate and reliable drought monitoring and prediction to aid operational drought managements.

In this chapter, we introduce the application of satellite remote sensing for drought monitoring and prediction over the globe. The overview of drought indices is first briefed in Section 15.2 along with remote sensing-based drought indices. The current global drought-monitoring efforts in different aspects are introduced in Section 15.3. An overview of drought prediction methods and satellite remote sensing-based drought prediction is introduced in Section 15.4, followed by the discussions and conclusions in Section 15.5.

15.2 REMOTE SENSING DROUGHT INDICES

15.2.1 OVERVIEW OF DROUGHT INDICES

Drought indices are commonly used to characterize the complicated drought condition. The past few decades have witnessed a flurry of developments of drought indices based on various climatic and hydrologic variables to characterize drought (Heim, 2002; Mishra and Singh, 2010; Niemeyer, 2008; Zargar et al., 2011). The Palmer Drought Severity Index (PDSI; Palmer, 1965) and the SPI (McKee et al., 1993) are among the most commonly used drought indices for drought characterization

(Guttman, 1998; Hayes et al., 2011). Drought indices can be broadly classified as the univariate drought indicator, which is generally based on the individual variable to measure a specific aspect of drought, and the multivariate (or integrated, composite) drought indicator, which essentially combines a variety of drought-related variables for comprehensive drought monitoring.

A suite of drought indicators or indices has been developed to characterize each type of drought. The SPI has been recommended to track the meteorological drought condition, along with a variety of other drought indices, such as the percent of normal precipitation, PDSI, Effective Drought Index (EDI), and Vegetation Condition Index (VCI), which have been commonly used for monitoring meteorological drought (Hayes et al., 2011; Quiring 2009). For reasons agricultural crops are sensitive to soil moisture and the soil moisture deficit in the root zone during the crop growth cycle will affect crop yields, a reliable agricultural drought index requires the proper consideration of factors including vegetation type, crop growth and root development, antecedent soil moisture condition, evapotranspiration, and temperature (Narasimhan and Srinivasan, 2005). For agricultural drought monitoring, a suite of drought indices, including Crop Moisture Index (CMI; Palmer, 1968), soil moisture percentile (SMP) or quantile (Mo, 2011; Sheffield et al., 2004), Standardized Soil moisture Index (SSI; Hao and AghaKouchak, 2013), and Normalized Soil Moisture (NSM; Dutra et al., 2008), has been developed and employed (Hayes et al., 2011). The hydrologic drought condition can be characterized by the reservoir level, Palmer Hydrologic Drought Index (PHDI), Surface Water Supply Index (SWSI), SPI, Standardized Runoff Index (SRI), and Standardized Streamflow Index (SSFI; Fernández et al., 2009; Hayes et al., 2011; Modarres, 2007; Shukla and Wood, 2008). As with agricultural drought, there has not been a consensus on the recommendation of the index to be used for hydrologic drought. Due to the limitations of observations networks for soil moisture and runoff, the land-surface simulations are commonly used for agricultural and hydrologic drought characterizations (Mo et al., 2011; Sheffield et al., 2014; Xia et al., 2014b). Certain drought indices can be used for depicting different types of drought. For example, the percent of normal precipitation can be used for all three types of drought conditions, and, although the SPI is used primarily for meteorological drought monitoring, it has also been used for agricultural drought characterization (Hayes et al., 2011; Narasimhan and Srinivasan, 2005). The standardization concept from the computation of SPI has been commonly used for a variety of drought-related variables, including soil moisture, runoff (or streamflow), groundwater, and snow melt, to develop the standardized drought index that is comparable in both time and space (Bloomfield and Marchant, 2013; Hao and Singh, 2015a; Hao et al., 2014; Núñez et al., 2014; Shukla and Wood, 2008; Staudinger et al., 2014).

Due to the complicated nature of drought, it is generally not sufficient to characterize drought with the individual variable for certain applications, especially when the overall condition of drought is desired. Along with the development of drought indices based on an individual variable, extensive efforts have been put in the past decade for the development of the integrated or Multivariate Drought Index (MDI) for comprehensive drought monitoring, as reviewed by Hao and Singh (2015a). The MDIs, such as US Drought Monitor (USDM; Svoboda et al., 2002), Reconnaissance Drought Index (RDI; Tsakiris et al., 2007), Standardized

Precipitation Evapotranspiration Index (SPEI; Vicente-Serrano et al., 2010), Optimal Blended NLDAS Drought Index (OBNDI; Xia et al., 2014a, b), Grand Mean Index (GMI; Mo and Lettenmaier, 2014b), can be developed based on the blending objective and subjective indicators, water balance models, and multivariate analysis methods, such as linear combination, principal component analysis (PCA), or joint distribution, which is mainly used to characterize various dependence structures among different variables (Hao and Singh, 2016). For example, PCA has been used for the development of the Aggregate Drought Index (ADI; Keyantash and Dracup, 2004) to comprehensively consider meteorological, agricultural, and hydrologic droughts through selection of a large number of variables that are related to each drought type, which may fall short in modeling the nonlinear dependence structure of variables. In addition, the joint distribution has also been used to develop several multivariate drought indices, such as Joint Drought Index (JDI; Kao and Govindaraju, 2010), Multivariate Standardized Drought Index (MSDI; Hao and AghaKouchak, 2013), Standardized Palmer Drought Index (SPDI)-based Joint Drought Index (SPDI-JDI; Ma et al., 2014), through the construction of the multivariate distribution with empirical or parametric copula methods (Beersma and Buishand, 2004; Hao and AghaKouchak, 2013; Hao and Singh, 2015b; Kao and Govindaraju, 2010; Ma et al., 2014), which is capable of modeling the nonlinear dependence but may be difficult to handle in high dimension when a large number of drought-related variables are involved. Note that the MDI is not superior to the univariate drought index and the univariate drought index may still be a popular choice, especially in characterizing a specific aspect of drought (Hao and Singh, 2015a).

15.2.2 Satellite Remote Sensing Drought Indices

Along with the traditional drought monitoring with hydroclimatic observations from the gauge-based observation networks and model simulations, recent decades have witnessed new opportunities for drought characterizations from satellite remote sensing products, which provide consistent observations of a variety of hydroclimatic variables of continuous temporal and spatial coverage at different resolutions. Theoretically, drought indices introduced before based on *in situ* observations or model simulations may also be computed for drought monitoring and assessment based on remote sensing products. For example, the SPI can also be employed based on the remote-sensed precipitation estimates for the monitoring of meteorological drought conditions (Ezzine et al., 2014; Naumann et al., 2012; Sahoo et al., 2015). However, remote sensing drought indices are subject to certain limitations, including the short data length, which hinders the development of certain drought indices, such as PDSI, for the drought characterization. A variety of satellite remote sensing-based drought indices have been developed for drought characterization based on the multiband information from optical remote sensing (radiation reflected from targets on the ground), thermal infrared (TIR) remote sensing (thermal radiation emitted from the target surface), and microwave remote sensing. Here, we briefly introduce commonly used drought indices based on satellite remote sensing for drought monitoring.

15.2.2.1 Indices from Optical Remote Sensing

Because droughts are associated with vegetation state and cover, vegetation indices (VIs) are commonly used for drought-monitoring purposes based on the optical remote sensing (Bannari et al., 1995; Silleos et al., 2006; Tucker and Choudhury, 1987). Drought indices derived from optical remote sensing typically are based on observations in multispectral bands including the visible red (R), near-infrared (NIR), and short wavelength infrared (SWIR), each of which provides different information about surface conditions (Karnieli et al., 2010). Various VIs have been developed based on the rationing, differencing, rationing differences, or linear combinations of two or more spectral bands, which enhances the vegetation signal while minimizing the influences from atmospheric and soil background effects (Hanes, 2013; Jackson and Huete, 1991). For the monitoring of the land-surface drought condition related to vegetation, the NDVI from AVHRR has been the forefront and the most commonly used drought index based on the normalized difference between the spectral reflectance measurements in the NIR and visible red region (R) (combined NIR–R band) (Tucker, 1979). The NDVI allows for the separation of vegetation from the bare soil, because the soil spectrum typically does not show distinct spectral difference between these bands (Karnieli et al., 2010). It has been shown in a variety of studies that NDVI correlates well with drought indices (e.g., SPI) or drought-related variables (e.g., soil moisture, rainfall) and, thus, can be used as an effective indicator of vegetation response to monitor drought (Bayarjargal et al., 2006; Di et al., 1994; Ji and Peters, 2003; Peled et al., 2010; Quiring and Ganesh, 2010; Wang et al., 2001a). A variety of NDVI-based drought indices have been developed in the past decades for drought monitoring, including Transformed Vegetation Index (TVI; Deering and Rouse, 1975), Corrected Transformed Vegetation Index (CTVI; Perry and Lautenschlager, 1984), VCI (Kogan, 1990; Kogan and Sullivan, 1993), the NDVI anomaly (Anyamba and Tucker, 2005), and Standardized Vegetation Index (SVI; Peters et al., 2002). Other vegetation drought indices from the NIR–R band have also been developed for characterizing vegetation conditions based on the so-called soil line concept (Baret et al., 1993), including the Ratio Vegetation Index (RVI; Pearson and Miller, 1972), Perpendicular Vegetation Index (PVI; Richardson and Weigand, 1977), Soil-adjusted Vegetation Index (SAVI; Huete, 1988) (and transformed, modified, or optimized SAVI [Baret and Guyot, 1991; Rondeaux et al., 1996; Qi et al., 1994]), Perpendicular Drought Index (PDI; Ghulam et al., 2007a), and Modified Perpendicular Drought Index (MPDI; Ghulam et al., 2007b). For example, to reduce the soil background effect and improve the NDVI in characterizing vegetation conditions, Huete (1988) proposed the SAVI using a soil-adjustment factor L ($= 0.5$) to account for first-order soil background variations.

Although the NDVI has been shown to be an effective indicator of vegetation and moisture condition, the NDVI-based indices are more suitable to serve as an after-effect drought indicator due to the lag between drought occurrences and NDVI changes, which may fall short in real time drought monitoring (Ghulam et al., 2007b). Because the short wavelength infrared (SWIR) reflectance is sensitive to leaf liquid water content (Ghulam et al., 2007b), drought indices based on the combination of NIR and SWIR have been developed recently, including

Leaf Water Content Index (LWCI; Hunt et al., 1987), Normalized Difference Infrared Index (NDII; Hunt and Rock, 1989), Normalized Difference Water Index (NDWI; Gao, 1996), Global Vegetation Moisture Index (GVMI; Ceccato et al., 2002), and Normalized Multiband Drought Index (NDMI; Wang and Qu, 2007). With more optical band from recent remote sensing instruments, such as MODIS, new opportunities have been provided in estimating the vegetation water content and soil moisture more accurately and robustly. For example, Wang and Qu (2007) developed the NMDI based on the vegetation and soil moisture spectral signatures (two SWIR bands and one NIR band) that responds to changes in both the water content (absorption of SWIR radiation) and spongy mesophyll (reflectance of NIR radiation) in vegetation canopies. Gu et al. (2007) developed the Normalized Difference Drought Index (NDDI) based on the NDVI and NDWI, which is a more sensitive indicator of drought in grasslands than NDVI alone due to the stronger response to summer drought condition. Spectral reflectance from other bands, such as blue band, has also been used to derive remote-sensed drought indicators, including Enhanced Vegetation Index (EVI; Huete et al., 2002, 2006) and Visible and Short wavelength Infrared Drought Index (VSDI; Zhang et al., 2013). For example, to address the disadvantage of saturated signals over high biomass conditions and sensitivity to canopy background variation in drought assessment from NDVI, the EVI was developed based on surface reflectance in the NIR, red, and blue bands as an optimized index to improve sensitivity in high biomass regions and vegetation monitoring through the de-coupling of canopy background signals and a reduction in atmosphere influences (Huete et al., 2002; Jiang et al., 2008).

15.2.2.2   Indices from Thermal Infrared Remote Sensing

Apart from the satellite remote sensing information from the optical domain, some drought indices are based on the land-surface temperature (LST) observations from the TIR region of the electromagnetic spectrum, which conveys information about the vegetation health condition and soil moisture state. The LST is widely used to formulate the energy and water budgets at the land surface–atmosphere interface between the earth's surface processes and is a biophysical factor sensitive to surface water stress (Ghulam et al., 2007b). In this context, the LST serves as a proxy for assessing evapotranspiration, vegetation water stress, soil moisture, and thermal inertia (Karnieli et al., 2010), which provide additional information on the drought condition (such as soil dryness, plant stress) not captured by the NDVI alone (Marshall et al., 2012). A variety of drought indices have been developed based on the LST, such as the Crop Water Stress Index (CWSI; Jackson et al., 1981), Temperature Condition Index (TCI; Kogan, 1995b), and ESI (Anderson et al., 2007, 2011).

The visible to NIR (VNIR) data and LST from TIR data have advantages and disadvantages of their own in the utility for drought detections, for which indices from VNIR data are more reliable in detecting vegetation condition and dynamics over intermediate levels of vegetation cover, while the LST-based assessment of land-surface conditions shows a better performance over sparse vegetation cover (Karnieli et al., 2010). As such, the LST has been used as additional sources along with VIs from optical remote sensing to aid drought characterizations by detecting vegetation and plant water condition. In the past few decades, several integrated or composite drought indices have been developed to combine the information from both the optical and thermal bands for drought characterizations by integrating both the VIs and LST, such as the Vegetation Health Index (VHI; Kogan, 1997), Vegetation Temperature Condition Index (VTCI; Wan et al., 2004; Wang et al., 2001b), Temperature-NDVI Ratio Index (McVicar and Bierwirth, 2001), and Temperature Vegetation Dryness Index (TVDI; Sandholt et al., 2002; Son et al., 2012). These drought indices are generally based on the existence of a strong negative (or inverse) relationship between NDVI and LST for the application of drought monitoring (Carlson et al., 1994; Nemani et al., 1993; Sun and Kafatos, 2007), and, thus, the existence of such a relationship is a fundamental basis. It has been shown that the LST–NDVI correlation is negative when water is the limiting factor for vegetation growth, whereas this correlation is positive when energy is the limiting factor for vegetation growth (in higher latitudes and elevations) (Karnieli et al., 2010). Thus, the application of remote sensing drought indices based on the negative LST–NDVI relationship, which is generally site- and time-specific, for drought monitoring should be restricted to areas and periods where negative correlations are identified.

15.2.2.3 Indices from Microwave Remote Sensing

VIs based on the VNIR band are often affected by the atmospheric condition or background soil conditions (Shi et al., 2008; Zhang and Jia, 2013). Compared with the optical remote sensors, the satellite microwave sensors are advantageous because they can penetrate clouds (all-weather capability), operate at day and night, and are less affected by atmospheric conditions (Petropoulos, 2013; Shi et al., 2008; Tao et al., 2008), and, thus, provide an important way for the vegetation and drought characterizations. Due to the all-weather working advantages, the satellite micro-wave remote sensing provides a great potential in drought monitoring, which is not investigated as common as that from optical or infrared bands (Zhang and Jia, 2013). A variety of observations or estimations of hydroclimatic variables can be obtained from microwave remote sensing, such as precipitation from TRMM, LST from AMSR-E (Holmes et al., 2009), and soil moisture from AMSR-E or European remote sensing (ERS) satellites (De Jeu et al., 2008), which can be used for the development of drought indices. A variety of drought indices based on microwave remote sensing products have been developed along these lines, such as Scaled Drought Condition Index (SDCI; Rhee et al., 2010), Microwave Integrated Drought Index (MIDI; Zhang and Jia, 2013), and Synthesized Drought Index (SDI; Du et al., 2013). For example, Rhee et al. (2010) developed the multisensor remote sensing drought index SDCI that combines LST and NDVI from the MODIS sensor and precipitation from TRMM. Zhang and Jia (2013) proposed the multisensor microwave remote sensing-based drought index, MIDI, to improve timely monitoring of short-term drought by integrating soil moisture and LST derived from AMSR-E and precipitation from TRMM.

15.2.2.4 Integrated Indices

A variety of remote sensing drought indices introduced before, such as NMDI, NDDI, VHI, SDCI, and MIDI, are developed based on the combination of different spectral bands or sensors, which can be regarded as the integrated (or multivariate, composite) remote sensing drought indices. In addition, integrated drought indices from remote sensing products can also be developed by combing remote sensing datasets with *in situ* observations or model simulations, such as the US Drought Monitor (USDM; Svoboda et al., 2002), North American Drought Monitor (NADM; Lawrimore et al., 2002), Combined Drought Indicator (CDI; Sepulcre et al., 2012), and Vegetation Drought Response Index (VegDRI; Brown et al., 2008; Tadesse et al., 2015). For example, a hybrid drought index VegDRI is developed to integrate satellite-based NDVI, climate-based drought indices (SPI and self-calibrated PDSI), and biophysical variables based on the linear regression model to characterize the drought-induced stress on vegetation, which addresses the limitations of individual satellite-based NDVI and climate indicators in drought assessment (Brown et al., 2008; Tadesse et al., 2015). In addition, an integrated soil moisture drought monitor is developed based on the triple collocation analysis (TCA) to merge three independent soil moisture products from the AMSR-E sensor, thermal remote sensing using the Atmosphere–Land Exchange Inverse (ALEXI) model, and physically based land-surface simulations using the Noah model (Anderson et al., 2012b). The development of the integrated drought indices to combine the remote sensing data products with *in situ* observations and modeling simulation for drought characterizations is still in its infancy and is expected to be further explored in the future for accurate drought assessments.

15.3 SATELLITE REMOTE SENSING DROUGHT MONITORING

Satellite remote sensing data have provided unique opportunities for global drought monitoring due to the continuous and consistent temporal and spatial coverage of land-surface variables related to drought. In the past decades, various remote-sensed products have been used to aid the drought monitoring of different aspects, including precipitation, evaporation, vegetation, soil moisture, groundwater, and snow cover of hydrologic components at the regional and global scales (Wardlow et al., 2012b). In this section, we will introduce recent developments of drought monitoring with remote sensing products over the globe for different components of the hydrologic cycle with focus on vegetation, precipitation, evaporation, soil moisture, and groundwater storage.

15.3.1 VEGETATION

Satellite remote sensing data have been widely used to track vegetation conditions to investigate the temporal and spatial evolution of drought, and a variety of VIs based on optical, thermal, and microwave remote sensing have been developed for this purpose as shown previously. In this section, we focus on the commonly used NDVI, VCI, TCI, and VHI for illustration purposes. The NDVI is defined based on the normalized difference between the NIR and visible red reflectance (R) (Rouse et al., 1974; Tucker, 1979):

$$\text{NDVI} = \frac{\rho_{\text{NIR}} - \rho_{\text{RED}}}{\rho_{\text{NIR}} + \rho_{\text{RED}}} \tag{15.1}$$

where ρ_{NIR} and ρ_{RED} are the spectral reflectance of NIR and red (visible) bands, respectively. The unhealthy or sparse plants (suffering from drought conditions) reflect more visible light and less NIR light, resulting in lower NDVI values. Although theoretically the NDVI may range from -1 to 1, the typical range of NDVI is between about -0.1 and 0.9 where high values indicate dense and green (or health) vegetation canopy. The NDVI from AVHRR and MODIS sensors have been commonly used for monitoring vegetation and drought conditions (Fensholt et al., 2012; Karnieli et al., 2010; Kogan, 1995a; Kogan and Sullivan, 1993; Ouyang et al., 2012, Peters et al., 2002; Son et al., 2012; Song et al., 2004). The disadvantage of NDVI is that it is affected by atmospheric influences (e.g., aerosols, cloud, and water vapor), only monitors the top of the canopy, and eventually saturates over high biomass conditions (Huete et al., 2002; Liu et al., 2011b).

The VCI is the normalization of NDVI to reflect relative changes in the vegetation condition from extremely poor (0) to excellent (100), which is defined as (Kogan, 1990, 1995a)

$$\text{VCI} = \frac{100\left(\text{NDVI} - \text{NDVI}_{\min}\right)}{\text{NDVI}_{\max} + \text{NDVI}_{\min}} \tag{15.2}$$

where:
NDVI is the smoothed weekly NDVI
$\text{NDVI}_{\max}$ and $\text{NDVI}_{\min}$ are the multiple-year maximum NDVI and minimum NDVI for a pixel with reference to a specific climatology, respectively

The VCI rescales the NDVI between the minimum and maximum values, with low VCI values indicating stressed vegetation due to unfavorable weather conditions and high VCI values indicating healthy vegetation conditions.

Based on the VCI normalization approach, TCI was developed based on LST observations from the TIR remote sensing for monitoring vegetation and drought conditions, which is defined as (Kogan, 1995b, 2000)

$$\text{TCI} = \frac{100(\text{BT}_{\max} - \text{BT})}{\text{BT}_{\max} + \text{BT}_{\min}} \tag{15.3}$$

where:
BT is the smoothed weekly thermal brightness temperatures
$\text{BT}_{\max}$ and $\text{BT}_{\min}$ are the pixel-specific multiple-year maximum and minimum thermal brightness temperatures, respectively

A drought-monitoring study in India showed that the utility of the VCI was improved when used together with the TCI (Singh et al., 2003).

It has been found that the TCI performed better than NDVI and VCI, especially in cases where soil moisture is excessive due to heavy rainfall or persistent cloudiness, in which depressed NDVI and low VCI values may be interpreted erroneously as drought (Liu and Kogan, 1996). To address the issue, the VHI was developed to combine the VCI and TCI to monitor drought and vegetation stress, which is expressed as (Kogan, 1995b, 1997)

$$VHI = \alpha VCI + (1 - \alpha)TCI \qquad (15.4)$$

where α is the coefficient to determine the contribution from each index. The VHI can be used as a proxy for characterizing vegetation health by combining the estimation of moisture and thermal condition through the integration of VCI and TCI, and has been used in a variety of studies for drought and vegetation condition assessments (Bhuiyan et al., 2006; Karnieli et al., 2006; Kogan, 2001, 2002).

The Global Vegetation Health Products, including NDVI, VCI, TCI, and VHI, are routinely produced and distributed by the NOAA Satellite and Information Service (NOAA-NESDIS), available at http://www.star.nesdis.noaa.gov/smcd/emb/vci/VH/vh_ftp.php. These vegetation health indices provide useful information for a variety of drought-related applications, including monitoring drought impacts on vegetation and crops, detecting early drought condition, and assessing drought duration, and intensity, and area coverage (Kogan, 1997). The 2012 US summer drought is among the most severe and extensive drought event to occur in the United States and has seriously affected the US agriculture (Hoerling et al., 2014; USDA, 2012; Yuan et al., 2015b). Here, the VCI and VHI are employed to assess the drought condition during this period. Figure 15.1 shows the VCI for the 34th week (20–26 August) of 2012 at the global scale. Generally, the VCI clearly shows the dryness in August 2012 in a large portion of the United States. Figure 15.2 shows the global VHI for the same period, in which the 2012 US drought is also partly shown.

15.3.2 Precipitation

Precipitation is the driver of the hydrologic cycle and is among the most commonly used satellite remote sensing products for meteorological drought monitoring. Satellite remote sensing precipitation is generally based on visible/infrared sensors,

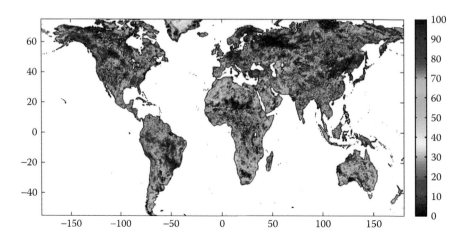

FIGURE 15.1   VCI for the 34th week (20–26 August) of 2012 at the global scale.

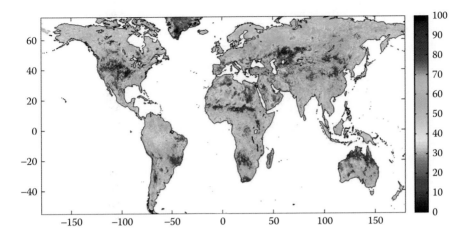

FIGURE 15.2   VHI for the 34th week (20–26 August) of 2012 at the global scale.

which measure precipitation indirectly based on the relationship between the cloud top temperature and rainfall rate, and microwave techniques, which measure rain directly and are relatively insensitive to cloud cover, or the integration of the visible/infrared and microwave observations (Tang et al., 2009; Yong et al., 2014). Several remote sensing products for precipitation estimation (Sapiano and Arkin, 2009), such as TRMM (Huffman et al., 2007), Precipitation Estimation from Remotely Sensed Information using Neural Networks (PERSIANN; Sorooshian et al., 2000), and Climate Prediction Center Morphing (CMORPH) method (Joyce et al., 2004) are available and have been commonly used for a variety of hydrologic applications such as flood forecasting. These remote sensing precipitation products can also be used directly for the meteorological drought monitoring based on the drought index SPI (Vernimmen et al., 2012). For example, the TRMM Multisatellite Precipitation Analysis (TMPA) (Version 7) provides near-global precipitation estimates (50°N–50°S) with a 0.25° spatial resolution based on multiple passive microwave and infrared satellite sensors (Zhou et al., 2014). The TMPA research quality product (TMPA-RP) includes corrections with *in situ* gauges information (available with time lag of few weeks to months), and the TMPA real-time product (TMPA-RT) is available in near real time without corrections, which have been used for drought monitoring at the regional and global scales (Ezzine et al., 2014; Naumann et al., 2012; Sahoo et al., 2015). In addition, the remote sensing precipitation products can also be used as forcing variables to drive land-surface models for agricultural and hydrologic drought characterization (Nijssen et al., 2014; Sheffield et al., 2008).

An impediment to using the remote-sensed precipitation products (~15 years) for drought monitoring is the lack of the relatively long record to form a baseline to compute the SPI, because generally a record of at least 30 years is required for the sufficient estimation of the distribution function. Although it is not optimal in computing the SPI for drought monitoring, studies have shown that the remote-sensed precipitation is feasible for monitoring the drought condition in different regions (Naumann et al., 2012, 2014; Sahoo et al., 2015). To address the issue of the short data record, recent efforts have been devoted to merge or assimilate the remote sensing precipitation product with products from other sources to provide a relatively long data record for drought monitoring (Adler et al., 2003; AghaKouchak and Nakhjiri, 2012; Sheffield et al., 2006).

In 2012, the Global Drought Monitoring System based on satellite remote sensing (http://eos.ou.edu/Global_Drought.html) was established at the Hydrometeorology and Remote Sensing (HyDROS) Laboratory of University of Oklahoma. The TMPA-RT monthly precipitation is used to compute the SPI for tracking meteorological drought conditions in the near real time. In addition, the precipitation estimation from TMPA is used to drive the hydrologic model, Coupled Routing and Excess Storage (CREST) (Wang et al., 2011b), to estimate the soil moisture and runoff for the agricultural and hydrologic drought characterization. Here, the global meteorological drought monitoring is illustrated based on the TMPA-RP and TMPA-RT (Version 7) products from 2001 to 2013 to characterize the 2012 US drought and northeastern Brazil drought. The northeastern Brazil drought is another serious drought event in 2012, which was the harshest drought in decades with potentially significant impacts on the vegetation and local livelihoods (Gutiérrez et al., 2014; Pereira et al., 2014; Rodrigues and McPhaden, 2014). It should be noted that because the data record is from 2001 to 2013, the drought severity calculated based on this period would differ from that based on a longer climatology from other products.

The global meteorological drought for the period August 2012 with TMPA-RP and TMPA-RT based on the 6 months SPI (SPI6) is shown in Figure 15.3. Results indicate severe drought conditions in large portions of the United States and eastern Brazil from TMPA-RP and TMPA-RT products. In spite of some discrepancies in the US data, two products show consistent drought severity and area coverage in northeastern Brazil. Overall, it is observed that drought conditions from TMPA-RP and TMPA-RT provide useful information for global meteorological drought monitoring,

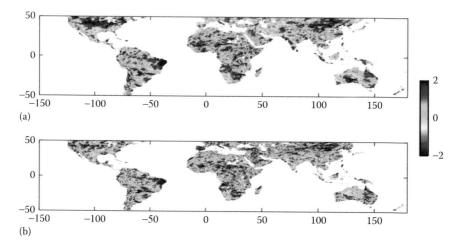

FIGURE 15.3 Monitoring of global meteorological drought based on SPI6 for August 2012 with (a) TMPA-RP and (b) TMPA-RT.

though discrepancies exist in certain regions, which is consistent with the previous studies (Sahoo et al., 2015; Zhou et al., 2014).

15.3.3 EVAPOTRANSPIRATION

ET is a measure of crop stress and, thus, can be used for drought monitoring related to agriculture. Although ET cannot be estimated with remote sensing directly, several approaches have been developed in an attempt to calculate ET from the remote-sensed estimation of surface energy fluxes, surface temperature, and VIs (Anderson et al., 2011; Tang et al., 2009). Unlike precipitation, actual evapotranspiration takes into account additional factors, such as vegetation type and phenology, and antecedent soil moisture conditions that contribute to crop stress, and, thus, ET-based indices are generally more sensitive than precipitation-based indices in characterizing the gradual changes in soil moisture and crop stress during agricultural droughts (Marshall et al., 2012). One of the unique features of ET in drought monitoring is its potential in detecting the flash drought, the rapid soil moisture depletion due to hot, dry, and windy atmospheric conditions, which the local precipitation anomalies may not be adequate to detect (Anderson et al., 2012a). The drought indices used for evaporation are generally based on the anomaly of ET or the ratio between ET and potential ET (or PET), including the Evapotranspiration Deficit Index (ETDI; Narasimhan and Srinivasan, 2005), ESI (Anderson et al., 2011), and DSI (Mu et al., 2013).

Monitoring of the "flash drought" event (with rapidly declining surface moisture and enhanced evaporative loss) requires suitable drought indicators that respond rapidly to the drought condition, for which the vegetation cover condition is a relatively slow response variable (Anderson et al., 2013). The ratio of actual ET and PET (denoted as f_{PET}) is generally used as an indicator of the terrestrial water availability, which can be expressed as (Anderson et al., 2007, 2011)

$$f_{PET} = \frac{ET}{PET} \tag{15.5}$$

The ESI is defined as the standardized anomalies of the f_{PET} and has been used as an alternative source to monitor the drought conditions in comparison with the precipitation-based drought indicator, such as the SPI and the USDM (Anderson et al., 2012a, 2013; Choi et al., 2013; Otkin et al., 2013). The real-time ESI maps over CONUS during the growing season can be obtained from http://hrsl.arsusda.gov/drought.

To incorporate the vegetation response in the drought index for characterizing the drought condition, the DSI was recently developed based on satellite products ET, PET, and NDVI, which is defined as (Mu et al., 2013)

$$DSI = \frac{z - \bar{z}}{\sigma_z} \tag{15.6}$$

where z is the summation of the standardized PET (z_{PET}) and NDVI (z_{NDVI}):

$$z = z_{PET} + z_{NDVI} \tag{15.7}$$

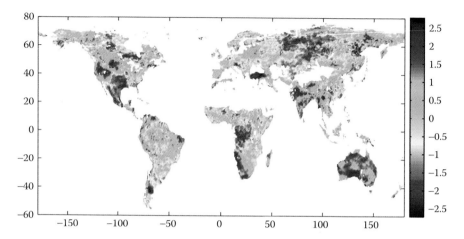

FIGURE 15.4   The annual DSI for the year 2011 at the global scale.

The DSI ranges as $[-\infty, \infty]$. The negative value of the DSI represents the dry condition, whereas the positive value represents the wet condition. Based on the MODIS ET/PET (Mu et al., 2007, 2011) and MODIS NDVI (Huete et al., 2002) products, the DSI products have been documented at 8-day, monthly, and annual timescales at the global scale (Mu et al., 2013). Figure 15.4 shows the global DSI results for the year 2011 (available from http://www.ntsg.umt.edu/data). The 2011 Texas drought is among the most extreme 1-year drought on record, resulting in agricultural losses up to $7.6 (Nielsen-Gammon, 2012). In Figure 15.4, the 2011 Texas drought is captured well with negative DSI values.

15.3.4 SOIL MOISTURE

Soil moisture modulates the moisture and energy exchange between the land surface and atmosphere and plays an important role in drought characterization. For reasons soil moisture is a fundamental link between global water and carbon cycles and is related to weather, climate, and society, accurate and timely measurements of soil moisture (and freeze/thaw state) with global coverage are critically important (Brown et al., 2013; Nghiem et al., 2012). Soil moisture can be obtained from measurement at *in situ* stations (Robock et al., 2000) or estimation from land-surface models. Due to the limitation of observation networks for soil moisture measurements, soil moisture estimation from land-surface simulations with relatively wide spatial coverage and long record has been commonly used for the drought monitoring with drought indices such as the soil moisture percentile (Mo, 2008; Mo and Lettenmaier, 2014b; Sheffield et al., 2004; Wang et al., 2011a). The potential limitation is that uncertainties from the model simulations may affect the accuracy of drought assessments from simulated soil moisture (Mo and Lettenmaier, 2014b).

Soil moisture can also be obtained from remote sensing, and microwave remote sensing at low frequencies is often considered most effective to characterize soil moisture from space (De Jeu et al., 2008; Lakshmi, 2013; Shen et al., 2013; Tang et al., 2009; Wang and Qu, 2009). Soil moisture has been derived based on microwave remote sensing with active sensors, including synthetic aperture radar (SAR) and scatterometers, and passive radiometers, such as TRMM Microwave Imager (TMI), AMSR-E, and Soil Moisture and Ocean Salinity Sensor (SMOS; Kerr et al., 2010), via various algorithms based on the bright temperature or backscatter information (dependent on soil dielectric properties) (Nghiem et al., 2012; Wang and Qu, 2009). There are several microwave remote sensing soil moisture products on the global coverage, such as European Space Agency Climate Change Initiative (ESA CCI; Dorigo et al., 2015; Liu et al., 2011a, 2012), AMSR-E, SMOS, and Advanced Scatterometer (ASCAT) on board of the Meteorological Operational (METOP) Satellite (Bartalis et al., 2007; Wagner et al., 2013).

A major limitation of microwave remote sensing of soil moisture is that it only provides information for only top few centimeters due to the limited penetration depth of the used signal. Due to the close relationship between surface soil moisture and top 10 cm (Albergel et al., 2008), surface soil moisture from microwave remote sensing can reflect the response of soil moisture dynamics to meteorological anomalies and can be used for the short-term drought monitoring, including the flash drought (Yuan et al., 2015a). The soil moisture products from microwave remote sensing, including ESA CCI (Yuan et al., 2015a), AMSR-E (Anderson et al., 2012b; Bolten et al., 2010; Chakraborty et al., 2013), SMOS (Scaini et al., 2015), and ASCAT (Zhang et al., 2015), have been used for drought characterizations mostly in recent years for different regions. Among these soil moisture products, the ESA CCI, which is a merged product consists of four passive (SMMR, SSM/I, TMI, and AMSR-E) and two active (ERS AMI and ASCAT) microwave sensors with a 0.25° spatial resolution, is the only product covering the minimum time period of 30 years (1979–2010) and, thus, is favorable for the drought monitoring at the regional and global scales (Dorigo et al., 2015). For example, Yuan et al. (2015a) employed the ESA CCI soil moisture product to characterize short-term droughts during crop growing seasons in China and found that passive and merged microwave soil moisture products performed well in drought monitoring over sparsely vegetated regions in northwestern China. The recently launched SMAP will provide soil moisture from combined radiometer/radar measurements at higher resolutions than radiometers alone can currently achieve (Brown et al., 2013; Entekhabi et al., 2010), which would be helpful to develop improved drought-monitoring capabilities.

15.3.5 GROUNDWATER STORAGE

Water stored deeply in the soil or groundwater is important for drought characterizations because it portends longer term weather trends and climate variations (Rodell, 2012). Groundwater provides relatively resilient water supplies with sustained surface flows through groundwater baseflow during the early stage of a drought event and, thus, is susceptible to persistent or prolonged drought (Bloomfield and Marchant, 2013; Hughes et al., 2012). Certain drought indices have been developed for drought characterizations based on groundwater, such as groundwater-level percentile, Groundwater Resource Index (GRI; Mendicino et al., 2008), and Standardized Groundwater level Index (SGI; Bloomfield and Marchant, 2013). Standard remote sensors generally fall short in monitoring groundwater because only the wet condition of the first few centimeters of the land surface can be detected due to the limitation of the penetration depth. Launched in 2002, GRACE estimates the Earth's gravity field to infer monthly changes in terrestrial water-storage (TWS), including soil moisture, groundwater, snow, and surface waters (Tapley et al., 2004). GRACE provides monthly estimates of total water-storage changes, which is an effective tool to be employed for various hydrologic applications, including drought characterizations in different regions (Chen et al., 2009; Houborg et al., 2012; Leblanc et al., 2009; Li and Rodell, 2015; Long et al., 2013; Thomas et al., 2014; Yirdaw et al., 2008). For example, Thomas et al. (2014) developed a quantitative approach for measuring hydrologic drought occurrence and severity based on terrestrial water-storage observations from GRACE.

GRACE terrestrial water-storage data are subject to several limitations (Famiglietti and Rodell 2013; Landerer and Swenson, 2012; Reager et al., 2015), including the aggregated observation of multiple water-storage components, latency in data product (2–4 months lag), and coarse spatial resolution, which hinder the hydrologic applications including drought monitoring. To address these issues, the GRACE record has been assimilated with land-surface models to facilitate near-real time analysis with disaggregated water storage of finer temporal and spatial resolutions (Houborg et al., 2012; Li et al., 2012; Reager et al., 2015; Zaitchik et al., 2008). For example, based on the Catchment Land Surface Model (CLSM), Houborg et al. (2012) applied the GRACE Data Assimilation System (DAS) to investigate the possibility of more comprehensive and objective identification of drought conditions by integrating spatially, temporally, and vertically disaggregated GRACE data into the US and North American Drought Monitors.

Based on terrestrial water-storage observations from GRACE satellite data integrated with other observations, the groundwater and soil moisture conditions from GRACE data assimilation have been produced routinely and are available from the National Drought Mitigation Center (http://drought.unl.edu/MonitoringTools/NASAGRACEDataAssimilation.aspx). These products provide useful information for drought characterizations across the United States. For example, the GRACE-based groundwater-storage percentile for November 28, 2011, in the United States is shown in Figure 15.5 (http://www.nasa.gov/topics/earth/features/tx-drought.html). It can be seen that the severe drought conditions of Texas in 2011 are captured well, which indicates a good performance of GRACE data in drought characterizations.

15.4 SATELLITE REMOTE SENSING DROUGHT PREDICTION

15.4.1 Overview of Drought Prediction Methods

One of the main challenges in reducing impacts of drought is the accurate drought prediction several months in advance. Accurate drought prediction would be of critical importance for drought early warning to reduce potential impacts of drought. Drought prediction methods can be broadly classified as dynamic methods based on the state-of-the-art climatic/hydrologic models and statistical methods based on inherent relationships among drought-related variables (Hao et al., 2015b;

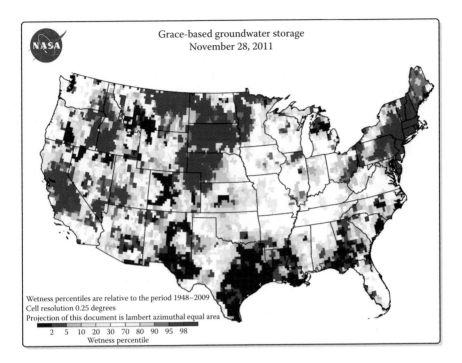

FIGURE 15.5 GRACE-based groundwater storage percentile for November 2011 in the United States. (Courtesy of NASA.)

Mishra and Singh, 2011). Significant advances have been achieved in the past decades for drought predictions.

The statistical methods for drought prediction generally rely on the statistical relationship in historical observations, including regression models, time series models, machine learning models, and the hybrid models (Hao et al., 2015b; Mishra and Singh, 2011). The regression model is commonly used to establish the relationship between dependent variables (drought indices to be predicted) with independent variables (or covariates), such as El Nino-Southern Oscillation (ENSO; Barros and Bowden, 2008; Liu and Juárez, 2001). In addition, the Autoregressive Integrated Moving Average Model (AMIMA) framework is the commonly used time series modeling technique to predict drought indices, such as SPI (Durdu, 2010; Mishra and Desai, 2005) or Palmer Drought Index (Rao and Padmanabhan, 1984). The machine learning method, such as the artificial neural network (ANN; Morid et al., 2007), support vector regression (SVR; Ganguli and Reddy, 2014), and Fuzzy logic (FL; Özger et al., 2012; Pongracz et al., 1999) can also be used to reveal complicated interactions among various variables for drought prediction. The hybrid models merge the advantages of individual models for drought prediction, such as the ARIMA-ANN model (Mishra et al., 2007). Other statistical methods based on historical observations have also been used for drought prediction, including the persistence or ensemble streamflow prediction (ESP) method (Day, 1985; Hao et al., 2015b; Lyon et al., 2012; Nicholls et al., 2005; Quan et al., 2012; Svensson, 2014; Van den Dool, 2006). For example, Lyon et al. (2012) developed the baseline method for meteorological drought prediction utilizing the inherence persistence by resampling from historical records, which was extended for the drought prediction based on SSI and MSDI (Hao et al., 2014). The traditional ESP method can be extended or revised to improve the performance of drought prediction through the selection of historical samples either based on the similarity of past situation (or the constructed analogue method) or based on conditioning on climate indices (Hamlet and Lettenmaier, 1999; Shukla et al., 2014; Trambauer et al., 2015; van den Dool et al., 2003; Werner et al., 2004; Yao and Georgakakos, 2001). The statistical methods are generally easy to implement; however, the performance may not be satisfactory if slow or even rapid changes in the climate are not adequately captured in historical records (Nicholls et al., 2005).

With an improved understanding of the ocean and atmosphere systems combined with the expanded range of data and the increase of computation capabilities, dynamic prediction systems based on computer models that represent the entire atmosphere–ocean–earth system have been developed recently (Kirtman et al., 2014; Nicholls et al., 2005; Saha et al., 2014). The dynamic model provides seasonal forecasts of climate variables, such as precipitation, which can be used to compute drought indices for meteorological drought prediction (Acharya et al., 2013; Quan et al., 2012; Yoon et al., 2012). For example, climate forecast systems, such as NCEP Climate Forecast System (CFS; Saha et al., 2014) and National Multi-Model Ensemble (NMME; Kirtman et al., 2014) have been assessed for meteorological drought prediction (Pan et al., 2013; Yoon et al., 2012; Yuan and Wood, 2013). Improvements in climate forecast have enhanced the capabilities of seasonal hydrologic predictions and, thus, for agricultural and hydrologic drought prediction. With forcing variables, such as precipitation and temperature, from dynamic model prediction, hydrologic models can be driven to predict soil moisture and streamflow, from which agricultural and hydrologic drought prediction can be achieved (Mo and Lettenmaier, 2014a; Luo and Wood, 2007; Wood et al., 2002). The statistical methods may provide the useful complementary prediction of the drought condition to that from dynamic methods in certain regions and seasons (Mo and Lyon, 2015; Quan et al., 2012).

15.4.2 Satellite Remote Sensing Drought Prediction

Drought prediction or forecast with a remote sensing product is still in its infancy. The commonly used method for remote sensing drought prediction is the statistical method based on historical observations, for which the multiple linear regression method (Dodamani et al., 2015; Liu and Juárez, 2001; Tadesse et al., 2014), the machine learning method (Jalili et al., 2014), and the ESP method (Hao et al., 2015a) are commonly used to predict drought indices. Various studies have shown the strong relationships between NDVI anomalies and El Niño/La Niña-Southern Oscillation (ENSO) phenomena on an interannual timescale for different regions (Anyamba and Eastman, 1996; Los et al., 2001; Lotsch et al., 2005; Martiny et al., 2006), and, thus, the NDVI anomalies can be used for the drought prediction based on the relationship between ENSO events and drought occurrence (Anyamba and Tucker, 2012). For example, Liu and Juárez (2001) employed regression models to predict the drought condition with NDVI anomaly as the dependent variable and various ENSO anomalies as independent variables, such as sea-surface temperature in the Pacific Ocean area and Southern Oscillation Index (SOI). Tadesse et al. (2014) developed the VegOut-Ethiopia to predict standardized values of NDVI at multiple time steps (weeks to months lead time) based on the analysis of historical patterns of satellite, climate, and oceanic data over historical records. In addition, the ANN has been used to predict the SPI value by incorporating a variety of features from the satellite imagery, including NDVI, VCI, and TCI extracted from NOAA-AVHRR (Jalili et al., 2014). Recently, Hao et al. (2015a) assessed the potential of the TRMM precipitation products for the global meteorological drought prediction with the ESP method and results show that the TMPA-RT precipitation products provide useful information to aid drought early warning.

Meanwhile, efforts have been devoted to the integration of *in situ* data, model simulations or reanalysis and remote sensing data products for the drought prediction, for which the remote sensing data sets are commonly used as the forcing data to drive hydrologic models to aid drought prediction (Nijssen et al., 2014; Sheffield et al., 2010, 2014). For example, Nijssen et al. (2014) introduced a GDIS based on multiple land-surface models, for which the satellite-based precipitation from TRMM is used as one of the forcing datasets.

15.4.3 GLOBAL METEOROLOGICAL DROUGHT PREDICTION WITH REMOTE SENSING PRODUCTS

As stated before, drought prediction based on remote sensing data is under development and a suite of drought prediction methods have been developed and employed for drought prediction. Although remote-sensed precipitation products play an important role for drought monitoring and have been assessed in a variety of studies (Naumann et al., 2012; Sahoo et al., 2015; Zhou et al., 2014), the drought prediction based solely on remote sensing precipitation data has seldom been explored. In this section, the global meteorological drought prediction is illustrated with the TMPA-RT and TMPA-RP products using the ESP method. The 1 month lead prediction of the SPIs based on the monthly TMPA-RT products for August 2012 is shown in Figure 15.6. As shown before, the SPI6 from TMPA-RP partly revealed the drought condition in the United States. From Figure 15.6, the drought condition in the United States is partly revealed from the prediction based on TMPA-RT data with certain discrepancies. Meanwhile, the 2012 Northeastern Brazil drought is captured well from the 1 month lead prediction.

To further show the performance of the drought prediction based on TRMM precipitation products, the drought prediction with SPI6 based on TMPA-RT for one grid (longitude: −95.1250, latitude: 33.3750) is compared with monitoring results with SPI6 based on TMPA-RP (regarded as observations), as shown in Figure 15.7. The observation from TMPA-RP shows the precipitation falls below the normal condition (SPI6 = 0) in the summer, which is exacerbated in the fall leading to drought conditions (with threshold −0.5). The prediction generally resembles the observation well. For example, the observed SPI6 for September is −1.3, while the predicted SPI6 is −1. These results indicate that the drought predictions from remote sensing

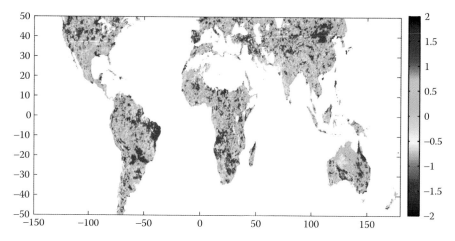

FIGURE 15.6 One month lead prediction of global meteorological drought based on SPI6 for August 2012 with TMPA-RT.

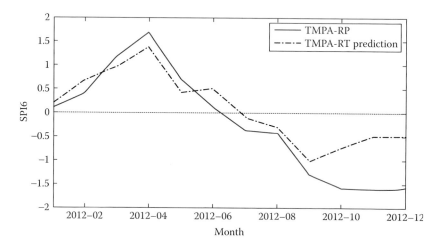

FIGURE 15.7 Comparison of the 1 month lead drought prediction based the TMPA-RT data with observations from TMPA-RP during 2012 for one grid.

precipitation products with relatively short record length provide useful information for the drought early warning.

15.5 DISCUSSIONS AND CONCLUSIONS

The remote sensing products provide continuous and consistent temporal and spatial observations of key land-surface variables for the regional or global coverage, which are invaluable for drought assessment, monitoring, and prediction at both regional and global scales, especially for regions with sparse observation networks. A variety of vegetation and drought indices, based on optical, TIR, and microwave remote sensing, developed in the past decades for drought monitoring were reviewed in this chapter. Various drought-monitoring efforts based on satellite remote-sensed variables, including vegetation, precipitation, evaporation, soil moisture, and groundwater, have been introduced with several recent drought events as examples, including the 2012 US drought, the 2011 Texas drought, and the 2012 northeastern Brazil drought. Drought early warning is important to reduce drought impacts and remote sensing products play an important role in this regard. The review of drought prediction methods based on remote sensing was also provided in this chapter, along with an example of global meteorological drought prediction based on the TRMM precipitation products. Drought monitoring and prediction based on satellite remote sensing would be essential to meet the demand of accurate, reliable, and integrated characterizations of drought condition.

For the drought monitoring with satellite remote sensing, a large number of vegetation and drought indices have been developed based on observations in multispectral bands including the visible red (R), NIR, and short wavelength infrared (SWIR) from the optical remote sensing or based on LST observations from the TIR remote sensing. Extensive efforts have been devoted to develop drought indices by integrating vegetation information from different bands from the optical remote sensing, along with the integration of VNIR data with LST from TIR data. The microwave remote sensing for drought assessments is less explored compared with those based on optical and thermal remote sensing. Recent advances in microwave remote sensing for drought monitoring mainly reside in the integration of various products, including precipitation, LST, and soil moisture, to develop the integrated (or composite, multivariate) drought indices (Rhee et al., 2010; Zhang and Jia, 2013). A variety of recently developed products from microwave remote sensing, such as soil moisture from ESA CCI, would play an important role for drought assessments in the future. A comprehensive drought-monitoring system generally requires to combine drought information from various sources, and efforts have been devoted to the integration of remote sensing products with expert inputs, *in situ* observations, and model simulations (Hao and Singh, 2015a; Svoboda et al., 2002). Overall, due to the complicated nature of drought, it is important to explore how to combine remote sensing information from different bands (or sensors) and with other products (e.g., *in situ* observations, model simulations) for accurate drought monitoring.

Drought prediction with remote sensing datasets is important to aid the drought early warning. Currently, statistical methods have been commonly used for remote sensing drought prediction based on the linear regression method, machine learning method, or ESP. Although the remote sensing product record is generally short with a certain bias or uncertainty, the statistical drought prediction still shows useful information for drought prediction in certain regions. The remote sensing precipitation products, such as TMPA-RP and TMPA-RT products are of critical importance in developing operational DIS (Nijssen et al., 2014). Due to continuous and consistent observations at the global scale, the remote sensing precipitation products are essential in the development of global drought monitoring and prediction systems. Integrating remote sensing products with *in situ* observations and model simulations to develop and improve operational DIS at the regional and global scales is a critical need to advance our capability of coping with drought.

A major limitation of remote sensing data for drought monitoring is the relatively short observational record. To facilitate the anomaly detection and maintain a consistent and reliable data input for operational DIS, it is required to generate products of long-term continuity (Wardlow et al., 2012a). Recently, extensive efforts have been devoted to merge or assimilate different products of land-surface variables or indices, including precipitation, soil moisture, groundwater storage, and NDVI, to extend the data record for drought assessments (AghaKouchak and Nakhjiri, 2012; Dorigo et al., 2015; Houborg et al., 2012; Liu et al., 2011a, 2011b, 2012; Sheffield et al., 2006; Tucker et al., 2005; Van Leeuwen et al., 2006). For example, data assimilations have been used to integrate *in situ* observations and remote sensing products to reduce input errors in land-surface simulations to provide a longer record of land-surface variables for drought modeling and assessment (Houborg et al., 2012; Reager et al., 2015; Sheffield et al., 2012). It should be born in mind that the representation of drought conditions based on remote sensing products of short record lengths is different from other drought indices based on a longer period. As such, it is necessary to rescale or normalize the drought indices for a valid comparison and assessment of drought conditions (Anderson et al., 2011; Hao and Singh, 2015a).

Assessments of the accuracy and utility of satellite remote sensing products for drought monitoring and prediction are still underway. With the increase of record length and temporal and spatial resolutions and advances in remote sensing accuracy, it is expected that more opportunities would unfold in the future for drought modeling and assessment. Apart from the efforts for drought characterizations presented in this chapter, there have been other emerging opportunities, such as using remote sensing for snow and ice cover (Kongoli et al., 2012) and streamflow retrieval (Andreadis et al., 2007), which can also be used to aid drought assessments. In addition, with the increase of EO missions, such as Soil Moisture Active Passive (SMAP), Surface Water & Ocean Topography (SWOT), and Global Precipitation Measurement (GPM), it is expected that unprecedented opportunities for drought modeling and assessment with satellite remote sensing will unfold in the coming decades. The advancement of remote sensing data products would enhance our capacity to cope with drought impacts and continue to play an important role in the operational drought monitoring and prediction at both regional and global scales.

ACKNOWLEDGMENTS

This work is supported by Youth Scholars Program of Beijing Normal University (Grant No. 2015NT02) and Supporting Program of the "Twelfth Five-year Plan" for Science & Technology Research of China (2012BAD15B05). We appreciate the NASA/National Drought Mitigation Center for providing the GRACE-based groundwater storage in the United States for the 2011 Texas drought (https://www.nasa.gov/topics/earth/features/tx-drought.html). We appreciate NOAA-NESDIS for providing the Global Vegetation Health Products (http://www.star.nesdis.noaa.gov/smcd/emb/vci/VH/vh_ftp.php). We thank the Numerical Terradynamic Simulation Group (NTSG) at University of Montana for providing the DSI products (http://www.ntsg.umt.edu/data).

APPENDIX A

TABLE A.1
List of Acronyms of Vegetation and Drought Indices

Abbreviation	Indices
ADI	Aggregate Drought Index
CMI	Crop Moisture Index
CTVI	Corrected Transformed Vegetation Index
CWSI	Crop Water Stress Index
DSI	Drought Severity Index
EDI	Effective Drought Index
ESI	Evaporative Stress Index
EVI	Enhanced Vegetation Index
GMI	Grand Mean Index
GRI	Groundwater Resource Index
GVMI	Global Vegetation Moisture Index
JDI	Joint Drought Index
LWCI	Leaf Water Content Index
MDI	Multivariate Drought Index
MIDI	Microwave Integrated Drought Index
MPDI	Modified Perpendicular Drought Index
NADM	North American Drought Monitor
NDII	Normalized Difference Infrared Index
NDVI	Normalized Difference Vegetation Index
NDWI	Normalized Difference Water Index
NMDI	Normalized Multiband Drought Index
NSM	Normalized Soil Moisture
OBNDI	Objective Blended NLDAS Drought Index
PDI	Perpendicular Drought Index
PDSI	Palmer Drought Severity Index
PHDI	Palmer Hydrologic Drought Index
PVI	Perpendicular Vegetation Index
RDI	Reconnaissance Drought Index
RVI	Ratio Vegetation Index
SAVI	Soil Adjusted Vegetation index
SDI	Synthesized Drought Index
SDCI	Scaled Drought Condition Index
SGI	Standardized Groundwater level Index
SMP	Soil Moisture Percentile
SPDI	Standardized Palmer Drought Index
SPEI	Standardized Precipitation Evapotranspiration Index
SPI	Standardized Precipitation Index
SRI	Standardized Runoff Index
SSFI	Standardized Streamflow Index
SSI	Standardized Soil Moisture Index

(Continued)

TABLE A.1 (*Continued*)
List of Acronyms of Vegetation and Drought Indices

Abbreviation	Indices
SVI	Standardized Vegetation Index
SWSI	Surface Water Supply Index
TCI	Temperature Condition Index
TVDI	Temperature–Vegetation Dryness Index
TVI	Transformed Vegetation Index
USDM	US Drought Monitor
VCI	Vegetation Condition Index
VegDRI	Vegetation Drought Response Index
VHI	Vegetation Health Index
VSDI	Visible and Short wavelength Infrared Drought Index
VTCI	Vegetation Temperature Condition Index

REFERENCES

Acácio, V. et al. 2013. Review of current drought monitoring systems and identification of (further) monitoring requirements. DROUGHT-R&SPI Technical Report No. 6.

Acharya, N. et al. 2013. Performance of general circulation models and their ensembles for the prediction of drought indices over India during summer monsoon. *Natural Hazards* 66(2):851–871.

Adler, R.F. et al. 2003. The version-2 global precipitation climatology project (GPCP) monthly precipitation analysis (1979–present). *Journal of Hydrometeorology* 4(6):1147–1167.

AghaKouchak, A. and N. Nakhjiri. 2012. A near real-time satellite-based global drought climate data record. *Environmental Research Letters* 7(4):044037.

Albergel, C. et al. 2008. From near-surface to root-zone soil moisture using an exponential filter: An assessment of the method based on in situ observations and model simulations. *Hydrology and Earth System Sciences Discussions* 12:1323–1337.

American Meteorological Society. 1997. Policy statement: Meteorological drought. *Bulletin of the American Meteorological Society* 78:847–849.

Anderson, M.C. et al. 2007. A climatological study of evapotranspiration and moisture stress across the continental United States based on thermal remote sensing: 1. Model formulation. *Journal of Geophysical Research: Atmospheres (1984–2012)* 112(D10):D10117.

Anderson, M.C. et al. 2011. Evaluation of drought indices based on thermal remote sensing of evapotranspiration over the continental United States. *Journal of Climate* 24(8):2025–2044.

Anderson, M.C. et al. 2012a. Thermal-based evaporative stress index for monitoring surface moisture depletion. In *Remote Sensing of Drought Innovative Monitoring Approaches*, Wardlow, B.D., Anderson, M.C., and Verdin, J.P. (eds.) Boca Raton, FL: CRC Press.

Anderson, W. et al. 2012b. Towards an integrated soil moisture drought monitor for East Africa. *Hydrology and Earth System Sciences* 16(8):2893–2913.

Anderson, M.C. et al. 2013. An intercomparison of drought Indicators based on thermal remote sensing and NLDAS-2 simulations with US Drought Monitor classifications. *Journal of Hydrometeorology* 14(4):1035–1056.

Andreadis, K.M. et al. 2007. Prospects for river discharge and depth estimation through assimilation of swath-altimetry into a raster-based hydrodynamics model. *Geophysical Research Letters* 34(10):L10403.

Anyamba, A. and J. Eastman. 1996. Interannual variability of NDVI over Africa and its relation to El Niño/Southern oscillation. *Remote Sensing* 17(13):2533–2548.

Anyamba, A. and C. Tucker. 2005. Analysis of Sahelian vegetation dynamics using NOAA-AVHRR NDVI data from 1981 to 2003. *Journal of Arid Environments* 63(3):596–614.

Anyamba, A. and C.J. Tucker. 2012. Historical perspectives on AVHRR NDVI and vegetation drought monitoring. In *Remote Sensing of Drought: Innovative Monitoring Approaches*, Wardlow, B.D., Anderson, M.C., and Verdin, J.P. (eds.) Boca Raton, FL: CRC Press.

Bannari, A. et al. 1995. A review of vegetation indices. *Remote Sensing Reviews* 13(1–2):95–120.

Baret, F. and G. Guyot. 1991. Potentials and limits of vegetation indices for LAI and APAR assessment. *Remote sensing of Environment* 35(2):161–173.

Baret, F. et al. 1993. The soil line concept in remote sensing. *Remote Sensing Reviews* 7(1):65–82.

Barros, A.P. and G.J. Bowden. 2008. Toward long-lead operational forecasts of drought: An experimental study in the Murray-Darling River Basin. *Journal of Hydrology* 357(3):349–367.

Bartalis, Z. et al. 2007. Initial soil moisture retrievals from the METOP-A Advanced Scatterometer (ASCAT). *Geophysical Research Letters* 34(20):L20401.

Bayarjargal, Y. et al. 2006. A comparative study of NOAA–AVHRR derived drought indices using change vector analysis. *Remote Sensing of Environment* 105(1):9–22.

Beersma, J.J. and T.A. Buishand. 2004. Joint probability of precipitation and discharge deficits in the Netherlands. *Water Resources Research* 40(12):W12508.

Bhuiyan, C. et al. 2006. Monitoring drought dynamics in the Aravalli region (India) using different indices based on ground and remote sensing data. *International Journal of Applied Earth Observation and Geoinformation* 8(4):289–302.

Bloomfield, J. and B. Marchant. 2013. Analysis of groundwater drought using a variant of the standardised precipitation index. *Hydrology and Earth System Sciences Discussions* 10(6):7537–7574.

Bolten, J.D. et al. 2010. Evaluating the utility of remotely sensed soil moisture retrievals for operational agricultural drought monitoring. *IEEE Journal of Selected Topics in Applied Earth Observations and Remote Sensing* 3(1):57–66.

Brown, J.F. et al. 2008. The vegetation drought response index (VegDRI): A new integrated approach for monitoring drought stress in vegetation. *GIScience & Remote Sensing* 45(1):16–46.

Brown, M.E. et al. 2013. NASA's soil moisture active passive (SMAP) mission and opportunities for applications users. *Bulletin of the American Meteorological Society* 94(8):1125–1128.

Carlson, T.N. et al. 1994. A method to make use of thermal infrared temperature and NDVI measurements to infer surface soil water content and fractional vegetation cover. *Remote Sensing Reviews* 9(1–2):161–173.

Ceccato, P. et al. 2002. Designing a spectral index to estimate vegetation water content from remote sensing data: Part 1: Theoretical approach. *Remote Sensing of environment* 82(2):188–197.

Chakraborty, A. et al. 2013. Assessing early season drought condition using AMSR-E soil moisture product. *Geomatics, Natural Hazards and Risk* 4(2):164–186.

Chen, J. et al. 2009. 2005 drought event in the Amazon River basin as measured by GRACE and estimated by climate models. *Journal of Geophysical Research: Solid Earth (1978–2012)* 114(B5):B05404.

Choi, M. et al. 2013. Evaluation of drought indices via remotely sensed data with hydrological variables. *Journal of Hydrology* 476:265–273.

Dai, A. 2011. Drought under global warming: A review. *Wiley Interdisciplinary Reviews: Climate Change* 2(1):45–65.

Day, G.N. 1985. Extended streamflow forecasting using NWSRFS. *Journal of Water Resources Planning and Management* 111(2):157–170.

De Jeu, R. et al. 2008. Global soil moisture patterns observed by space borne microwave radiometers and scatterometers. *Surveys in Geophysics* 29(4–5):399–420.

Deering, D. and J. Rouse. 1975. Measuring 'forage production' of grazing units from Landsat MSS data. *International Symposium on Remote Sensing of Environment, 10 th*, Ann Arbor, Michigan.

Deng, M. et al. 2013. Web-service-based monitoring and analysis of global agricultural drought. *Photogrammetric Engineering & Remote Sensing* 79(10):929–943.

Di, L. et al. 1994. Modelling relationships between NDVI and precipitation during vegetative growth cycles. *International Journal of Remote Sensing* 15(10):2121–2136.

Dodamani, B. et al. 2015. Agricultural drought modeling using remote sensing. *International Journal of Environmental Science and Development* 6(5):326.

Dorigo, W. et al. 2015. Evaluation of the ESA CCI soil moisture product using ground-based observations. *Remote Sensing of Environment* 162:380–395.

Du, L. et al. 2013. A comprehensive drought monitoring method integrating MODIS and TRMM data. *International Journal of Applied Earth Observation and Geoinformation* 23:245–253.

Durdu, Ö.F. 2010. Application of linear stochastic models for drought forecasting in the Büyük Menderes river basin, western Turkey. *Stochastic Environmental Research and Risk Assessment* 24(8):1145–1162.

Dutra, E. et al. 2008. ERA-40 reanalysis hydrological applications in the characterization of regional drought. *Geophysical Research Letters* 35(19):L19402.

Entekhabi, D. et al. 2010. The soil moisture active passive (SMAP) mission. *Proceedings of the IEEE* 98(5):704–716.

Ezzine, H. et al. 2014. Seasonal comparisons of meteorological and agricultural drought indices in Morocco using open short time-series data. *International Journal of Applied Earth Observation and Geoinformation* 26:36–48.

Famiglietti, J.S. and M. Rodell. 2013. Water in the balance. *Science* 340(6138):1300–1301.

Fensholt, R. et al. 2012. Greenness in semi-arid areas across the globe 1981–2007—An earth observing satellite based analysis of trends and drivers. *Remote Sensing of Environment* 121:144–158.

Fernández, C. et al. 2009. Streamflow drought time series forecasting: A case study in a small watershed in North West spain. *Stochastic Environmental Research and Risk Assessment* 23(8):1063–1070.

Funk, C. 2009. New satellite observations and rainfall forecasts help provide earlier warning of African drought. *The Earth Observer* 21(1):23–27.

Funk, C. et al. 2012. Mapping recent decadal climate variations in precipitation and temperature across Eastern Africa and the Sahel. In *Remote Sensing of Drought: Innovative Monitoring Approaches*, Wardlow, B.D., Anderson, M.C., and Verdin, J.P. (eds.) Boca Raton, FL: CRC Press.

Ganguli, P. and M.J. Reddy. 2014. Ensemble prediction of regional droughts using climate inputs and the SVM–copula approach. *Hydrological Processes* 28(19):4989–5009.

Gao, B.-C. 1996. NDWI—A normalized difference water index for remote sensing of vegetation liquid water from space. *Remote Sensing of Environment* 58(3):257–266.

Ghulam, A. et al. 2007a. Designing of the perpendicular drought index. *Environmental Geology* 52(6):1045–1052.

Ghulam, A. et al. 2007b. Modified perpendicular drought index (MPDI): A real-time drought monitoring method. *ISPRS Journal of Photogrammetry and Remote Sensing* 62(2):150–164.

Gu, Y. et al. 2007. A five-year analysis of MODIS NDVI and NDWI for grassland drought assessment over the central Great Plains of the United States. *Geophysical Research Letters* 34(6):L06407.

Gutiérrez, A.P.A. et al. 2014. Drought preparedness in Brazil. *Weather and Climate Extremes* 3:95–106.

Guttman, N.B. 1998. Comparing the palmer drought index and the standardized precipitation index. *JAWRA Journal of the American Water Resources Association* 34(1):113–121.

Hamlet, A.F. and D.P. Lettenmaier. 1999. Columbia River streamflow forecasting based on ENSO and PDO climate signals. *Journal of Water Resources Planning and Management* 125(6):333–341.

Hanes, J. 2013. *Biophysical Applications of Satellite Remote Sensing.* Berlin, Germany: Springer.

Hao, Z. and A. AghaKouchak. 2013. Multivariate standardized drought index: A multi-index parametric approach for drought analysis. *Advances in Water Resources* 57:12–18.

Hao, Z. and A. AghaKouchak. 2014. A multivariate multi-Index drought monitoring framework. *Journal of Hydrometeorology* 15:89–101.

Hao, Z. and V.P. Singh. 2013. Entropy-based method for bivariate drought analysis. *Journal of Hydrologic Engineering* 18(7):780–786.

Hao, Z. and V.P. Singh. 2015a. Drought characterization from a multivariate perspective: A review *Journal of Hydrology* 527:668–678.

Hao, Z. and V.P. Singh. 2015b. Integrating entropy and copula theories for hydrologic modeling and analysis. *Entropy* 17(4):2253–2280.

Hao, Z. and V.P. Singh. 2016. Review of dependence modeling in hydrology and water resources. *Progress in Physical Geography* (in press).

Hao, Z. et al. 2014. Global integrated drought monitoring and prediction system. *Scientific Data* 1:1–10.

Hao, Z. et al. 2015a. On the potential of remote sensing precipitation products for real time drought prediction. *Submitted manuscript.*

Hao, Z. et al. 2015b. A statistical method for categorical drought prediction based on NLDAS-2. *Journal of Applied Meteorology and Climatology* (in press).

Hao, Z. et al. 2016. Probabilistic drought characterization in the categorical form using ordinal regression. *Journal of Hydrology* 535:331–339.

Hayes, M. et al. 2011. The Lincoln declaration on drought indices: Universal meteorological drought index recommended. *Bulletin of the American Meteorological Society* 92(4):485–488.

Hayes, M.J. et al. 2012. Drought monitoring historical and current perspectives. In *Remote Sensing of Drought: Innovative Monitoring Approaches*, Wardlow, B.D., Anderson, M.C., and Verdin, J.P. (eds.) Boca Raton, FL: CRC Press, pp. 1–19.

Heim, R.R. 2002. A review of twentieth-century drought indices used in the United States. *Bulletin of the American Meteorological Society* 83(8):1149–1166.

Heim, R.R. and M.J. Brewer. 2012. The global drought monitor portal: The foundation for a global drought information system. *Earth Interactions* 16(15):1–28.

Hoerling, M. et al. 2014. Causes and predictability of the 2012 Great Plains drought. *Bulletin of the American Meteorological Society* 95(2):269–282.

Holmes, T. et al. 2009. Land surface temperature from Ka band (37 GHz) passive microwave observations. *Journal of Geophysical Research: Atmospheres (1984–2012)* 114(D4).

Houborg, R. et al. 2012. Drought indicators based on model-assimilated gravity recovery and climate experiment (GRACE) terrestrial water storage observations. *Water Resources Research* 48, W07525.

Huete, A.R. 1988. A soil-adjusted vegetation index (SAVI). *Remote Sensing of Environment* 25(3):295–309.

Huete, A. et al. 2002. Overview of the radiometric and biophysical performance of the MODIS vegetation indices. *Remote Sensing of Environment* 83(1):195–213.

Huete, A.R. et al. 2006. Amazon rainforests green-up with sunlight in dry season. *Geophysical Research Letters* 33(6):L06405.

Huffman, G.J. et al. 2007. The TRMM multisatellite precipitation analysis (TMPA): Quasi-global, multiyear, combined-sensor precipitation estimates at fine scales. *Journal of Hydrometeorology* 8(1):38–55.

Hughes, J. et al. 2012. Drought, groundwater storage and stream flow decline in southwestern Australia. *Geophysical Research Letters* 39(3):L03408.

Hunt, E.R. and B.N. Rock. 1989. Detection of changes in leaf water content using near-and middle-infrared reflectances. *Remote Sensing of Environment* 30(1):43–54.

Hunt, E.R. et al. 1987. Measurement of leaf relative water content by infrared reflectance. *Remote Sensing of Environment* 22(3):429–435.

Jackson, R.D. and A.R. Huete. 1991. Interpreting vegetation indices. *Preventive Veterinary Medicine* 11(3):185–200.

Jackson, R. et al. 1981. Canopy temperature as a crop water stress indicator. *Water Resources Research* 17(4):1133–1138.

Jackson, T.J. et al. 2010. Validation of advanced microwave scanning radiometer soil moisture products. *IEEE Transactions on Geoscience and Remote Sensing* 48(12):4256–4272.

Jalili, M. et al. 2014. Nation-wide prediction of drought conditions in Iran based on remote sensing data. *IEEE Transactions on Computers* 63(1):90–101.

Ji, L. and A.J. Peters. 2003. Assessing vegetation response to drought in the northern Great Plains using vegetation and drought indices. *Remote Sensing of Environment* 87(1):85–98.

Jiang, Z. et al. 2008. Development of a two-band enhanced vegetation index without a blue band. *Remote Sensing of Environment* 112(10):3833–3845.

Joyce, R.J. et al. 2004. CMORPH: A method that produces global precipitation estimates from passive microwave and infrared data at high spatial and temporal resolution. *Journal of Hydrometeorology* 5(3):487–503.

Justice, C.O. et al. 1998. The moderate resolution imaging spectroradiometer (MODIS): Land remote sensing for global change research. *IEEE Transactions on Geoscience and Remote Sensing* 36(4):1228–1249.

Kao, S.C. and R.S. Govindaraju. 2010. A copula-based joint deficit index for droughts. *Journal of Hydrology* 380(1–2):121–134.

Karnieli, A. et al. 2006. Comments on the use of the vegetation health index over Mongolia. *International Journal of Remote Sensing* 27(10):2017–2024.

Karnieli, A. et al. 2010. Use of NDVI and land surface temperature for drought assessment: Merits and limitations. *Journal of Climate* 23(3):618–633.

Kawanishi, T. et al. 2003. The advanced microwave scanning radiometer for the earth observing system (AMSR-E), NASDA's contribution to the EOS for global energy and water cycle studies. *IEEE Transactions on Geoscience and Remote Sensing* 41(2):184–194.

Kerr, Y.H. et al. 2010. The SMOS mission: New tool for monitoring key elements of the global water cycle. *Proceedings of the IEEE* 98(5):666–687.

Keyantash, J.A. and J.A. Dracup. 2004. An aggregate drought index: Assessing drought severity based on fluctuations in the hydrologic cycle and surface water storage. *Water Resources Research* 40(9):W09304.

Kirtman, B. et al. 2014. The north american multimodel ensemble: Phase-1 seasonal-to-interannual prediction; phase-2 toward developing intraseasonal prediction. *Bulletin of the American Meteorological Society* 95(4):585–601.

Kogan, F. 1990. Remote sensing of weather impacts on vegetation in non-homogeneous areas. *International Journal of Remote Sensing* 11(8):1405–1419.

Kogan, F. 1995b. Application of vegetation index and brightness temperature for drought detection. *Advances in Space Research* 15(11):91–100.

Kogan, F. 2002. World droughts in the new millennium from AVHRR-based vegetation health indices. *Eos, Transactions American Geophysical Union* 83(48):557–563.

Kogan, F. and J. Sullivan. 1993. Development of global drought-watch system using NOAA/ AVHRR data. *Advances in Space Research* 13(5):219–222.

Kogan, F.N. 1995a. Droughts of the late 1980s in the United States as derived from NOAA polar-orbiting satellite data. *Bulletin of the American Meteorological Society* 76(5):655–668.

Kogan, F.N. 1997. Global drought watch from space. *Bulletin of the American Meteorological Society* 78(4):621–636.

Kogan, F.N. 2000. Satellite-observed sensitivity of world land ecosystems to El Nino/La Nina. *Remote Sensing of Environment* 74(3):445–462.

Kogan, F.N. 2001. Operational space technology for global vegetation assessment. *Bulletin of the American Meteorological Society* 82(9):1949–1964.

Kongoli, C. et al. 2012. Snow cover monitoring from remote sensing satellites: Possibilities for drought assessment. In *Remote Sensing of Drought: Innovative Monitoring Approaches*, Wardlow, B.D., Anderson, M.C., and Verdin, J.P. (eds.) Boca Raton, FL: CRC Press.

Lakshmi, V. 2013. Remote sensing of soil moisture. *ISRN Soil Science* 2013:33.

Landerer, F. and S. Swenson. 2012. Accuracy of scaled GRACE terrestrial water storage estimates. *Water Resources Research* 48(4):4531.

Lawrimore, J. et al. 2002. Beginning a new era of drought monitoring across North America. *Bulletin of the American Meteorological Society* 83(8):1191–1192.

Leblanc, M.J. et al. 2009. Basin-scale, integrated observations of the early 21st century multiyear drought in southeast Australia. *Water Resources Research* 45(4):W04408.

Li, B. and M. Rodell. 2015. Evaluation of a model-based groundwater drought indicator in the conterminous US. *Journal of Hydrology* 526:78–88.

Li, B. et al. 2012. Assimilation of GRACE terrestrial water storage into a land surface model: Evaluation and potential value for drought monitoring in western and central Europe. *Journal of Hydrology* 446:103–115.

Liu, W. and R.N. Juárez. 2001. ENSO drought onset prediction in northeast Brazil using NDVI. *International Journal of Remote Sensing* 22(17):3483–3501.

Liu, W. and F. Kogan. 1996. Monitoring regional drought using the vegetation condition index. *International Journal of Remote Sensing* 17(14):2761–2782.

Liu, Y. et al. 2011a. Developing an improved soil moisture dataset by blending passive and active microwave satellite-based retrievals. *Hydrology and Earth System Sciences* 15(2):425–436.

Liu, Y. et al. 2012. Trend-preserving blending of passive and active microwave soil moisture retrievals. *Remote Sensing of Environment* 123:280–297.

Liu, Y.Y. et al. 2011b. Global long-term passive microwave satellite-based retrievals of vegetation optical depth. *Geophysical Research Letters* 38(18):L18402.

Lloyd-Hughes, B. and M. A. Saunders. 2007. University College London Global Drought Monitor. http://drought.mssl.ucl.ac.uk.

Long, D. et al. 2013. GRACE satellite monitoring of large depletion in water storage in response to the 2011 drought in Texas. *Geophysical Research Letters* 40(13):3395–3401.

Los, S.O. et al. 2001. Global interannual variations in sea surface temperature and land surface vegetation, air temperature, and precipitation. *Journal of Climate* 14(7):1535–1549.

Lotsch, A. et al. 2005. Response of terrestrial ecosystems to recent Northern Hemispheric drought. *Geophysical Research Letters* 32(6):L06705.

Luo, L. and E.F. Wood. 2007. Monitoring and predicting the 2007 US drought. *Geophysical Research Letters* 34(22):L22702.

Lyon, B. et al. 2012. Baseline probabilities for the seasonal prediction of meteorological drought. *Journal of Climate and Applied Meteorology* 51(7):1222–1237.

Ma, M. et al. 2014. New variants of the Palmer drought scheme capable of integrated utility. *Journal of Hydrology* 519:1108–1119.

Marshall, M.T. et al. 2012. Agricultural drought monitoring in Kenya using evapotranspiration derived from remote sensing and reanalysis data. *In Remote Sensing of Drought: Innovative Monitoring Approaches*, Wardlow, B.D., Anderson, M.C., and Verdin, J.P. (eds.) Boca Raton, FL: CRC Press.

Martiny, N. et al. 2006. Compared regimes of NDVI and rainfall in semi-arid regions of Africa. *International Journal of Remote Sensing* 27(23):5201–5223.

McKee, T.B. et al. 1993. The relationship of drought frequency and duration to time scales. *Eighth Conference on Applied Climatology, at American Meteorological Society*, Anaheim, CA, USA.

McVicar, T. and P. Bierwirth. 2001. Rapidly assessing the 1997 drought in Papua New Guinea using composite AVHRR imagery. *International Journal of Remote Sensing* 22(11):2109–2128.

Mendicino, G. et al. 2008. A groundwater resource index (GRI) for drought monitoring and forecasting in a mediterranean climate. *Journal of Hydrology* 357(3):282–302.

Mishra, A. and V. Desai. 2005. Drought forecasting using stochastic models. *Stochastic Environmental Research and Risk Assessment* 19(5):326–339.

Mishra, A. et al. 2007. Drought forecasting using a hybrid stochastic and neural network model. *Journal of Hydrologic Engineering* 12(6):626–638.

Mishra, A.K. and V.P. Singh. 2010. A review of drought concepts. *Journal of Hydrology* 391(1–2):202–216.

Mishra, A.K. and V.P. Singh. 2011. Drought modeling-A review. *Journal of Hydrology* 403(1–2):157–175.

Mitchell, K.E. et al. 2004. The multi-institution North American land data assimilation system (NLDAS): Utilizing multiple GCIP products and partners in a continental distributed hydrological modeling system. *Journal of Geophysical Research: Atmospheres (1984–2012)* 109(D7):D07S90.

Mo, K.C. 2008. Model-based drought indices over the United States. *Journal of Hydrometeorology* 9(6):1212–1230.

Mo, K.C. 2011. Drought onset and recovery over the United States. *Journal of Geophysical Research* 116(D20):D20106.

Mo, K.C. and B. Lyon. 2015. Global meteorological drought prediction using the North American multi-model ensemble. *Journal of Hydrometeorology* 16(3):1409–1424.

Mo, K.C. and D.P. Lettenmaier. 2014a. Hydrologic prediction over Conterminous US using the National Multi Model ensemble. *Journal of Hydrometeorology* 15(4):1457–1472.

Mo, K.C. and D.P. Lettenmaier. 2014b. Objective drought classification using multiple land surface models. *Journal of Hydrometeorology* 15:990–1010.

Mo, K.C. et al. 2011. Drought indices based on the climate forecast system reanalysis and ensemble NLDAS. *Journal of Hydrometeorology* 12(2):181–205.

Modarres, R. 2007. Streamflow drought time series forecasting. *Stochastic Environmental Research and Risk Assessment* 21(3):223–233.

Morid, S. et al. 2007. Drought forecasting using artificial neural networks and time series of drought indices. *International Journal of Climatology* 27(15):2103–2111.

Mu, Q. et al. 2007. Development of a global evapotranspiration algorithm based on MODIS and global meteorology data. *Remote Sensing of environment* 111(4):519–536.

Mu, Q. et al. 2011. Improvements to a MODIS global terrestrial evapotranspiration algorithm. *Remote Sensing of Environment* 115(8):1781–1800.

Mu, Q. et al. 2013. A remotely sensed global terrestrial drought severity index. *Bulletin of the American Meteorological Society* 94(1):83–98.

Narasimhan, B. and R. Srinivasan. 2005. Development and evaluation of soil moisture deficit index (SMDI) and evapotranspiration deficit index (ETDI) for agricultural drought monitoring. *Agricultural and Forest Meteorology* 133(1):69–88.

Naumann, G. et al. 2012. Monitoring drought conditions and their uncertainties in Africa using TRMM data. *Journal of Applied Meteorology & Climatology* 51(10):1867–1874.

Naumann, G. et al. 2014. Comparison of drought indicators derived from multiple data sets over Africa. *Hydrology and Earth System Sciences* 18(5):1625–1640.

Nemani, R. et al. 1993. Developing satellite-derived estimates of surface moisture status. *Journal of Applied Meteorology* 32(3):548–557.

Nghiem, S.V. et al. 2012. Microwave remote sensing of soil moisture science and applications. In *Remote Sensing of Drought: Innovative Monitoring Approaches*, Wardlow, B.D., Anderson, M.C., and Verdin, J.P. (eds.) Boca Raton, FL: CRC Press.

Nicholls, N. et al. 2005. The challenge of climate prediction in mitigating drought impacts. In *Drought and Water Crises: Science, technology, and management issues*, Wilhite, D.A. (ed.) Boca Raton, FL: CRC Press.

Nielsen-Gammon, J.W. 2012. The 2011 Texas Drought. *Texas Water Journal* 3(1):59–95.

Niemeyer, S. 2008. New drought indices. *Options Méditerranéennes. Série A: Séminaires Méditerranéens* 80:267–274.

Nijssen, B. et al. 2014. A prototype global drought information system based on multiple land surface models. *Journal of Hydrometeorology* 15:1661–1676.

Njoku, E.G. et al. 2003. Soil moisture retrieval from AMSR-E. *IEEE Transactions on Geoscience and Remote Sensing* 41(2):215–229.

Núñez, J. et al. 2014. On the use of standardized drought indices under decadal climate variability: Critical assessment and drought policy implications. *Journal of Hydrology* 517:458–470.

Otkin, J.A. et al. 2013. Examining rapid onset drought development using the thermal infrared based evaporative stress index. *Journal of Hydrometeorology* 14(4):1057–1074.

Ouyang, W. et al. 2012. Integration of multi-sensor data to assess grassland dynamics in a Yellow River sub-watershed. *Ecological Indicators* 18:163–170.

Özger, M. et al. 2012. Long lead time drought forecasting using a wavelet and fuzzy logic combination model: A case study in Texas. *Journal of Hydrometeorology* 13(1):284–297.

Palmer, W. 1965. Meteorological drought. *Research Paper No. 45*. U.S. Weather Bureau. Washington, D.C.

Palmer, W.C. 1968. Keeping track of crop moisture conditions, nationwide: The new crop moisture index. *Weatherwise* 21(4):156–161.

Pan, M. et al. 2013. A probabilistic framework for assessing drought recovery. *Geophysical Research Letters* 40(14):3637–3642.

Pearson, R.L. and L.D. Miller. 1972. Remote mapping of standing crop biomass for estimation of the productivity of the short-grass prairie, Pawnee National Grasslands, Colorado. In: *Proceedings of the 8th International Symposium on Remote Sensing of Environment*. Michgan, Ann Arbor.

Peled, E. et al. 2010. Technical note: Comparing and ranking soil drought indices performance over Europe, through remote-sensing of vegetation. *Hydrology and Earth System Sciences* 14(2):271–277.

Pereira, M.P.S. et al. 2014. The influence of oceanic basins on drought and ecosystem dynamics in Northeast Brazil. *Environmental Research Letters* 9(12):124013.

Perry, C.R. and L.F. Lautenschlager. 1984. Functional equivalence of spectral vegetation indices. *Remote Sensing of Environment* 14(1):169–182.

Peters, A.J. et al. 2002. Drought monitoring with NDVI-based standardized vegetation index. *Photogrammetric Engineering and Remote Sensing* 68(1):71–75.

Petropoulos, G.P. 2013. *Remote Sensing of Energy Fluxes and Soil Moisture Content*. Boca Raton, FL: CRC Press.

Pongracz, R. et al. 1999. Application of fuzzy rule-based modeling technique to regional drought. *Journal of Hydrology* 224(3):100–114.

Qi, J. et al. 1994. A modified soil adjusted vegetation index. *Remote Sensing of Environment* 48(2):119–126.

Quan, X.-W. et al. 2012. Prospects for dynamical prediction of meteorological drought. *Journal of Climate and Applied Meteorology* 51(7):1238–1252.

Quiring, S.M. 2009. Monitoring drought: An evaluation of meteorological drought indices. *Geography Compass* 3(1):64–88.

Quiring, S.M. and S. Ganesh. 2010. Evaluating the utility of the vegetation condition index (VCI) for monitoring meteorological drought in Texas. *Agricultural and Forest Meteorology* 150(3):330–339.

Rao, A.R. and G. Padmanabhan. 1984. Analysis and modeling of Palmer's drought index series. *Journal of Hydrology* 68(1):211–229.

Reager, J.T. et al. 2015. Assimilation of GRACE terrestrial water storage observations into a land surface model for the assessment of regional flood potential. *Remote Sensing* 7(11):14663–14679.

Rhee, J. et al. 2010. Monitoring agricultural drought for arid and humid regions using multi-sensor remote sensing data. *Remote Sensing of Environment* 114(12):2875–2887.

Richardson, A.J. and C. Weigand. 1977. Distinguishing vegetation from soil background information. *Photogrammetric Engineering and Remote Sensing* 43(12):1541–1552.

Robock, A. et al. 2000. The global soil moisture data bank. *Bulletin of the American Meteorological Society* 81(6):1281–1299.

Rodell, M. 2012. Satellite gravimetry applied to drought monitoring. *In Remote Sensing of Drought: Innovative Monitoring Approaches*, Wardlow, B.D., Anderson, M.C., and Verdin, J.P. (eds.) Boca Raton, FL: CRC Press.

Rodrigues, R.R. and M.J. McPhaden. 2014. Why did the 2011–2012 La Niña cause a severe drought in the Brazilian Northeast? *Geophysical Research Letters* 41(3):1012–1018.

Rondeaux, G. et al. 1996. Optimization of soil-adjusted vegetation indices. *Remote Sensing of Environment* 55(2):95–107.

Rouse, J. et al. 1974. Monitoring vegetation systems in the Great Plains with ERTS. *NASA special publication* 351:309.

Saha, S. et al. 2014. The NCEP climate forecast system version 2. *Journal of Climate* 27:2185–2208.

Sahoo, A.K. et al. 2015. Evaluation of the tropical rainfall measuring mission multi-satellite precipitation analysis (TMPA) for assessment of large-scale meteorological drought. *Remote Sensing of Environment* 159:181–193.

Sandholt, I. et al. 2002. A simple interpretation of the surface temperature/vegetation index space for assessment of surface moisture status. *Remote Sensing of Environment* 79(2):213–224.

Sapiano, M. and P. Arkin. 2009. An intercomparison and validation of high-resolution satellite precipitation estimates with 3-hourly gauge data. *Journal of Hydrometeorology* 10(1):149–166.

Scaini, A. et al. 2015. SMOS-derived soil moisture anomalies and drought indices: A comparative analysis using in situ measurements. *Hydrological Processes* 29(3):373–383.

Sepulcre, G. et al. 2012. Development of a combined drought indicator to detect agricultural drought in Europe. *Natural Hazards and Earth System Sciences* 12(11):3519–3531.

Shah, R.D. and V. Mishra. 2015. Development of an experimental near-real time drought monitor for India. *Journal of Hydrometeorology* 16:327–345.

Sheffield, J. et al. 2004. A simulated soil moisture based drought analysis for the United States. *Journal of Geophysical Research: Atmospheres (1984–2012)* 109(D24):D24108.

Sheffield, J. et al. 2006. Development of a 50-year high-resolution global dataset of meteorological forcings for land surface modeling. *Journal of Climate* 19(13):3088–3111.

Sheffield, J. et al. 2008. Experimental drought monitoring for Africa. *GEWEX News* 8(3):4–6.

Sheffield, J. et al. 2010. *Global drought monitoring and forecasting based on satellite data and land surface modeling.* AGU Fall Meeting, San Francisco, CA.

Sheffield, J. et al. 2012. North American land data assimilation system-a framework for merging model and satellite data for improved drought monitoring. In *Remote Sensing of Drought: Innovative Monitoring Approaches*, Wardlow, B.D., Anderson, M.C., and Verdin, J.P. (eds.) Boca Raton, FL: CRC Press.

Sheffield, J. et al. 2014. A drought monitoring and forecasting system for sub-sahara African water resources and food security. *Bulletin of the American Meteorological Society* 95:861–882.

Shen, X. et al. 2013. Bare surface soil moisture estimation using double-angle and dual-polarization L-band radar data. *IEEE Transactions on Geoscience and Remote Sensing* 51(7):3931–3942.

Shi, J. et al. 2008. Microwave vegetation indices for short vegetation covers from satellite passive microwave sensor AMSR-E. *Remote Sensing of Environment* 112(12):4285–4300.

Shukla, S. et al. 2014. Using constructed analogs to improve the skill of national multi-model ensemble March–April–May precipitation forecasts in equatorial East Africa. *Environmental Research Letters* 9(9):094009.

Shukla, S. and A.W. Wood. 2008. Use of a standardized runoff index for characterizing hydrologic drought. *Geophysical Research Letters* 35(2):L02405.

Silleos, N.G. et al. 2006. Vegetation indices: Advances made in biomass estimation and vegetation monitoring in the last 30 years. *Geocarto International* 21(4):21–28.

Singh, R.P. et al. 2003. Vegetation and temperature condition indices from NOAA AVHRR data for drought monitoring over India. *International Journal of Remote Sensing* 24(22):4393–4402.

Son, N. et al. 2012. Monitoring agricultural drought in the Lower Mekong Basin using MODIS NDVI and land surface temperature data. *International Journal of Applied Earth Observation and Geoinformation* 18:417–427.

Song, X. et al. 2004. Early detection system of drought in East Asia using NDVI from NOAA/AVHRR data. *International Journal of Remote Sensing* 25(16):3105–3111.

Sorooshian, S. et al. 2000. Evaluation of PERSIANN system satellite-based estimates of tropical rainfall. *Bulletin of the American Meteorological Society* 81(9):2035–2046.

Staudinger, M. et al. 2014. A drought index accounting for snow. *Water Resources Research* 50(10):7861–7872.

Sun, D. and M. Kafatos. 2007. Note on the NDVI-LST relationship and the use of temperature-related drought indices over North America. *Geophysical Research Letters* 34(24):L24406.

Svensson, C. 2014. Seasonal river flow forecasts for the United Kingdom using persistence and historical analogues. *Hydrological Sciences Journal* 61(1):19–35.

Svoboda, M. et al. 2002. The drought monitor. *Bulletin of the American Meteorological Society* 83:1181–1190.

Tadesse, T. et al. 2014. Satellite-based hybrid drought monitoring tool for prediction of vegetation condition in Eastern Africa: A case study for Ethiopia. *Water Resources Research* 50(3):2176–2190.

Tadesse, T. et al. 2015. Assessing the vegetation condition impacts of the 2011 drought across the U.S. Southern Great Plains using the vegetation drought response index (VegDRI). *Journal of Applied Meteorology and Climatology* 54(1):153–169.

Tang, Q. et al. 2009. Remote sensing: Hydrology. *Progress in Physical Geography* 33(4):490–509.

Tao, J. et al. 2008. Monitoring vegetation water content using microwave vegetation indices. *IEEE Geoscience and Remote Sensing Symposium (IGARSS 2008)*, Boston, MA.

Tapley, B.D. et al. 2004. The gravity recovery and climate experiment: Mission overview and early results. *Geophysical Research Letters* 31(9):L09607.

Thomas, A.C. et al. 2014. A GRACE-based water storage deficit approach for hydrological drought characterization. *Geophysical Research Letters* 41(5):1537–1545.

Trambauer, P. et al. 2015. Hydrological drought forecasting and skill assessment for the Limpopo River basin, Southern Africa. *Hydrology and Earth System Sciences* 19(4):1695–1711.

Trenberth, K. et al. 2004. Exploring drought and its implications for the future. *Eos, Transactions American Geophysical Union* 85(3):27–27.

Trenberth, K.E. et al. 2003. The changing character of precipitation. *Bulletin of the American Meteorological Society* 84(9):1205–1217.

Tsakiris, G. et al. 2007. Regional drought assessment based on the reconnaissance drought index (RDI). *Water Resources Management* 21(5):821–833.

Tucker, C.J. 1979. Red and photographic infrared linear combinations for monitoring vegetation. *Remote Sensing of Environment* 8(2):127–150.

Tucker, C.J. and B.J. Choudhury. 1987. Satellite remote sensing of drought conditions. *Remote Sensing of Environment* 23(2):243–251.

Tucker, C.J. et al. 2005. An extended AVHRR 8-km NDVI dataset compatible with MODIS and SPOT vegetation NDVI data. *International Journal of Remote Sensing* 26(20):4485–4498.

USDA. 2012. U.S. drought 2012: Farm and food impacts. http://www.ers.usda.gov/topics/in-the-news/us-drought-2012-farm-and-food-impacts.aspx.

Van den Dool, H. 2006. *Empirical Methods in Short-Term Climate Prediction*: Oxford, New York: Oxford University Press.

van den Dool, H. et al. 2003. Performance and analysis of the constructed analogue method applied to US soil moisture over 1981–2001. *Journal of Geophysical Research* 108(D16):8617.

Van Leeuwen, W.J. et al. 2006. Multi-sensor NDVI data continuity: Uncertainties and implications for vegetation monitoring applications. *Remote Sensing of Environment* 100(1):67–81.

Vernimmen, R. et al. 2012. Evaluation and bias correction of satellite rainfall data for drought monitoring in Indonesia. *Hydrology and Earth System Sciences* 16(1):133–146.

Vicente-Serrano, S.M. et al. 2010. A multiscalar drought index sensitive to global warming: The standardized precipitation evapotranspiration index. *Journal of Climate* 23(7):1696–1718.

Vogt, J. et al. 2011. *Developing a European drought observatory for monitoring, assessing and forecasting droughts across the European continent*. AGU Fall Meeting, San Francisco, CA.

Wagner, W. et al. 2013. The ASCAT soil moisture product: A review of its specifications, validation results, and emerging applications. *Meteorologische Zeitschrift* 22(1):5–33.

Wan, Z. et al. 2004. Using MODIS land surface temperature and normalized difference vegetation index products for monitoring drought in the southern Great Plains, USA. *International Journal of Remote Sensing* 25(1):61–72.

Wang, A. et al. 2009. Multimodel ensemble reconstruction of drought over the continental United States. *Journal of Climate* 22(10):2694–2712.

Wang, A. et al. 2011a. Soil moisture drought in China, 1950–2006. *Journal of Climate* 24(13):3257–3271.

Wang, J. et al. 2011b. The coupled routing and excess storage (CREST) distributed hydrological model. *Hydrological Sciences Journal* 56(1):84–98.

Wang, J. et al. 2001a. Spatial patterns of NDVI in response to precipitation and temperature in the central Great Plains. *International Journal of Remote Sensing* 22(18):3827–3844.

Wang, L. and J.J. Qu. 2007. NMDI: A normalized multi-band drought index for monitoring soil and vegetation moisture with satellite remote sensing. *Geophysical Research Letters* 34(20):1–5.

Wang, L. and J.J. Qu. 2009. Satellite remote sensing applications for surface soil moisture monitoring: A review. *Frontiers of Earth Science in China* 3(2):237–247.

Wang, P. X. et al. 2001b. Vegetation temperature condition index and its application for drought monitoring. *Proceedings of International Geoscience and Remote Sensing Symposium,* Sydney, Australia.

Wardlow, B.D. et al. 2012a. Future opportunities and challenges in remote sensing of drought. In *Remote Sensing of Drought: Innovative Monitoring Approaches*, Wardlow, B.D., Anderson, M.C., and Verdin, J.P. (eds.) Boca Raton, FL: CRC Press.

Wardlow, B.D. et al. 2012b. *Remote Sensing of Drought: Innovative Monitoring Approaches*: Boca Raton, FL: CRC Press.

Werner, K. et al. 2004. Climate index weighting schemes for NWS ESP-based seasonal volume forecasts. *Journal of Hydrometeorology* 5(6):1076–1090.

Wood, A.W. 2008. The University of Washington surface water monitor: An experimental platform for national hydrologic assessment and prediction. *Proceedings of the AMS 22nd Conference on Hydrology*, at New Orleans, LA.

Wood, A.W. and D.P. Lettenmaier. 2006. A test bed for new seasonal hydrologic forecasting approaches in the western United States. *Bulletin of the American Meteorological Society* 87(12):1699–1712.

Wood, A.W. et al. 2002. Long-range experimental hydrologic forecasting for the eastern United States. *Journal of Geophysical Research: Atmospheres* 107(D20):4429.

Xia, Y. et al. 2012. Continental-scale water and energy flux analysis and validation for the North American Land Data Assimilation System project phase 2 (NLDAS-2): 1. Intercomparison and application of model products. *Journal of Geophysical Research: Atmospheres (1984–2012)* 117(D3):D03110.

Xia, Y. et al. 2014a. Uncertainties, correlations, and optimal blends of drought indices from the NLDAS multiple land surface model ensemble. *Journal of Hydrometeorology* 15:1636–1650.

Xia, Y. et al. 2014b. Application of USDM statistics in NLDAS-2: Optimal blended NLDAS drought index over the continental United States. *Journal of Geophysical Research: Atmospheres* 119:2947–2965.

Yao, H. and A. Georgakakos. 2001. Assessment of Folsom Lake response to historical and potential future climate scenarios: 2. Reservoir management. *Journal of Hydrology* 249(1):176–196.

Yirdaw, S.Z. et al. 2008. GRACE satellite observations of terrestrial moisture changes for drought characterization in the Canadian Prairie. *Journal of Hydrology* 356(1):84–92.

Yong, B. et al. 2014. Global view of real-time TRMM multi-satellite precipitation analysis: Implication to its successor global precipitation measurement mission. *Bulletin of the American Meteorological Society* 96:283–296.

Yoon, J.-H. et al. 2012. Dynamic-model-based seasonal prediction of meteorological drought over the contiguous United States. *Journal of Hydrometeorology* 13(2):463–482.

Yuan, X. et al. 2015a. Microwave remote sensing of short-term droughts during crop growing seasons. *Geophysical Research Letters* 42(11):4394–4401.

Yuan, X. et al. 2015b. Seasonal forecasting of global hydrologic extremes: System development and evaluation over GEWEX basins. *Bulletin of the American Meteorological Society* 96:1895–1912.

Yuan, X. and E.F. Wood. 2013. Multimodel seasonal forecasting of global drought onset. *Geophysical Research Letters* 40(18):4900–4905.

Zaitchik, B.F. et al. 2008. Assimilation of GRACE terrestrial water storage data into a land surface model: Results for the Mississippi River basin. *Journal of Hydrometeorology* 9(3):535–548.

Zargar, A. et al. 2011. A review of drought indices. *Environmental Reviews* 19:333–349.

Zhang, A. and G. Jia. 2013. Monitoring meteorological drought in semiarid regions using multi-sensor microwave remote sensing data. *Remote Sensing of Environment* 134:12–23.

Zhang, J. et al. 2015. Evaluation of the ASCAT surface soil moisture product for agricultural drought monitoring in USA. *IEEE International Geoscience and Remote Sensing Symposium,* Milan, Italy.

Zhang, N. et al. 2013. VSDI: A visible and shortwave infrared drought index for monitoring soil and vegetation moisture based on optical remote sensing. *International Journal of Remote Sensing* 34(13):4585–4609.

Zhou, T. et al. 2014. Evaluation of real-time satellite precipitation data for global drought monitoring. *Journal of Hydrometeorology* 15(4):1651–1660.

16 Capacity-Building Efforts in Hydrologic Modeling for Africa
Workshops about CREST Model

Robert A. Clark III

CONTENTS

16.1 INTRODUCTION

Capacity building harnesses the human resources of a country to improve development outcomes. Although "capacity building" is sometimes treated as a synonym for "training," it is far more wide ranging. In the hydrologic modeling context, capacity-building efforts in developing countries seek to enhance the technical skills and scientific knowledge of hydrologists, meteorologists, and engineers as well as to improve the resilience of communities to natural disasters, such as flood and drought. Capacity-building efforts should also provide assistance in the development, implementation, and maintenance of the technical infrastructure necessary to link expert knowledge with its end users. The end goal is to enable developing countries to understand and react to natural disasters in a way that reduces or mitigates harmful impacts to their citizenry. By its very definition, capacity building is a lengthy and involved process and it cannot succeed without sufficient dedication from both local stakeholders and outside experts.

Hydrologic models enable users to predict, monitor, and study the movement of water within the water cycle. As computing power grows ever more ubiquitous and inexpensive, the range of plausible uses for hydrologic models also grows. Hydrologic modeling and capacity building across the African continent are a natural fit. Hydrologic models hold the promise of improving forecasts and monitoring of droughts and flooding, which are together among the most devastating and most common environmental disasters to afflict African nations. Hydrologic models also permit governments, nongovernmental organizations (NGOs), and citizens to plan around water surpluses and water shortages. In the agricultural realm, model outputs like soil moisture can guide irrigation decisions, while surveys to collect soil properties cannot only increase the fidelity of hydrologic model outputs but also inform the choice of which crop to plant, where, and at what time. The choice of hydrologic model requires brief consideration. Models can be statistical (stochastic), in which the model inputs and outputs are linked via mathematical equations or statistical relationships, or physical, where the modeler attempts to characterize the physical processes linking the inputs and outputs; these can also be described via the terms "empirical" model and "conceptual" model, respectively. Models can also be characterized by whether they use lumped parameters (where the model settings are the same across the entire modeled domain) or distributed parameters (where the model settings vary from model cell to model cell). These parameters, or model settings, can be thought of as tuning knobs. Consider a stereo where turning the base or treble knob alters the output sound. In a hydrologic model, changing a parameter changes the model output: streamflow, soil moisture, or some other modeled variable of interest.

Hydrologic models exist in a two-dimensional continuum where one dimension represents ease of understanding and stretches from complex to simple and the other dimension describes the required computing power and stretches from small to large. The location of models along these dimensions is generally correlated (i.e., simple-to-use models typically require less computing power). As capacity-building efforts begin, models should be selected based on the available computing power. If the model user interface is appropriately modular, only the simplest, most critical aspects of the model need to be made available initially. Then, as user understanding grows, users can be trained to use more powerful and complex model components until the upper limit of available computing power is reached.

Since 2010, the University of Oklahoma (OU) and the National Aeronautics and Space Administration (NASA) have developed hydrologic models designed for use *by* (and not just *in*) developing countries. These models CREST (Coupled Routing and Excess Storage) and EF5 (Ensemble Framework for Flash Flood Forecasting) have now been set up and run in multiple river basins across Africa by national governments, NGOs, and other interested parties, and plans for additional model training and deployments are on the cards, as shown in Figure 16.1. Most of the joint OU-NASA training and capacity-building workshops have focused on the use of hydrologic models to monitor and forecast streamflow in large rivers, but stream

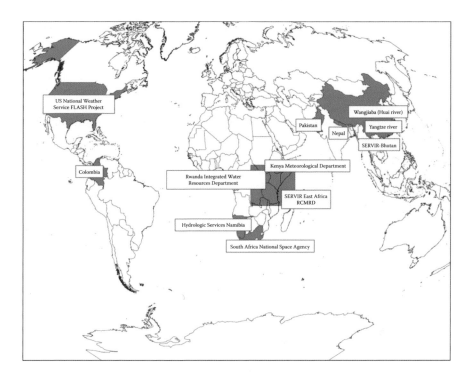

FIGURE 16.1   Map of CREST and EF5 Implementations.

flow in small streams and catchments, drought, and flood inundation can be forecast and monitored using the same principles. The CREST model, developed by scientists at the NASA Goddard Space Flight Center and OU, has been used in most of these training workshops. CREST has many advantages: it is simple to set up and run, is fast even on old or inexpensive computer hardware, is highly scalable, and is free. Users may select the grid cell resolution of the simulation and so the model is equally usable over small flash flood scale catchments or entire continents. CREST is a physical or process model and can be run with either lumped or distributed parameters. Water entering the model is either infiltrated into the soil layer of the cell via a variable infiltration curve (VIC) or turned into surface runoff and routed downstream via linear reservoirs. Since 2015, the CREST infiltration component has been incorporated into the EF5 system, which enables the user to select from multiple hydrologic modeling cores, including CREST-like infiltration with kinematic wave routing and Sacramento-like infiltration with kinematic wave routing. Both CREST and EF5 are cross-platform and compatible with the vast majority of modern personal computers; neither requires specialized equipment except in extremely complicated and unusual situations.

Capacity-building exercises in this realm usually involve participants with a wide range of knowledge of hydrologic modeling and hydrologic science. In the past, government officials, university students, academics, community leaders, technicians, and others have completed the CREST or EF5 model training courses.

The courses are designed in a modular fashion so that modules can be added or removed as needed, based on the baseline knowledge of the participants. A successful training session involves heavy doses of hands-on work and repeated run-throughs to practice and refine core skills so that the model can be run independently of assistance from the workshop facilitator. The successful course will be taught by flexible facilitators who plan ahead for technical contingencies, such as the lack of reliable Internet access, slower-than-expected personal computing resources, and the like. Finally, it is critical to remember that capacity building does not end when the workshop participants go home. Facilitators should be willing to engage with participants remotely after the conclusion of the workshop and to return and conduct additional follow-on workshops, if necessary.

16.2   EFFORTS PRIOR TO 2015

Hydrologic capacity building was baked right into the CREST project from the beginning, as efforts to build hydrologic modeling capacity via the CREST model began in 2010 soon after the initial release of the CREST model itself (Wang et al., 2011). The primary objective of the model developers was to create a scalable hydrologic model where the user could easily select his or her desired grid cell resolution. Initial testing of the model took place over the Nzoia River Basin of eastern Kenya, which drains into Lake Victoria. This early work demonstrated that the model would be useful for flood monitoring in Africa, despite the fact that precipitation forcing to hydrologic models over most of the continent tends to comes only from satellite estimates that possess coarse temporal and spatial resolution. Eventually, the original CREST model, as a result of several rounds of minor source code changes, became known as CREST version 1.6.

16.2.1   2010: KENYA

In April 2010, the first CREST version 1.6 training workshop was conducted at the Regional Centre for Mapping of Resources for Development (RCMRD) facility in Nairobi, Kenya. RCMRD is an intergovernmental group consisting of 20 African countries in the eastern and southern parts of the continent. During this training session, a University of Oklahoma undergraduate student explained the structure of the CREST model and taught RCMRD personnel how to compile CREST and apply it in additional catchments beyond the Nzoia. Figure 16.2 shows the environment in which this first CREST capacity-building workshop was conducted. Note that, in this case, the workshop organizers, not the facilitators, provided own computing resources so that participants could complete the hands-on portions of the training.

As a result of this effort, in August 2011, CREST was adopted by the NASA Marshall Spaceflight Center in Huntsville, Alabama, for deployment by RCMRD. In this implementation, the model is forced by near-real-time remotely sensed precipitation estimates from the Tropical Rainfall Measurement Mission (TRMM). Gridded streamflow and soil moisture estimates from CREST are produced over Kenya, Tanzania, Uganda, and adjacent areas of surrounding countries. OU scientists

FIGURE 16.2 2010 CREST 1.6 Training workshop in Nairobi, Kenya. (Courtesy of Zac Flamig.)

assisted with the initial CREST calibration for several gauges in RCMRD's area of interest, but since this preliminary set up the model has been run at RCMRD and is the responsibility of their personnel. Over time, RCMRD has built their own web interface for displaying and analyzing outputs from the CREST model. The CREST version 1.6 training was heavily hands-on and primarily consisted of several iterations of running model examples over the Nzoia River Basin. Although general model concepts were introduced in the training, they were not a major focus of the workshop.

16.2.2 2012: NAMIBIA

In March 2012, OU scientists traveled to Namibia to begin discussions of implementing hydrologic models, including CREST, over that country. Namibia is highly susceptible to floods and droughts and monitoring and predicting water resources are of critical importance to most of the country. Conversion of the output of hydrologic models to flood inundation maps would increase the lead time of flood predictions and result in better outcomes for flood-prone citizenry. In an effort to achieve this, the Open Cloud Consortium (OCC) and NASA's Goddard Space Flight Center have developed a cloud-computing framework called Project Matsu that, among other things, can process satellite imagery into useful outputs. The OCC is a nonprofit group of universities, companies, and government agencies that supports large-scale research projects by sharing costs of cloud-computing infrastructure and data acquisition.

One of Project Matsu's web applications is the Namibia Flood Dashboard, which contains water information for Namibia and the surrounding area. This information is obtained from ground-based stream gauges operated primarily by the Namibian government and from space-borne sensors operated primarily by NASA. In 2012, the Hydrologic Services Namibia (HSN) hosted the Flood Dashboard Workshop to introduce stakeholders in Namibia and the surrounding countries to the principles of hydrologic modeling and to define what role model outputs could play in forecasting and monitoring flood and drought conditions across the country. Workshop participants also traveled to the Okavango River Basin in the northern part of the country to have first-hand experience on how modeling along a flood-prone stretch of river could work in Namibia.

These activities illustrate an important component of capacity-building work. Although workshops and training are important, discussion with local stakeholders and the elucidation of a long-term vision always come first. Additionally, first-hand observations of the conditions requiring capacity-building activities (in this case, susceptible river basins) are invaluable.

16.2.3   2012: KENYA

After the initial 2010 training workshop at RCMRD, OU released a new version of CREST (version 2.0) with new features and improvements to the user experience and computational efficiency (Xue et al., 2013). In concert with this, CREST got a new set of training modules. This April 2012 training was designed to teach the new and improved CREST model to RCMRD experts so that these experts could upgrade their CREST implementations to version 2.0. Additionally, participants from national hydrologic services or universities in 13 other Asian and African counties attended.

This workshop, completed over 4 days, consists of nine modules as shown in Table 16.1. Each of these nine modules is a Microsoft PowerPoint presentation of varying length. The general strategy of this workshop is to gradually introduce to the participants the concepts underlying the CREST model. Although less focused on hands-on examples than the 2010 version of the CREST training, the last five modules of the CREST 2.0 training each contain opportunities for hands-on modeling work. In the development of the CREST 2.0 model, the Wangchu River Basin in Bhutan (see Figure 16.3) was used for initial testing, calibration, and validation, and, therefore, this example is used extensively throughout the training, with the Nzoia River Basin example appearing only in figures and diagrams.

16.2.4   2013: NAMIBIA AND RWANDA

In March 2013, NASA and OU took another step toward training HSN hydrologists and technicians to use CREST. At this workshop, discussion once again focused on the Okavango River Basin but now took on a more specific tone regarding exactly how and when the CREST 2.0 model could be implemented on the basin. Figure 16.4

TABLE 16.1

List of Modules Included in 2010 CREST 2.0 Training Workshop

Day	Title	Description of Contents
Day 1: 1	Introduction to CREST model	Model schematics, flowcharts, organization, current and future features
Day 1: 2	Overview of data for CREST model	Topography, precipitation, PET, soil texture, and land cover data and sources
Day 1: 3	CREST applications at local, regional, and global scale	Real-time global CREST (Earth Observing System), US flash flood demonstration system, SERVIR East Africa, SERVIR Bhutan, Pakistan capacity building, China streamflow forecasts
Day 1: 4	Input and output data of CREST model	Organization of files and folder, options in.Project file, model outputs and post processing
Day 1: 5	Running the example basin	Move example files into their appropriate folders, modify. Project file, run model from command line, view results
Day 2: 1	CREST data preparation	Downloading HydroSHEDS data, clipping to basin extent, converting to ASCII files, downloading precipitation and PET data, matching.Project file and DEM derivative files
Day 2: 2	CREST calibration	Manual calibration, automatic calibration, evaluation indices, meaning of parameters
Day 3: 1	Visualization of model results	Process time series results, create hydrographs in Excel, view distributed streamflow in ArcMap, create files to view in Google Earth
Day 4: 1	Comments for setting up CREST model	General checklist for producing independent CREST examples

shows OU PhD student Zachary Flamig and HSN hydrologist McCloud Katjizeu presenting plans for CREST and the Okavango River to representatives of the Namibian government and private sector.

NASA and OU personnel also conducted field exercises with HSN hydrologists. A team traveled along portions of the river in order to map the edge of the Okavango River via Global Positioning System (GPS) and detailed field notes. Another team used a helicopter to map the same areas from the air using GPS and cameras. The goal of these exercises was to improve satellite estimates of inundated areas in the Okavango River Basin. Unfortunately, only a few months after these activities, the head of HSN passed away unexpectedly. Luckily, the new head of HSN already had an excellent working relationship with the NASA and OU group of modeling experts and so, in this case, capacsity-building efforts could continue. However, this illustrates the need of outside experts engaging in capacity-building activities to develop strong working relationships across multiple offices and through several levels of management.

After his work in Namibia, Flamig traveled to Kigali, Rwanda, to conduct a CREST 2.0 training workshop for that nation's Integrated Water Resources Department. Here,

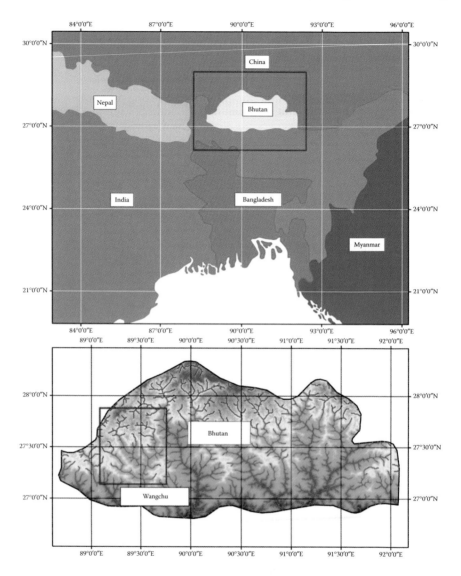

FIGURE 16.3 Map of Wangchu River Basin in Bhutan. (After Xue, X. et al., *J. Hydrol.*, 499, 91–99, 2013.)

20 government-employed Rwandan meteorologists and hydrologists completed the CREST 2.0 training workshop outlined in Table 16.1. Figure 16.5 is a picture of Flamig at the workshop in Kigali. As shown in Figure 16.2, each participant has access to a personal computer. When possible, these workshops are completed at no cost to the local participants. NASA's SERVIR project funds the vast majority of the projects and OCC's grants from the National Science Foundation pay for travel for workshop facilitators in some cases.

FIGURE 16.4 2013 Capacity Building Meeting in Windhoek, Namibia. (Courtesy of Zac Flamig.)

FIGURE 16.5 2013 CREST 2.0 Training workshop in Kigali, Rwanda. (Courtesy of Zac Flamig.)

16.2.5 2013: NIGERIA

OU sent a delegation of scientists and students to Nigeria in December 2013. This meeting can be considered roughly analogous to the 2012 Namibia meetings, so activities were primarily focused around high-level initial discussions of Nigeria's needs and what shape a mooted early warning system might take in that country. Although CREST 2.0 was discussed, the process of deciding on, training to use, and then implementing a particular system can be lengthy, and the 2013 meeting is just the first tentative step to implementing the model in Nigeria.

16.2.6 2014: NAMIBIA

Two OU graduate students traveled to Namibia in February 2014. This was the third visit by the OU and NASA group to the country and the first when CREST training was made an explicit part of the visit with HSN personnel. This workshop largely followed the outline in Table 16.1, but with a different schedule and some additional content. OU scientists had previously calibrated and validated initial CREST model runs on the Okvango River between 2013 and 2014 and so results from that work were presented during the training as an example, where workshop participants could see the requirements for getting the model working on one of Namibia's large rivers. Most of the remaining workshop, however, still relied on the older Wangchu Basin example. Table 16.2 outlines the training workshop conducted in Namibia in 2014.

The astute reader will note that the facilitators had roughly 5 or 6 h per day to engage in training activities. This illustrates one important consideration when planning training in countries that frequently receive visits from outside experts. In many cases, local technical experts spend so much of their workdays in training activities that they find they have little time left to do their actual work. In the 2014 training, HSN management requested frequent breaks (termed "health breaks") between each training module and 90–120 min lunch breaks. Additionally, computer issues are bound to arise in any capacity building that relies heavily on personal computing resources. Workshop facilitators learned to adjust when necessary, usually by reserving significantly more time for hands-on exercises than originally planned. One particularly thorny problem arose during the morning of the third day, where the hands-on activities required a working install of ArcMap that is a very expensive and specialized program that was not available to all workshop participants, so the instructors modified the lesson on the fly as much as possible to reduce the number of steps for which this program was required. Finally, not all participants had access to mobile or laptop computing and so time at the beginning and end of each session was needed for some participants to set up, use, and take down desktop computers. Figure 16.6 shows OU graduate student Race Clark discussing streamflow measurements in the Kuiseb River Basin with HSN hydrologists.

At the conclusion of the training workshop, HSN, NASA, and OU traveled into the Namib Desert and surrounding areas to gain the first-hand knowledge of HSN's current streamflow monitoring capabilities and to geo-locate HSN monitoring

TABLE 16.2

Outline of 2014 CREST 2.0 Model Training Workshop

Day	Title	Description of Contents	Time Required
Day 1: Afternoon	Introduction to CREST model	Model schematics, flowcharts, organization, current and future features	1 h
Day 1: Afternoon	Overview of data for CREST model	Topography, precipitation, PET, soil texture, and land cover data and sources	1 h
Day 2: Morning	CREST applications at local, regional, and global scale	Real-time global CREST (Earth Observing System), US flash flood demonstration system, SERVIR East Africa, SERVIR Bhutan, Pakistan capacity building, China streamflow forecasts	30 min
Day 2: Morning	Input and output data of CREST model	Description of each folder in a CREST project	30 min
Day 2: Morning	Running the example basin	Organization of files, modifying the. Project file	30 min
Day 2: Morning	Hands-on	Run CREST 2.0 on Wangchu basin	1 h
Day 2: Afternoon	CREST calibration	Parameters, SCE-UA, evaluation indices	30 min
Day 2: Afternoon	Hands-on	Manual calibration on Wangchu, SCE-UA calibration on Wangchu	1 h
Day 2: Afternoon	Additional CREST options	Relative file paths, batch files	15 min
Day 2: Afternoon	Hands-on	Run Wangchu with relative file paths and using a bath file	1 h
Day 3: Morning	Data preparation	Obtain PET, DEM, and precipitation	30 min
Day 3: Morning	Hands-on	Process DEM in ArcMap; correctly locate precipitation, PET, and DEM derivatives in CREST project; run Wangchu	1.5 h
Day 3: Morning	CREST visualization	Create hydrographs in Microsoft Excel	30 min
Day 3: Afternoon	Hands-on	Create hydrograph in Microsoft Excel for Wangchu basin	30 min
Day 3: Afternoon	How to create your own CREST example	General comments	15 min
Day 3: Afternoon	Okavango/Rundu example	Discussion of Okavango results and calibration, future uses of CREST	30 min
Day 3: Afternoon	Hands-on	Run Okavango example, create hydrograph of results	30 min

FIGURE 16.6 Kuiseb river gauge site at the Gogabeb research and training centre in Namibia. (Courtesy of Race Clark.)

FIGURE 16.7 2014 CREST 2.0 workshop participants at Schleisenweir on the Kuiseb river in Namibia. (Courtesy of Race Clark.)

stations via GPS to the Earth Observatory-1 satellite operated by NASA. This field trip also provided HSN an opportunity to share their local hydrologic knowledge with outside experts and to help these outsiders to better tailor future training activities to HSN's needs. Figure 16.7 shows HSN personnel and workshop facilitators near the Schleisenweir station on the Kuiseb River.

16.2.7   2014 AND BEYOND: LECTURE VIDEOS

Although in-person training is almost always desirable, security, cost, and other considerations sometimes require remote training to be used as an alternative. The developers of CREST 2.0 have, thus, created narrated PowerPoint presentations that explain the basic features and needs of the CREST model for regions where travel is difficult or impossible. As of 2015, these presentations are roughly equivalent to the first three training modules in either Table 16.1 or 16.2 and, therefore, lack the hands-on components present in most other versions of CREST or EF5 training.

16.3   CURRENT EFFORTS

Between 2010 and 2014, CREST, in its various forms, was deployed operationally at one supra-national center (RCMRD), tested on dozens of basins across three continents, and taught to hydrologists and meteorologists from over a dozen countries. This modeling software continues to evolve, expand, and improve. Throughout much of CREST's history, a similar model, but with distinct cell-to-cell routing, user interface, and codebase, evolved in parallel: EF5. This model offers multiple advantages over CREST 1.6 and CREST 2.0: it enables ensemble streamflow forecasts by using multiple model cores at once, incorporates the kinematic wave assumption for cell-to-cell routing, offers improved cross-platform compatibility, and supports a wider range of precipitation, topographical, and potential evapotranspiration (PET) file formats as inputs. Most importantly, EF5 underlies a pre-operational flash flood forecast system used by the US National Weather Service, which ensures continued code development, improvement, and support for years to come.

16.3.1   EF5 TRAINING

As a result of lessons learned over several years of capacity-building workshops, EF5 also comes with a new set of training modules distinct from those provided with CREST 2.0. General goals of the new training are the following: use only open-source software packages, maintain small file sizes for the easier distribution, include multiple examples from across the world representing different terrain and hydroclimatic regimes, plan for frequent breaks and computer compatibility, and include hands-on components or checklists in every training module. Additionally, the new training provides companion documents for each hands-on process so that trainees can more easily replicate those processes without an experienced facilitator present. An EF5 training course should require 4 days to complete, but its modular structure means that modules can be removed to decrease the required time, or additional model examples added if trainees want more experience. A total of seven examples, as shown in Figure 16.8, are part of the training; these have been selected to include different types of terrain and hydroclimatic regimes across different continents.

Day 1 begins with a 90-min welcome session. In this welcome session, participants learn about OU and NASA-SERVIR's involvement in hydrologic modeling and capacity building. Even in the welcome session, participants are expected to engage with the facilitators interactively and will be taught how to install and configure

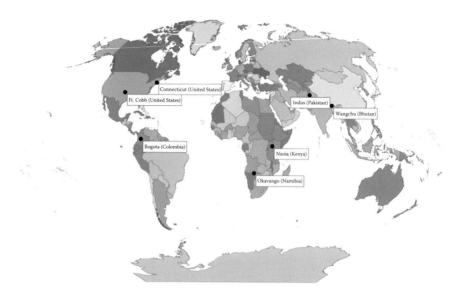

FIGURE 16.8   Map of EF5 training examples.

QGIS (also known as Quantum Geographic Information System, an open-source alternative to Esri's ArcGIS suite) and TauDEM (an open-source alternative to the ArcMap Hydro Toolbox). QGIS allows participants to clip or crop topographical data to the appropriate extent, to sample it to the desired resolution, and to visualize and verify the result. TauDEM corrects, conditions, and further processes topographical data for use in hydrologic models. Like the entire EF5 training course, the welcome session can be completed and the appropriate software package installed without Internet connectivity. In general, a training facilitator will give the PowerPoint presentation and then answer questions before entering the hands-on part of the training module. For remote workshops, the PowerPoint could be presented with pre-recorded narration and then the companion checklist used for the corresponding hands-on portion of the module.

Following a morning health break, the second module of Day 1 takes around 2 h to complete. Titled "Introduction to Hydrologic Models," this section of the workshop covers the basic concepts of hydrologic modeling. Models are further discussed by their categories such as those that can use distributed parameter sets and those that use only lumped parameter sets. The water cycle processes usually included in hydrologic models are described theoretically, and then model implementations of these processes are discussed. Concepts such as soil moisture, rainfall forcing, infiltration, routing, storage, and runoff generation are all explained. The hands-on portion of this module is intended to give participants early confidence in their abilities as it requires no modeling expertise and only basic computer skills. Example 1 is on the Wangchu Basin of Bhutan and is completely prepackaged from start to finish; the participants need only to double-click the EF5 executable and let it run to completion. After completing the welcome module described before, participants will have basic QGIS knowledge, so this is built on by asking participants to visualize

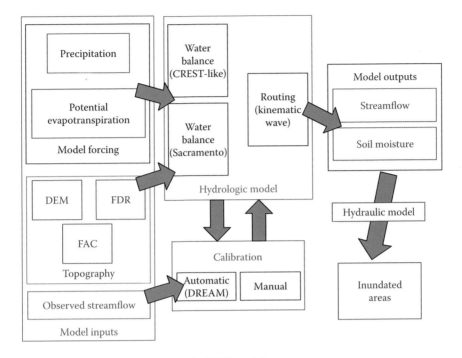

FIGURE 16.9   Block diagram of typical EF5 workflow.

the model inputs (topographical information, precipitation forcing, and PET forcing) in that program. The model outputs (streamflow, in this case) are visualized in Microsoft Excel, with which it is presumed that workshop participants will have basic familiarity.

After a lunch break, the training starts to focus more specifically on the implementation of hydrologic theory within EF5 in the module entitled "EF5 Overview." This includes a brief discussion of the kinematic wave equations used for routing in EF5 and the VIC equations used to separate surface water flow and interflow (Liang et al., 1994). Participants are asked to visualize a typical EF5 workflow via block diagrams, flowcharts, and other schematics. These diagrams define EF5 inputs, outputs, and processes (see Figure 16.9). The presentation in this module summarizes the requirements and options relating to model inputs and outputs, while accompanying documents fully explain these models' characteristics. Other features of EF5 are described, focusing primarily on the line-by-line explanation of the EF5 control file, which permits the user to tell the model what to do and how to do it. This module does not include its own hands-on component, but the training facilitator should make frequent references back to Example 1 (Wangchu Basin) as EF5 features and components are fully explained.

After the afternoon health break, the facilitator will discuss "DEM Derivatives," a roughly 2-h-long module. EF5, like most hydrologic models, requires three basic topographical components. The first of these is the DEM, or digital elevation model, which is a digital model representing the surface of the earth over the area of interest.

The model is stored as a grid (also called a "raster") and the value stored for each grid cell corresponds to the elevation of the terrain within that grid cell. Secondly, the model requires a flow direction raster (FDR) that can be calculated from a DEM. The FDR is a grid at the same resolution as the DEM and in each FDR grid cell a number is stored that corresponds to the direction of flow out of that cell. In EF5, a strategy called the eight-direction (or D8) flow model is used (Jenson and Domingue, 1988), where water can exit a cell in one of eight directions: the cardinal and inter-cardinal directions. The final piece needed by the model is a flow accumulation grid (FAC). The FAC is a raster where the value of each cell corresponds to the number of cells that flow into that cell; the FDR is required to produce the FAC raster. These three rasters can be created in most GIS programs and in a few other standalone hydrologic software packages; in the EF5 training, brief mention is given to the procedure in ArcMap but to ensure that open-source software is used when-ever possible, step-by-step instructions are restricted to QGIS and TauDEM. The training also discusses the importance of DEM resolution and gives some advice to the trainees regarding how to select the appropriate DEM resolution for a given river basin. DEMs can be freely obtained from multiple sources; the default in the EF5 training comes from the HydroSHEDS (*Hydro*logic data and maps based on *SH*uttle *E*levation *D*erivatives at multiple *S*cales) project at the World Wildlife Fund (Lehner et al., 2008). HydroSHEDS consists of 30-arcsec hydrologically conditioned DEMs processed from NASA's SRTM (Shuttle Radar and Topography Mission). Other DEMs produced from satellites, radars, or lidars are available at varying resolutions for some locations around the world. By the end of the module, workshop par-ticipants should learn how and why to select DEM data from the various sources depending on their requirements. In the hands-on part of this module, participants create their own DEM, FAC, and FDR in QGIS for the Okavango River at Rundu, Namibia from the original HydroSHEDS dataset; this is "Example 2" in the training workshop. Once the participants have created the three necessary topographical files for Example 2, the first training day is complete.

Day 2 begins with a discussion of the various "Uses of EF5," which usually lasts about an hour. This module reinforces why the participants are learning how to use EF5; the model's wide and growing user base makes capacity-building efforts easier by reassuring trainees that the model will continue to be developed and improved. These uses fall into three general categories: activities of RCMRD in east Africa, global CREST simulations, and the FLASH (Flooded Locations and Simulated Hydrographs) project in the United States. These can be individually preferentially emphasized or deemphasized, depending on the workshop audience. This module is short enough that a local facilitator could add additional information about EF5 or CREST implementations in other relevant areas of the world and slightly shorten the Day 2 lunch break. The hands-on component of this module involves exploring real-time CREST and EF5 outputs from the ongoing global CREST, FLASH, and RCMRD projects. In the event that the workshop venue is not equipped with Internet access, screenshots from the web interfaces of these projects can be displayed and discussed in a group setting instead.

In Figure 16.9, a block diagram of a typical EF5 workflow was presented. In Day 1, the model itself and the topographical inputs to the model were explained.

In the second module of Day 2, the precipitation and PET forcing to the model are described in the module "Rainfall and PET." Rainfall data typically used in African capacity-building applications come from NASA satellites, particularly those from the TRMM (Liu et al., 2012). TRMM rainfall estimates are available in multiple versions, but those of central interest in EF5 training workshops are TRMM V7, which is a post-processed ground gage-corrected rainfall estimate, and TRMM RT (real-time), which is available in near real time and does not include any gage correction. V7 can be used in retrospective applications, like calibrating a hydrologic model, while RT would be used in ongoing real-time modeling applications. Other satellite-based rainfall estimates are available, but are generally less widely used. One is CMORPH (The National Oceanic and Atmospheric Administration's Climate Prediction Center Morphing Technique) that combines TRMM data with that from other NASA spacecraft (Joyce et al., 2004). Another option is PERSIANN (Precipitation Estimation from Remotely Sensed Information using Artificial Neural Networks) that is a solution that incorporates information from American and European weather satellites along with the TRMM satellites (Hsu and Sorooshian, 2008). As PERSIANN is available only with a 2 days delay, it is usually unsuitable for real-time modeling, but not for model calibration. The module includes the discussion of the different resolutions of available precipitation forcing grids, along with the pros and cons of each. Although not a major focus of the module, participants are told that it is possible to use rain gage data objectively analyzed to a grid to force the hydrologic model. It is also possible to use quantitative precipitation forecasts, like those available from the American GFS (Global Forecasting System) to extend streamflow predictions farther into the future. The final portion of this module traces the process that results in PET grids. The EF5 materials recommend the use of United States Agency for International Development Famine Early Warning Systems Network (USAID FEWS NET) PET estimates that are calculated using the Penman–Monteith equation as outlined in Shuttleworth (1992). Participants will then complete the hands-on component of this module, where they are tasked with downloading, processing, and visualizing the TRMM V7 rainfall and FEWS NET PET data needed to properly force EF5 over the Okavango River Basin (which is part of Example 2 in the training). As previously stated, if Internet access is not available, unprocessed rainfall and PET data should be provided to participants. At the close of this module, participants have successfully learned about the model and its required input information and have produced their own versions of each.

After a lunch break, participants return to learn about "Organizing Input and Output Data." This session begins with an explanation of the various possible file formats for use in EF5 for topographical and forcing files. There is a brief review of how to create DEM, FDR, and FAC files and then a more in-depth review of the EF5 control file. Concepts like batch file development (for running EF5 in Microsoft Windows) and relative versus absolute file paths are explained. The hands-on portion of this module involves using the user-created topographical information and user-processed rainfall and PET data together with an appropriate EF5 control file to run a preliminary simulation over the Okavango River Basin. Participants will visualize selected model inputs in QGIS and develop a hydrograph comparing simulated streamflow and observed streamflow in Microsoft Excel.

Of course, in most cases, a physically based hydrologic model requires adjustments to its parameters before skillful simulations are produced. The concluding lesson of Day 2 teaches workshop attendees the theoretical basis for this calibration process in "Parameters and Manual Calibration." Each of the parameters in EF5's water balance and routing sections are explained with equations and diagrams, and their effects on simulated streamflow are described. Standard model skill evaluation indices are introduced, including correlation coefficient (CC), model bias (or "bias"), and the Nash–Sutcliffe coefficient of efficiency (NSCE) model (NSCE, colloquially called "Nash") (Nash and Sutcliffe, 1970). Best practices for the manual calibration of lumped parameters are presented, along with the concept of distributed parameters. Due to time limitations and the conceptual difficulties involved, no hands-on development of distributed parameter sets is introduced, but advanced EF5 training materials that will eventually be developed will include this functionality. The hands-on component of this module requires participants to re-run Example 2 from the previous module with adjusted parameters in an effort to improve the bias, NSCE, or CC of the simulation. If time permits, it is suggested that participants visualize their new calibrated hydrographs in Microsoft Excel to explore more fully how the calibration process affects the simulated hydrographs. With this process complete, the second day of the training workshop ends.

On Day 3, participants move on to studying "Automatic Calibration." EF5 includes one algorithm for automatically selecting parameter sets with improved simulation skill. This algorithm is the Differential Evolution Adaptive Metropolis, or DREAM for short (Vrugt et al., 2009). CREST 1.6 and 2.0 included a different algorithm: the Shuffled Complex Evolution (SCE-UA) Method (Duan et al., 1992). This first module of Day 3 explains the similarities and differences between the two algorithms and details how and why each works so that participants do not think of automatic calibration as a simple "black box." In this module, the difference between model calibration and model validation periods is also explained, as well as the theory behind the production of *a priori* model parameters based on observable physical attributes of the land surface over which the model is running. In the hands-on portion of the module, participants use DREAM within EF5 to automatically calibrate the Okavango River Basin example. They are also asked to visualize the output in Microsoft Excel, to improve their hydrograph-building skills. The module then concludes with a discussion of what the model parameter sets looked like before and after the calibration process and why certain parameters changed in the ways they did.

After a morning health break, participants return to practice processing and manipulating topographical information in the "DEM Practice" module. The lecture portion of this module simply reviews materials already taught to the participants regarding obtaining DEM data and producing FDR and FAC data from them. Participants will also briefly review rainfall and PET forcing data. As the workshop progresses, the amount of time devoted to hands-on experiences increases dramatically, and this module is no exception to that trend. Participants will process and use DEM and forcing data for Example 3, which is sited on the Bogota River near the city of Bogota, Colombia. This basin is much smaller and has more complex terrain than the Okavango River in Example 2. These vastly different basins provide a real-world example of why the resolution of forcing and topographical data matters for model skill while simultaneously demonstrating that EF5 can be made to work in all different types of topography and hydroclimate.

After a lunch break, trainees will continue their work on Example 3 by reviewing the initial lesson on parameters and calibration strategies. They immediately transition into hands-on work by calibrating Example 3 in a competition to see whose simulations can achieve the best NSCE, CC, and bias, and how quickly each can do so. Each will visualize her or his results in Microsoft Excel and specific pupils could be asked to discuss their calibration and modeling strategy during the example, if desired by the workshop facilitators.

While the mechanics of hydrologic simulation are important, model output does not exist in a vacuum. Like any simulation, hydrologic model has sources of error and uncertainty that must be considered. Day 3 wraps up with a series of "Comments on Interpreting and Using Model Output." In this session, participants will hear more about how to interpret model hydrographs, especially in a forecasting context. They will also hear about how to determine one's confidence in a simulation as well as the sources of uncertainty inherent in hydrologic modeling. Finally, the facilitators will describe the theory of linking a hydrologic model like EF5 to a hydraulic model to enable forecasting and monitoring of flood inundation.

The morning of Day 4 is entirely hands-on. The facilitators will point to the Nzoia River Basin (Example 4) on a map and tell the trainees to set up the required data, calibrate the model, run the model, and post-process the output. The next module will offer a brief break from hands-on work with an exploration of the "Operational Uses of EF5" which will offer some comments on how EF5 could be set up operationally in an ongoing, real-time environment. This topic is best addressed by discussing the current state of such EF5 implementations, but for most workshops, the content of this module is highly dependent on the workshop facilitators.

In the afternoon, participants will again be asked to run the model entirely on their own for a new example basin: the Fort Cobb River in the United States. This second chance at independently running the model gives participants a chance to learn from any mistakes or hiccups they encountered in the Day 4 morning session while still having the benefit of the workshop facilitators nearby to answer any lingering questions. After a health break, the workshop wraps up with a module entitled "The Future." Facilitators will briefly address using forecast rainfall, like that from the GFS, as model forcing. They will also explore the Namibia Flood Dashboard with workshop participants, discuss the principles of model data assimilation, how model output can be used for drought monitoring, and a reminder of how distributed parameter grids work in EF5. Many of these topics are addressed in a planned "Advanced EF5 Training Workshop" that remains under development.

At the end of the workshop, participants will have successfully installed and used QGIS, TauDEM, and EF5. They will have access to NASA satellite rainfall data, FEWS NET PET, and HydroSHEDS topographical data. With appropriate observed streamflow, participants will be able to independently set up, run, calibrate, and interpret output from a hydrologic model. To practice these skills, they can continue to play with the Wangchu, Okavango, Bogota, Nzoia, and Fort Cobb examples. There are two additional examples that can be used during a workshop, if more time is needed, or after the workshop at the participants' leisure: the Connecticut River in the United States and the Indus River in Pakistan. Participants will be able to look at any presentation given by the facilitators, at the accompanying hands-on worksheets, or at video recordings of past workshops (though this last source may vary significantly from workshop to workshop). The training course also includes the EF5 user manual, documentation on all other pieces of included software and datasets, and a handful of scientific papers relevant to the workshop principles. Table 16.3 summarizes the EF5 training workshop in an outline form.

TABLE 16.3
Outline of EF5 Model Training Workshop

Lesson	Title	Description of Contents	Time Required
1.1	Welcome	Introductions, organization of training, installing software	90 min
1.2	Introduction to hydrologic models	Uses of models, modeling concepts, Example 1	2 h
1.3	EF5 overview	Model features, control file, discussion of Example 1	90 min
1.4	DEM derivatives	QGIS instructions, DEM, FDR, and FAC descriptions, create files for Example 2	2 h
2.1	Uses of EF5	Description of projects using EF5 around the world; Internet or static examples	1 h
2.2	Rainfall and PET	TRMM, FEWS NET, process and visualize forcing files for Example 2	90 min
2.3	Organizing input and output data	Review of control file, file paths, batch files, run Example 2 and visualize outputs	2 h
2.4	Parameters and manual calibration	Parameter definitions, model equations, parameter effects, manual calibration of Example 2	2 h
3.1	Automatic calibration	SCE-UA, DREAM, calibration and validation periods, automatically calibrate Example 2, discussion of results	2 h
3.2	DEM practice	Review of module 1.4, create topographical files for Example 3	90 min
3.3	Example 3	Review module 2.4, compete to produce best NSCE, CC, or bias on Example 3	2 h
3.4	Comments on interpreting and using model output	Confidence, uncertainty, meaning of hydrographs, hydraulic modeling	1 h
4.1	Example 4	Run Example 4 from start to finish	3 h
4.2	Operational uses of EF5	Operational applications of EF5, how to set up	1 h
4.3	Example 5	Run Example 5 from start to finish	2 h
4.4	The future	Forecast rainfall, Namibia flood dashboard, data assimilation, drought, distributed parameters	30 min

16.4 FUTURE EFFORTS AND SUMMARY

As the community using EF5 grows and matures, the current EF5 training course will inevitably be modified. The initial course is comprehensive for someone just learning the principles of hydrologic modeling, but for a user with significant modeling experience, it is likely too simplistic. As laid out in Table 16.3, the course requires 26 and a half hours of instruction time to complete, which would be roughly equivalent to a quarter of a semester in a college or university setting, assuming a three credit hour course. With the additional examples and advanced training modules in development, the course, in its final state, will be the equivalent of a three credit hour college course, hopefully in the form of a MOOC (massively open online course). This MOOC would beef up the content explaining hydrologic principles and practices in the early stages of the course, while adding significant content on drought monitoring, distributed parameters, *a priori* parameters, forecasting techniques and model interpretation, ensemble modeling, and operationalizing EF5.

Capacity building is a lengthy and involved process that achieves great results when all stakeholders share a similar vision of success. As the experience of NASA and OU personnel involved in the CREST and EF5 projects demonstrates, great results do not develop in days or weeks but rather over months and years. Hydrologic modeling is a natural fit for capacity-building activities: it hones computing skills, transfers technology to those well placed to benefit, and harnesses the scientific knowledge of local experts in ways extremely useful to their constituents. The vibrant user community associated with CREST and EF5 after just a handful of years is a strong testament to the successes that can be achieved by applying the principles of capacity building to the domain of hydrologic modeling.

REFERENCES

Duan, Q., S. Sorooshian, and V. Gupta. 1992. Effective and efficient global optimization for conceptual rainfall-runoff models. *Water Resour. Res.* 28:1015–1031.

Hsu, K.-L. and S. Sorooshian. 2008. Satellite-based precipitation measurement using PERSIANN system. In *Hydrological Modelling and the Water Cycle*, eds. S. Sorooshian, K.-L. Hsu, E. Coppola et al., Berlin: Springer, pp. 27–48.

Jenson, S. K. and J. O. Domingue. 1988. Extracting topographic structure from digital elevation data for geographic information system analysis. *Photogramm. Eng. Rem. S.* 54:1593–1600.

Joyce, R. J., J. E. Janowiak, P. A. Arkin et al. 2004. CMORPH: A method that produces global precipitation estimates from passive microwave and infrared data at high spatial and temporal resolution. *J. Hydromet.* 5:487–503.

Lehner, B., K. Verdin, and A. Jarvis. 2008. New global hydrography derived from spaceborne elevation data. *Eos, Trans. Amer. Geophys. Union.* 89:93–104.

Liang, X., D. P. Lettenmaier, E. F. Wood et al. 1994. A simple hydrologically based model of land surface water and energy fluxes for GSMs. *J. Geophys. Res.* 99:14415–14428.

Liu, Z., D. Ostrenga, W. Teng et al. 2012. Tropical rainfall measuring mission (TRMM) precipitation data and services for research and applications. *Bull. Amer. Meteor. Soc.* 93:1317–1325.

Nash, J. E. and J. V. Sutcliffe. 1970. River flow forecasting through conceptual models part I—A discussion of principles. *J. Hydrol.* 10:282–290.

Shuttleworth, J. 1992. Evaporation. In *Handbook of Hydrology*, ed. D. Maidment, New York: McGraw-Hill, pp. 4.1–4.53.

Vrugt, J. A., C. J. F. ter Braak, C. Diks et al. 2009. Accelerating Markov chain Monte Carlo simulation by differential evolution with self-adaptive randomized subspace sampling. *Int. J. Nonlin. Sci. Num.* 10:273–290.

Wang, J., Y. Hong, L. Li et al. 2011. The coupled routing and excess storage (CREST) distributed hydrological model. *Hydrolog. Sci. J.* 56:84–98.

Xue, X., Y. Hong, A. S. Limaye et al. 2013. Statistical and hydrological evaluation of TRMM-based Multi-satellite precipitation analysis over the Wangchu Basin of Bhutan: Are the latest satellite precipitation products 3B42V7 ready for use in ungauged basins? *J. Hydrol.* 499:91–99.

17 Assessment of Shallow Landslides Induced by Mitch Using a Physically Based Model
A Case Study in Honduras

*Zonghu Liao, Kun Yang, Yang Hong,
and Dalia Kirschbaum*

CONTENTS

17.1 INTRODUCTION

Rainfall-induced landslides pose significant threats to human lives and property worldwide (Hong et al., 2006). Overpopulation, deforestation, mining, and uncontrolled land-use for agricultural and transportation purposes, increasingly put large numbers of people at risk from landslides (Boebel et al., 2006).

Landslide disaster preparedness and forecasting systems require knowledge of the physical mechanisms influencing landsliding, information on geotechnical composition, and landslide inventory data.

Recent examinations of remote sensing datasets offer an opportunity to enhance the sparse field-based landslide inventory datasets for hazard forecasting and mapping at larger spatial scales. Liao et al. (2010) develop an early warning system by incorporating a physically based model to assess landslide hazard in Indonesia experimentally. The model, SLIDE (SLope-Infiltration-Distributed Equilibrium),

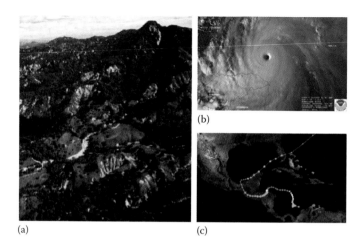

(a)   (c)

FIGURE 17.1 (a) High concentrations of debris flows in Honduras following the Hurricane Mitch (From Harp, EL et al., Digital inventory of landslides and related deposits in Honduras Triggered by Hurricane Mitch, U.S. Geological Survey Open-File Report 02–61, 2002.), (b) aerial view, and (c) path of Hurricane Mitch in the Caribbean. (Courtesy of NOAA, October 26, 1998.)

has been further developed and implemented to identify the spatial and temporal distribution of landslides in Honduras. This chapter describes the approach for landslide forecasting using Hurricane Mitch as a case study within two areas in Honduras. Hurricane Mitch made landfall in late October, 1998, and dropped historic amounts of rainfall in Central America. The hurricane and related landslides killed an estimated 5657 people, injuring another 12,272 according to the National Emergency Cabinet. Figure 17.1 shows the storm path and an example of the shallow landslides triggered in Honduras. This chapter describes the framework of the SLIDE model and remotely sensed and in situ datasets employed; it then describes the application of the model in Honduras; finally, it quantitatively evaluates the forecasting results and discusses the advantage and limitations of the model.

17.2 METHODOLOGY

17.2.1 APPROACHES FOR LANDSLIDE FORECAST

Three major approaches to developing a forecasting model are discussed: (1) data collection and parameter initialization; (2) model development based on infinite slope equilibrium equations; and (3) test and evaluation of the model. The flow chart of the forecasting model is shown in Figure 17.2.

17.2.2 AVAILABLE DATA AND PARAMETER INITIALIZATION

Landslide occurrence depends on complex interactions among a large number of factors, mainly including slope, soil properties, lithology, and land cover. With increasing frequency, the remote sensing data sets have been used to develop susceptibility map regionally and globally. Hong et al. (2007) have used some of the aforementioned remote

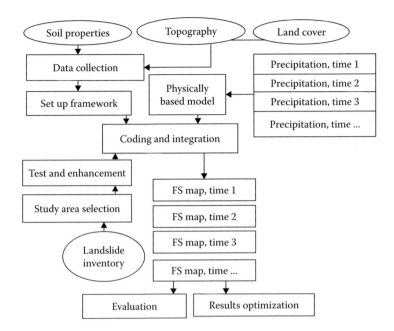

FIGURE 17.2 Conceptual framework of a physical model for forecasting of rainfall-induced shallow landslides.

sensing data to derive a global landslide susceptibility map. In this research, elevation, land cover, and precipitation data sets were all derived from satellites.

> *Elevation*: Elevation data were retrieved from the ASTER (Advanced Spaceborne Thermal Emission and Reflection Radiometer) Global Digital Elevation Model, which was developed jointly by the Ministry of Economy, Trade, and Industry of Japan and the NASA, USA (http://asterweb.jpl.nasa.gov/data.asp). Topographic properties were derived from a 30-m ASTER DEM.
>
> *Soil*: Soil parameter values were determined from soil types provided by the Food and Agriculture Organization of the United Nations (FAO; http://www.fao.org/AG/agl/agll/dsmw.htm) and the Moderate Resolution Imaging Spectroradiometer land classification map. There are 16 soil types with referenced values provided. However, only the initial 12 types have been listed in Table 17.1. The remaining four soil types are not related to soil cohesion or friction and are therefore not listed. These not included are organic materials, water, and bedrock.
>
> *Land Cover*: The University of Maryland's 1 km global land cover classification is produced by using data for 1992–1993 from the Advanced Very High Resolution Radiometer (Hansen et al., 2000). There are 12 classes in total and a certain cohesion value of vegetation will be added to soil cohesion in slope stability calculation. Schmidt et al. published the method to obtain values of root cohesion for different species of vegetation and a summary table of selected root cohesion from different sources.

TABLE 17.1

Table of Soil Parameter Values

USDA Class	Soil Type	Bulk Density (g/cm³)	Field Capacity (cm³/cm³)	Porosity (Fraction)	Saturated Hydraulic Conductivity (cm/h)	Cohesion[a] (KPa)	Friction Angle (Degree)
1	Sand	1.49	0.08	0.43	38.41	2–8	38–42
2	Loamy loam	1.52	0.15	0.42	10.87	5–10	27–30
3	Sandy loam	1.57	0.21	0.4	5.24	4–8	28–36
4	Silt loam	1.42	0.32	0.46	3.96	6–12	23–25
5	Silt	1.28	0.28	0.52	8.59	6–12	23–30
6	Loam	1.49	0.29	0.43	1.97	7–40	21–24
7	Sandy clay loam	1.6	0.27	0.39	2.4	8–50	18–22
8	Silt clay loam	1.38	0.36	0.48	4.57	36–82	16–17
9	Clay loam	1.43	0.34	0.46	1.77	20–50	18–22
10	Sandy clay	1.57	0.31	0.41	1.19	7–42	21–24
11	Silty clay	1.35	0.37	0.49	2.95	20–40	17–20
12	Clay	1.39	0.36	0.47	3.18	30–50	15–18

[a] The columns 1–6 referenced to FAO and columns 7 and 8 referenced to the geotechnical literatures.

This method was referenced after we obtained the land cover classification validated by public vegetation information.

Precipitation: The spatial distribution, duration, and intensity of precipitation play an important role in triggering landslides. The precipitation data used in this study are obtained from the NASA Tropical Rainfall Measurement Mission (TRMM) Multisatellite Precipitation Analysis (TMPA) (Huffman et al., 2007), which provides a provides a calibrated sequential scheme for combining precipitation from multiple satellite and gauge analyses at a resolution of $0.25° \times 0.25°$ over the latitude band 50°N–50°S every 3 h. The real-time rainfall is available on the NASA TRMM web site (http://trmm.gsfc.nasa.gov).

Landslide inventory data: This event-based inventory was compiled in the months following the Hurricane Mitch in November, 1998, by USGS. The hurricane triggered more than 500,000 landslides throughout the Honduras, 95% of which were debris flows with various soil depths including shallow landslides and deep-seated landslides (Harp et al., 2002). It is a supreme mass movement event of the century and cast disastrous impact over the whole country.

17.2.3 SLIDE MODEL

The SLIDE model integrates the contribution of apparent cohesion to the shear strength of the soil and the soil depth influenced by infiltration (Figure 17.3). Factor of Safety (FS) is expressed as the ratio of shear strength to shear stress to calculate slope stability. A slope is considered stable when FS > 1 and a landslide is predicted

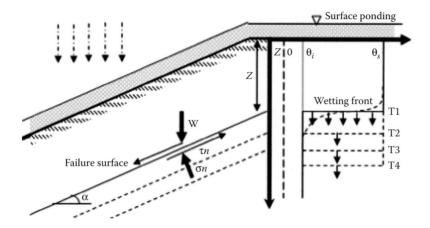

FIGURE 17.3 Infiltration processes and infinite landslide model.

when FS nears or drops below 1. In this study of shallow landslides, an infinite-slope equation is translated as the cohesion and frictional components:

$$FS(Z_t, t) = \frac{c' + c_\phi(t)}{\gamma_s Z_t \sin \alpha \cos \alpha} + \frac{\tan \varphi}{\tan \alpha} \quad (17.1)$$

where:
  c' is soil cohesion, incorporating a value for root zone cohesion
  γ_s is the unit weight of soil
  α is the slope angle
  φ is the soil friction angle

$c_\phi(t)$ represents the apparent cohesion related the matric suction, which in turn, depends on the degree of saturation of the soil (Montrasio and Valentino, 2008), written as

$$c_\phi(t) = AS_r (1 - S_r)^\lambda (1 - m_t)^\partial \quad (17.2)$$

where:
  A is a parameter depending on the kind of soil and is linked to the peak shear stress at failure
  λ and ∂ are numerical parameters that allow to translate the peak of apparent cohesion related to S_r, the degree of saturation of the soil
  m_t represents the dimensionless thickness of the infiltrated layer, which is a fractional parameter between 0 and 1:

$$m_t = \frac{\sum_{t=1}^{T} I_t}{nZ_t (1 - S_r)} \quad (17.3)$$

in which I_t is rain intensity, n is the porosity, and Z_t is the soil depth at time t, which is determined by the infiltration process (Rui, 2004):

$$Z_t = \sqrt{\frac{2K_sH_ct}{\theta_n - \theta_0}}$$ (17.4)

where:
   K_s is hydrologic conductivity
   H_c is capillary pressure
   t is time
   θ_n is water content of the saturated soil
   θ_0 is initial water content of the soil

17.3   CASE STUDY

Intense rainfall from Hurricane Mitch from October 27–31, 1998, exceeded 900 mm within some regions of Honduras and triggered in excess of 50,000 landslides through-out the country (Harp et al., 2002). Two study areas were selected to compute the FS maps, each of which covers approximately 600 km² over the south coast of Honduras (Figure 17.4). In our selected study areas, the highest recorded rainfall intensity was 12 mm/h, with an accumulation of close to 200 mm in 5 days (see Figure 17.5). This region was selected for analysis because it currently has one of the largest and most regionally extensive landslide inventory dataset complied following Hurricane Mitch.

Parameter values for the model are summarized in Table 17.2. As soil values are the most difficult parameter to assess, we used the general soil information from Table 17.1 after the validation. Furthermore, as Honduras has a tropical climate and the study areas are covered by pine woods, we assume the root cohesion values added to the total cohesion range from 3.7 to 6.4 (Schmidt et al., 2001; Waldron et al., 1983). Because of the large geographic area considered as well as the limited high-resolution datasets for this region, we make several assumptions to evaluate the SLIDE model in this area:

- Detailed geological information was not included in this model due to the limitations of the simplified physical model and lack of homogeneous information at realistic spatial resolutions over the large study area. However, this information may be incorporated into the model when data become available.
- The layer of the soil subject to sliding is generally characterized by a certain degree of heterogeneity and can be affected by holes in the soil caused by various living organisms (Montrasio and Valentino, 2008). In this study, we consider that the soil properties in the layer are homogeneous.
- Run-off and evapotranspiration were neglected in the water balance. Therefore, we assume that all rainfall infiltrates into the soil, which is not physically realistic.

All assumptions in this section were made to make the model easily applicable over larger areas and able to employ various remotely sensed datasets.

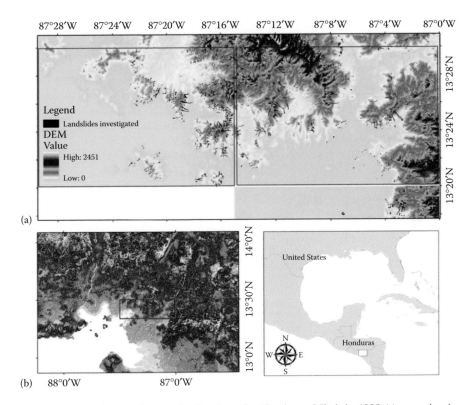

(a)

(b)

FIGURE 17.4 Two study areas in Honduras for Hurricane Mitch in 1998 (a), zooming in from the Central America (b).

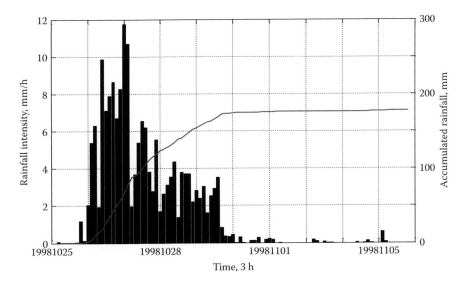

FIGURE 17.5 Precipitation of Hurricane Mitch in 1998 for the selected study areas.

TABLE 17.2

Parameters, Symbols, and Values Are Used for SLIDE Application

Property	Symbol	Unit	Value
Slope angle	α	Degree	Varies
Soil depth	Z	m	Varies
Soil type	1–16	Unit less	Sandy clay/Clay
Land cover	LC	Unit less	6, 7, 11
Coefficients	λ, ∂	Unit less	0.4, 3.4
Friction	ϕ	Degree	17–22
Cohesion (root included)	c'	KPa	15–24
Coefficient	A	Unit less	20
Unite weight of soil	γ_s	KN/m^3	20
Porosity	n	1	0.40–0.60
Water content	θ_0, θ_n	1	0.2, 0.6
Degree of saturation	S_r	%	20
Hydrologic conductivity	K_s	cm/h	1.00–4.00
Capillary	H_c	mm	50

17.3.1 RESULTS AND EVALUATION

Using TMPA data as the rainfall input, the parameterized SLIDE model was run at 3-hourly resolution and a FS value was generated throughout the Hurricane Mitch event from November 25, 1998 to December 5, 1998 at every 30-m grid cell. Figure 17.6a and b illustrate the unstable areas computed by the SLIDE model for the two study areas over the Hurricane Mitch event. Unstable areas (FS ≤ 1) are shown in red and observed landslides are shown in black.

First, we evaluate the results by comparing the unstable maps with observation by four indices from Fawcett (2006). We use labels {Y, N} for the class forecasts produced by the physical model in 30 m grids. The labels {p, n} were used for the class of observation in the field. In the class of observation, landslide areas were computed within a radius of 500 m. There are four possible outcomes when classifying a grid from the unstable map. If a computed unstable grid is inside the observed landslide area, it is counted as true positive (*tp*; also called hit); if it is outside the observed landslide area, it is counted as false positives (*fp*; also called false alarm). If a computed stable grid matches an observed landslide grid, it is counted as false negative; otherwise, it is called true negatives. Figure 17.7 shows the classification matrix and the equations of several indices. The true positives rate defines how well the predicted results agree with the observations. The *tp* rates are 78% for Figure 17.6a, and 75% for Figure 17.6b. The false positives rate indicates the tradeoff between the predicted results and observations. The *fp* rates are 1% and 6%, respectively. The error rate is defined as the portion of the computed unstable grids that did not contain observed landslides (Sorbino et al., 2010). The SLIDE model provides a value of error rate of 35% in Figure 17.6a while error rate is 49%

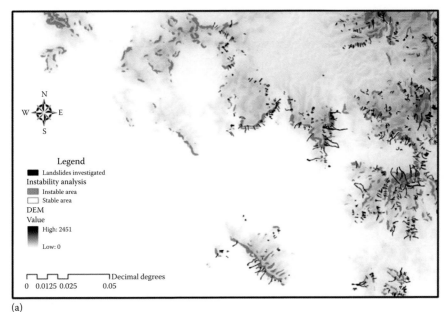

(a)

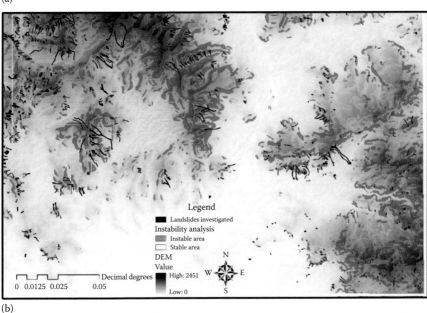

(b)

FIGURE 17.6 Instability maps of landslides produced by SLIDE for the study areas (a) and (b).

in Figure 17.6b. Accuracy represents the level of agreement between the forecast and the observations of landslides. The accuracy values are 98% for the study area a and 92% for the study area b. The precision means positive predictive value. The precisions for two areas are 65% in Figure 17.6a and 51% in Figure 17.6b. All the statistics values are listed in Table 17.3.

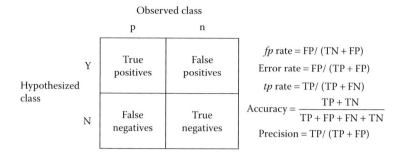

FIGURE 17.7   Four quantitative evaluation indices. (From Fawcett T, *Pattern Recognition Letters*, 27, 861–874, 2006.)

TABLE 17.3
Evaluation Statistics for Instability Maps of Landslides Produced by SLIDE

Indices	True Positive Rate (%)	False Positive Rate (%)	Accuracy (%)	Precision (%)	Error Rate (%)
Study area (a)	78	1	98	65	35
Study area (b)	75	6	92	51	49
Perfect values	100	0	100	100	0

The model forecasts show good agreement with the mapped inventory over the two study areas. However, the model overpredicts the landslide areas, particularly in the west part of Figure 17.6a. In Figure 17.6b, the predicted landslide areas by the model are much larger than observed areas. Generally, the model forecast results in larger hazard zones than what is observed, resulting in an overestimation of hazard areas. This is consistent with other studies performed at regional scales (Nadim et al., 2006).

17.4   DISCUSSION AND CONCLUSIONS

This case study tests the model performance for a specific event to test the forecast capabilities of the forecast system. Quantitative evaluation of the landslide Hurricane Mitch case study indicates that the model demonstrates good predictive skill when compared to the landslide inventory over the two study areas. The *tp* rates for the two study areas are 78% and 75%, respectively, while the *fp* rates are as low as 1% and 6%. The low *fp* rates come from the accurate prediction of stable areas by a physical model. The high *tp* rates and low *fp* rates demonstrate the potential to employ a physically based model for the forecast of landslide disasters using remote sensing and geospatial data sets over larger regions; particularly in regions where comprehensive field data and fine-scale precipitation observations and forecasts are available. The remotely sensed data products used in the SLIDE model describe the most important factors influencing the slope movement, which serves to simplify the model calculations and decrease the computational load.

However, overprediction is clearly indicated by the error rates which are 35% and 49%, though it is possibly due to a number of model assumptions and the limitation of physical models. Furthermore, there are several limitations of simplifying the physically based relationships and employing remotely sensed, rather than in situ data products. These shortcomings can limit model accuracy and should be improved for future applications, these include

1. Uniform geological structures of slopes are assumed as initial conditions, which could lead to potential errors of modeling results. Neglecting various geological features within the model calculations greatly limits the forecast ability of the physical models. In the future application of the other study area, information of geological structures from high risk-prone area could better enhance the hazard assessments.

2. The negative effects that decrease the accuracy of modeling results by applying same soil values to the regional scale modeling cannot be neglected. A same type of soil from different places reveals various physical and mechanical properties in a certain range, regardless of the vegetation roots and the rainfall infiltration. In this study, the same soil under different stratigraphic conditions was set to be homogeneous by inputting a same set of physical and mechanical values. This hypothesized procedure would reduce the spatial variances. A study of the empirical soil properties would improve this problem.

3. Higher spatiotemporal resolution of satellite rainfall, DEMs, and accurately validated remote sensing soil information are expected to better account for landslide-prone regions. In situ tests of hotspot areas would provide better soil values for input and improve the performance of model forecast. The higher spatial resolution of DEM would be used to derive better topographical features of slope, for example, slope, resulting in reducing the overprediction of the model.

4. Overprediction and error rates of landslides are difficult to reduce due to the geomorphologic variations, different triggering mechanisms, and unpredictable mass volume of sliding. Furthermore, landslide-prone areas can also largely be affected by anthropogenic impacts such as improper building and roads, which have not been incorporated within the physically based model approach. As only an extreme event is presented, the areas identified as "unstable" and computed by error indices could still be unstable and triggered in future events.

While there are several limiting factors affecting the accuracy of model forecasting, the SLIDE model shows potential as a landslides forecasting tool over larger regions. The model may be suitable to be incorporated into future real-time early warning system when high-quality data sets are available.

ACKNOWLEDGMENTS

The computing for this project was performed at the OU Supercomputing Center for Education & Research (OSCER) at the University of Oklahoma (OU). This research was supported by an appointment to the NASA Postdoctoral Program at the *Goddard Space Flight Center*, administered by Oak Ridge Associated Universities through a contract with NASA. The authors would also like to extend the appreciations to USGS scientists make the landslide inventory data available for research community. Support fund was partially provided by Science Foundation of China University of Petroleum, Beijing (No. 2462014YJRC013).

REFERENCES

Boebel O, Kindermann L, Klinck H, Bornemann H, Plotz J, Steinhage D, Riedel S, Burkhardt E (2006) Satellite remote sensing for global landslide monitoring. *EOS, Transactions, American Geophisical Union*, 87(37):357–358.

Fawcett T (2006) An introduction to ROC analysis. *Pattern Recognition Letters*, 27:861–874.

Hansen MC, DeFries RS, Townshend JRG, Sohlberg R (2000) Global land cover classification at 1 km spatial resolution using a classification tree approach. *International Journal of Remote Sensing*, 21(6–7):1331–1364.

Harp EL, Hagaman KW, Held MD, McKenna JP (2002) Digital inventory of landslides and related deposits in Honduras Triggered by Hurricane Mitch, U.S. Geological Survey Open-File Report 02–61.

Hong Y, Adler R, Huffman G (2006) Evaluation of the potential of NASA multi-satellite precipitation analysis in global landslide hazard assessment. *Geophysical Research Letters*, 33:L22402, doi: 10.1029/2006GRL028010.

Hong Y, Adler R, Huffman G, Negri A (2007) Use of satellite remote sensing data in mapping of global shallow landslides susceptibility. *Journal of Natural Hazards*, 43(2), doi: 10.1007/s11069-006-9104-z.

Huffman GJ, Adler RF, Bolvin DT, Gu G, Nelkin EJ, Bowman KP, Hong Y, Stocker EF, Wolff DB (2007) The TRMM multisatellite precipitation analysis (TMPA): Quasi-global, multiyear, combined-sensor precipitation estimates at fine scales. *Journal of Hydrometeorology* 8:38–55. doi: 10.1175/JHM560.1.

Liao Z, Hong Y, Wang J, Fukuoka H, Sassa K, Karnawati D, Fathani F (2010) Prototyping an experimental early warning system for rainfall-induced landslides in Indonesia using satellite remote sensing and geospatial datasets. *Landslides*, 7(3):317–324.

Montrasio L, Valentino R (2008) A model for triggering mechanisms of shallow landslides. *Natural Hazards and Earth System Sciences*, 8:1149–1159.

Nadim F, Kjekstad O, Peduzzi P, Herold C, Jaedicke C (2006) Global landslide and avalanche hotspots. *Landslides*, 3:159–173.

Rui X (2004) *Theory of hydrology*, vol. 386. Waterpub, Beijing, pp. 90–91.

Schmidt KM, Roering JJ, Stock JD, Dietrich WE, Montgomery DR, Schaub T (2001) The variability of root cohesion as an influence on shallow landslide susceptibility in the Oregon Coast Range. *Canadian Geotechnical Journal*, 38:995–1024.

Sorbino G, Sica C, Cascini L (2010) Susceptibility analysis of shallow landslides source areas using physically based models. *Natural Hazards*, 53:313–332.

Waldron LJ, Dakessian S, Nemson JA (1983) Shear resistance enhancement of 1.22-meter diameter soil cross sections by pine and alfalfa roots. *Soil Science Society of America Journal*, 47:9–14.

18 Applied Research and Future of Flood Monitoring in Indus River Basin

Sadiq Ibrahim Khan, Thomas E. Adams III, and Zachary L. Flamig

CONTENTS

18.1 INTRODUCTION

In recent years, extreme hydrometeorological hazards (e.g., floods and droughts) have caused major disasters in human-dominated regions as well as less-intensively managed natural eco-systems. Of all natural hazards, floods are the most impactful disasters that cause more fatalities and more property damage than any other type of weather-related hazard. In terms of occurrence, economic loss, and mortality, floods are the world's leading natural disaster (Smith, 1996; Guha-Sapir et al., 2012), and costs and loss of life are increasing (Brissette et al., 2003; Shepherd et al., 2011). Some of this change may be related to societal, population, economic factors, or combinations (Barredo, 2009; Bouwer, 2011). But as a warming climate system accelerates the water cycle, the likelihood of extreme events such as floods increases (Huntington, 2006; Trenberth et al., 2007; Dirmeyer, 2011; Andersen and Shepherd, 2013).

In South Asia, the summer monsoon not only brings enough rainfall to subdue scorching temperatures and rejuvenates vegetation, but also delivers intense storms, strong winds, and some of the highest lightning flash densities in the world (Albrecht et al., 2009; Christian et al., 2003). Every summer, these heavy downpours often produce severe floods and landslides that have devastating impacts on lives, livelihoods, and infrastructure. The 2010 extreme monsoon in Pakistan resulted in the worst flooding with destruction throughout the low-lying regions of the Indus River Valley that stretches from the Himalayas to the Arabian Sea. Anomalously heavy precipitation fell during late July and early August resulting in one of the worst hydrometeorological disasters in history, causing inconsolable human casualties and heavy economic destruction. By the end of August, 20% of Pakistan was submerged under the flood waters, there were close to 2000 deaths, and more than 40 billion dollars' worth of damage (WMO, 2011). This is not an isolated incident; Pakistan is one of the top five disaster-prone countries that account for nearly 80% of the total population exposed to recurring flood risks every years.

To achieve reliable hydrologic forecast instantaneous and seamless high spatio-temporal data and models are employed to estimate precipitation, soil moisture, evapo-transpiration, groundwater, snowmelt, and runoff. Improved hydrologic modeling, real-time data access, and engaging key stakeholders are the common denominators for flood forecasting and water resources management. Another important part of the equation is collaboration among different stakeholders to achieve the single goal of water extreme predictions. The utility of forecasts for hazard events can further be optimized by incorporating societal factors into the warning systems to immediately improve societal resilience to multihazards at regional and global scales.

The advanced early warning systems with close collaboration and inputs from local stakeholders and decision makers are critical to enhance preparedness and mitigation of water-related extremes and disasters. International collaboration through multidisciplinary research and education activities can strengthen linkages and help both government and nongovernmental agencies to formulate impact based disaster risk reduction strategies. This chapter is focused on an applied research project on flood monitoring and forecasting over Indus River Basin. The research and collaborative effort initiated knowledge sharing activities among stakeholders on water-related hazards through an advanced educational and training program. Specifically, Section 18.2 discusses flood modeling research over Indus River Basin. Section 18.3 is focused on capacity building in flood monitoring and last but not least Section 18.4 discussed some future directions to advance operational flood monitoring and prediction strategies in developing countries.

18.2   FLOOD MONITORING RESEARCH OVER INDUS BASINS

18.2.1   Geospatial Data for Flood Monitoring

Hydrologic modeling and geospatial data assimilation has provided new insights into high resolution rainfall patterns and runoff response. The near real-time availability of in situ and remote sensing data over vast regions, many of which are entirely ungauged, has spawned the capability for systematic rainfall monitoring and subsequent flood and drought monitoring. However, assimilation of remote sensing data into distributed hydrologic models is marred with uncertainties related to data estimates and numerical models. In this regard, two recent studies on flood monitoring have argued about the issue of hydrologic model recalibration by using satellite rainfall data (Nikolopoulos et al., 2013; Vergara et al., 2013). These studies showed that, although, the coarse rainfall resolution associated with some of the main satellite products (e.g., 3 h and 25 km) reduces the skill of the hydrologic model particularly at decreasing drainage areas and high values of streamflow, calibrating model parameters at rainfall resolutions representative to satellite products could significantly improve the hydrologic prediction skill of satellite-based rainfall estimates. Others examined the impact of satellite pixel resolution on hydrologic model calibration (Gourley et al., 2011; Maggioni et al., 2013). They also found that hydrologic model parameters were sensitive to pixel resolution, and it was recommended to use a reference rainfall dataset that corresponds to the resolution of the satellite rainfall if these data are to be used for flood forecasting purposes.

Geospatial data are uniquely suited to provide the timely and uniform information of the Earth's surface and atmosphere that are needed to evaluate flood hazards over large areas. For example, remote sensing-based precipitation products from ground-based weather radars and space radars can assist in hydrologic monitoring in ungauged watersheds. Among the key precipitation products that are used globally comes from the Tropical Rainfall Measurement Mission (TRMM) Precipitation Radar (PR), and both accuracy as well as spatiotemporal coverage of which is further improved under the latest Global Precipitation Measurement (GPM) program (http://pmm.nasa.gov) program.

Reliable precipitation estimates are critically important for flood modeling and hazard assessment. Therefore, research and training activities were conducted to evaluate the advanced version of satellite-based rainfall estimation. The latest TRMM (3B42V7) requires validation before it is used to study the rainfall diurnal variation over the study domain, therefore the spatial error structure of surface precipitation are systematically studied by comparing with rain gauge observations. The spatial and temporal characteristics of the Asian monsoon over with the climate prediction center morphing technique product and the latest TRMM Version 7 product (3B42RT) are quantitatively studied (Khan et al., 2014).

Observed daily precipitation data of 24 gauged stations was obtained from the Pakistan Meteorological Department (PMD) (Figure 18.1). The gauge network data recordings are subject to significant sources of error, both from the recording process and the instruments. The original data provided by local meteorological department, the PMD, were manually inputted. Therefore, first, the gauge observations were subjected to a quality control assessment through visual inspection and second, systematic evaluation to detect and correct any outliers in the data record. Daily rainfall accumulation time series for each gauge was plotted to screen out the corrupt data.

Some of the geomorphological data sets include NASA's Shuttle Radar and Topography Mission based *Hydrologic data and maps based on Shuttle Elevation Derivatives at multiple Scales* (HydroSHED) data that provides critical information of topography, watersheds, and river channel networks. Satellites such as Earth Observing-1 (EO-1) Mission (a 10 m panchromatic band plus 30 m

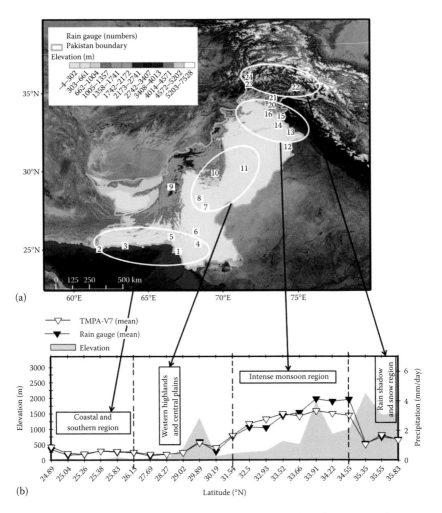

FIGURE 18.1 (a) Study domain with topography (m) and the rain gauge stations (numbered), used to compare the TRMM 3B42RT version 7 satellite precipitation product. Study regions based on precipitation variation over coastal, Western highlands, intense monsoon, and rain shadow region. (b) Gauge and TRMM 3B42RT mean precipitation (mm/day) variability over 24°N–36°N and 60°E–75°E.

multispectral bands), along with data from the Advanced Spaceborne Thermal Emission and Reflection Radiometer (ASTER) instrument onboard Terra and other commercial satellites such as GeoEye, provides important land surface imagery used to evaluate floods/landslide events around the world.

This Indus river flood monitoring project facilitated practical applications of geospatial data products and models to advance flood monitoring system. The applied research activities that included flood monitoring using satellite data and cross-validation using ground-based rain gauges and streamflow stations along the Indus River, Pakistan. The optical imagery from the Moderate Resolution Imaging

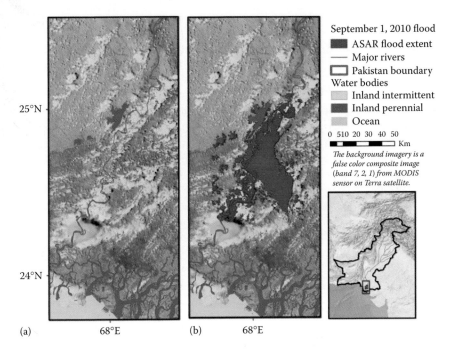

FIGURE 18.2   (a) MODIS false color composite (bands 7, 2, 1) for September 1, 2010, and (b) lower Indus river and delta flood inundation delineated from the ASAR image.

Spectroradiometer (MODIS) was employed to delineate the extent of the 2010 flood along the Indus River, Pakistan. Moreover, the all-weather all-time capability of higher resolution imagery from the advanced synthetic aperture radar (ASAR) is used to monitor flooding in the lower Indus river basin (Figure 18.2). The project utilized a significant amount of NASA's Earth Science results, including remote sensing data and CREST models for evaluation, of particular importance will be the multi-spectral sensors (MODIS, ASTER, etc.) based flood maps produced.

The research work comprised the following main datasets to calibrate the hydrologic model as shown in Table 18.1.

In addition to satellite imagery-based flood inundation, a proxy for river discharge from the Advanced Microwave Scanning Radiometer (AMSR-E) aboard NASA's Aqua satellite and rainfall estimates from the TRMM are used to study streamflow time series and precipitation patterns. The AMSR-E detected water surface signal was cross-validated with ground-based river discharge observations at multiple streamflow stations along the main Indus River. A high correlation was found as indicated by a Pearson correlation coefficient of above 0.8 for the discharge gauge stations located in the southwest of Indus River basin. In addition to remote sensing products, access and collection of in situ observations are keys to success of this project. Six years (January 2005–December 2010) of rainfall observations from the rain gauge stations (Figure 18.3) were compared with the satellite precipitation product to characterize the spatiotemporal error characteristics.

TABLE 18.1

Summary of the Hydrologic Model and Remote Sensing Data Used for the Project

Type	Period	Resolution	Description	Available From
		Geospatial Datasets for Hydrologic Modeling		
TRMM	1997–	3 h/0.25°	Gridded rainfall	NASA–GSFC
GPM Day-1	2011.7–	3 h/0.1°	Global precipitation	NASA–GSFC
HydroSHED DEM	2002–	30 m–1 km	Topography/Watershed	USGS
ASTER/MODIS	2000–	1 day/30 m–1 km	Flood inundation maps	NASA–GSFC
EO-1/GeoEye	2009–	10–30 m	Landcover/inundation	NASA/USGS
AMSR-E/SMAP	2002–	3 h–1 day/9–25 km	Gridded soil moisture	NSIDC/future mission
		Distributed Hydrologic Model		
CREST model	2009–	A gridded/cell-to-cell routing with coupled runoff generation module.		

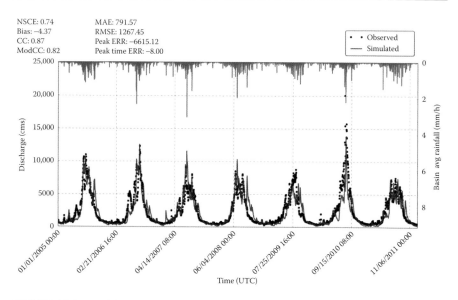

FIGURE 18.3   EF5/CREST calibration at Tarbela, with the Nash–Sutcliffe efficiency coefficient model, Pearson correlation coefficient, bias and other model evaluation statistics on top left corner.

18.2.2   FLOOD MODELING RESEARCH AND DEVELOPMENT

Numerical models have been used for operational flood forecasting since the development of the first watershed hydrologic model—Stanford Watershed model in 1966 (Crawford and Linsley, 1966). Hydrologic models have evolved from lumped-process models (Williams and Hann, 1978) to semidistributed model (Beven and Kirkby, 1979; Zhao et al., 1980) and to fully distributed models (Abbott et al., 1986a, 1986b; Wigmosta et al., 1994; Wang et al., 2011). However, studies on dynamically coupling hydrologic processes predicted by distributed hydrologic models with soil physics and mechanics determining slope stability are still in a very early stage (Bogaard and Greco, 2014) due to lack of knowledge of interactions between these processes and difference in the spatiotemporal scales of the flood events.

In the context of flood monitoring, operational agencies in emerging countries should focus on the optimal spatiotemporal resolution for operational decision-support system (DSS). In this section, we introduce some of the efforts to achieve this objective by implementing a distributed hydrologic modeling framework, for example the ensemble framework for flash flood forecasting (EF5) that implement the *Coupled Routing and Excess Storage* (CREST) distributed hydrologic model (Wang et al., 2011; Xue et al., 2013). The CREST model developed by the OU HyDROS Lab (http://hydro.ou.edu) together with collaborators at NASA Marshall and Goddard centers has been used to predict streamflow, soil moisture, and evapotranspiration using input from gridded meteorological forcing fields.

This distributed hydrologic model has been described by authors in some of the other chapter of this book. In this chapter, the same CREST hydrologic model is discussed. The datasets required to set up the model (i.e., digital elevation model, soil types, land use, evapotranspiration, etc.) are all derived from near real-time remote sensing data or archival data in public domains. Detailed description on the CREST model structure and training material can be found at http://hydro.ou.edu/research/crest/. The hydrologic model was tested for five stations namely Tarbela (Figure 18.3), Chashma, Kalabagh, Taunsa, and Guddu over the Indus River Basin.

18.2.3 COUPLING ATMOSPHERIC FORECASTS AND HYDROLOGIC MODELING

Hydrometeorological hazards are multifaceted in that they are not only an atmospheric phenomenon, but are also influenced by static and dynamic hydrologic conditions (e.g., soil types and their volumetric water contents) and the intersection of the natural hazard with the locations of infrastructure and modifications by city planning and societal values and behaviors. Floods are caused by slow-moving or large-areal-extent convective storms producing extreme precipitation over a specific area for a long duration. Therefore, improved accuracy, higher spatial, and temporal resolution of quantitative precipitation estimation (QPE) and quantitative precipitation forecast (QPF) can provide the greatest potential to benefit the hydrologic community because rainfall is the primary factor affecting the timing and stage of river discharge, which is critical for flood prediction. The impacts of improved QPE and QPF can only be realized through close coupling between the atmospheric and land-surface model components, data assimilation, and the development of combined ensemble forecasting systems to provide uncertainty information for early warning and decision making.

Advances in sensor technology and computers are opening the door for models to be coupled dynamically, namely, atmospheric and hydrologic models, where the terrestrial movement of water impacts the heat fluxes and mass balances. For a holistic description of the system, this two-way coupled model system should be extended to include soil moisture models, urban systems, and agricultural and natural vegetative growth models. In this context, applied research activities were conducted to couple global forecast system model with distributed hydrologic model for high-impact hazard prediction such as large scale floods.

18.3   CAPACITY BUILDING IN FLOOD MONITORING

The local government departments in Pakistan that are mandated to provide critical information on these hazards lack funds, expertise, real-time data, and management capacity to plan and implement well targeted disaster risk reduction strategies. The primary goal of the project discussed in this chapter is to disseminate the research related activities that are carried out under this project and future projects for the Pakistan–US Science and Technology Cooperation Program. Applied research activities were conducted that included flood monitoring using satellite data and cross-validation using ground-based rain gauges and streamflow stations along the Indus River, Pakistan. The archived river flow and rainfall data was compiled, cleaned, and re-organized. Gaps in data were identified. Moreover, linkages with the Pakistan Meteorological Department (PMD) and the Federal Flood Commission were developed. The two organizations were requested to provide the gauge-based daily rainfall and river flow data. Some of the research work comprised the main activities already discussed earlier in Sections 18.2.2 and 18.2.3 of this chapter.

On capacity building front, two training workshops titled "HydroMeteorological Modeling and Simulation Using GIS and RS Technologies" were conducted at the Institute of Geographical Information Systems (IGIS) at National University of Science and Technology (NUST) (Figure 18.4). University faculty member completed the proposed training at the National Weather Center at University of Oklahoma (OU). Weekly seminar and lectures provided ample opportunities to collaborate with other researchers on environmental observation and modeling research. Moreover, some of the specific activities are listed below:

- Geospatial processing of in situ and remote sensing data for hydrologic model.
- Compilation of CREST model using the control files, data files, parameter files.
- Making adjustments in hydrologic model to output selected parameters.
- Interpretation of CREST model outputs.

A functional website is online (http://flash.ou.edu/pakistan/) and will provide up to date information on current flood situation and future flood forecast over Pakistan.

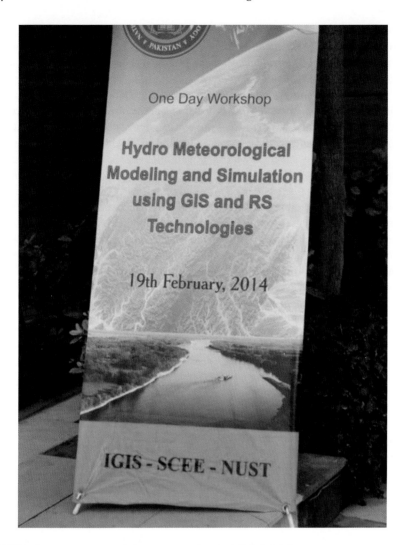

FIGURE 18.4   One day workshop organized at the IGIS, NUST.

18.4   FUTURE IMPROVEMENTS IN HYDROMETEOROLOGICAL FORECAST AND WARNING SYSTEM

The most significant problems facing any research to operations endeavor is the lack of a full understanding of important operational issues by the researchers and the resistance to or lack of commitment to change by the operational staff. Without a thorough understanding of operational issues by researchers and the commitment to change within the operational environment, any research to operations program will fail. To succeed the program must involve full cooperation and equal partnership between the research and operational bodies. Too often researchers do not understand the importance of operational details and constraints, while those in operations may not fully understand the need for scientific advancement; change can be threatening and operational concerns can be wrongly viewed as naive and passé. The words *research to operations* has become a convenient catch-phrase that is aimed at capturing the process of modernizing an operational enterprise with the infusion of modern or up-to-date scientific models or methodologies, while preserving operational integrity and proven best practices. One does not want to throw the baby out with the bath water. The development of, successful testing, evaluation, and implementation of improved models and methodologies is critical to making significant improvements to hydrometeorological forecast and warning systems. But, the old dictum of *garbage in, garbage out* still rings true. That is, reliable, un-biased, quality controlled observational data must be used as model forcings. Consequently, the foundation of sound forecast and warning systems is the observational data network.

Experience has shown in many different national and regional settings, that properly deployed advanced hydrometeorological forecast and warning systems can succeed provided the fundamental observational network, telecommunications system, and data acquisition and processing systems infrastructure has been implemented. Moreover, the overall success of these systems depends on long-term commitment to routine system maintenance and meaningful training of operational staff. The latter is a key element of national capacity building that assures sustainable hydrometeorological forecast and warning operations.

18.4.1 HYDROMETEOROLOGICAL OBSERVATION STATIONS AND WEATHER RADAR

Hydrometeorological observation stations or automated weather stations (AWS) form the backbone of hydrologic data collection. It is recommended that AWS specifications follow those currently used and recommended by the FFD and PID, which is attractive because of their familiarity with the maintenance of these data collection platforms. Many AWS locations in the northern mountainous regions are needed and that is where the bulk of the new stations should be located based on guidance from the FFD and PID. What is critical is that these stations are needed for bias correction of satellite-based and radar remotely sensed precipitation estimates. Also, the gauge-only estimates serve as a backup when individual radars may go off-line during periods with data transmission outages or if the radar itself is shutdown for maintenance or technical problems. An often overlooked aspect of observational networks is the need for a long-term commitment to the maintenance of the observation platforms and data transmission network. It is better to have a well-maintained, less dense network that is reliable than a denser unreliable observation network.

From a hydrologic perspective, radar-based QPE is a fundamental remote sensing tool to provide hydrologic model forcings for the detection and prediction of floods and flash floods. Table 18.2* lists characteristics of commonly used radars

---

* From WMO, Joint meeting of CBS expert team on surface-based remotely-sensed observations, Geneva, Switzerland, November 23–27, 2009—Ercan Buyukbas, Turkish State Meteorological Service (TSMS), assess the current and potential capabilities of weather radars for the use in wmo integrated global observing system (WIGOS).

TABLE 18.2
Comparison of Three Types of Weather Radars Most Commonly Used in European Union Countries

Characteristic	S-Band Radars	C-Band Radars	X-Band Radars
Frequency	2–4 GHz (2.9 GHz)	4–8 GHz (5.6 GHz)	8–12 GHz (9.3 GHz)
Wave length	15–7.5 cm (10.3 cm)	7.5–3.8 cm (5.3 cm)	3.8–2.5 cm (3.2 cm)
Typical range	300–500 km	120–240 km	50–100 km
Peak power	500 kW–1 MW (750 kW)	250–500 kW (250 kW)	50–200 kW (200 kW)
Measuring sensitivity	Rain, snow, hail (The bigger particles as compared to C band)	Rain, snow, hail, drizzle (The bigger particles as compared to X-band)	Rain, snow, hail, light drizzle (The smaller particles as compared to S-band and C-band)
Atmospheric attenuation	Less attenuation as compared to C-band and X-band	Less attenuation as compared to X-band while much attenuation as compared to S-band	Much attenuation as compared to C-band and S-band
Antenna size	7.5 m	4.2 m	2.5 m
Cost	1.5× C-band, 2× X-band	0.7× S-band, 1.3× X-band	0.5× S-band, 0.8× C-band

Source: World Meteorological Organization (WMO), Joint meeting of CBS expert team on surface-based remotely-sensed observations, Geneva, Switzerland, November 23–27, 2009.

by countries in the European Union (EU). While S-band Doppler radars are more expensive, they are widely used in EU member Nations and in the United States by the NOAA/National Weather Service, which has nearly 20 years of operational experience using S-band Doppler weather radar systems. For this reason, the use of S-band Doppler radar systems is recommended in Pakistan. Based on local discussions with PMD staff, it is clear that the two existing systems must be upgraded and additional radar systems must be added to provide coverage for the economically critical Punjab region and upstream headwater areas. However, enough additional radars must be deployed to provide full coverage for Pakistan.

Research results presented in scientific literature shows unequivocally that even well-calibrated radars produce large uncertainties in the estimation of rainfall rates due to both natural atmospheric effects and human sources that lead to estimation bias. In order for any hydrologic model to properly simulate streamflow and river stages with accuracy, rainfall measurements used as hydrologic model inputs must be optimal, unbiased estimates of actual precipitation. Errors in precipitation estimation, in turn, contribute to large errors in hydrologic forecasting. Best practices in precipitation estimation utilize gridded precipitation estimation from radars, which are adjusted by rain gauge estimates from the precipitation gauge network, and, when necessary, are supplemented by satellite precipitation estimates. The combined multisensor precipitation field measurements can and should be integrated into a national precipitation grid for the country at various temporal scales. The recommended strategy would be the generation of 1-h precipitation grids, which can be aggregated to longer duration grids as needed. The national precipitation grid products are needed to significantly advance hydrometeorological forecasting and warning services in Pakistan beyond current capabilities.

It is recommended that real-time national precipitation grid estimates be produced with sufficient temporal and spatial resolution to be suitable as inputs to hydrologic models used in Pakistan as well as meeting the precipitation data needs for the establishment of a flash flood guidance system. The first phase of this task must include an international survey to identify the characteristics and suitability of available radar precipitation processing systems currently used operationally for potential use in a modernize Pakistan Hydrometeorological forecasting system. Excellent precipitation processing systems are in place in the United States. There are two such precipitation processing systems developed and used by the United States. National Oceanic and Atmospheric Administration (NOAA) National Weather Service (NWS) that utilizes radar, rain gauge precipitation data, and satellite estimates to produce rainfall estimates at a 4 km resolution grid at 1-h intervals for the United States. The *Multisensor Precipitation Estimator (MPE)* application runs within the *Advanced Weather Interactive Processing System (AWIPS, version 2)* and provides direct input to the *Community Hydrologic Prediction System* that was adapted from the Deltares (The Netherlands), *Flood Early Warning System (FEWS)*. The support contractor has proposed that FEWS be established as a demonstration system used by VHMS to build the full operational hydrometeorological forecasting system for the country.

The other precipitation processing system developed and used by the NOAA/NWS is the multiradar/multisensor (MRMS) system, which is run at the National Centers for Environmental Prediction. The MRMS system utilizes automated algorithms that quickly integrate data streams from multiple radars, surface and upper air observations, lightning detection systems, and satellite and forecast models. Numerous two-dimensional multiple-sensor products offer assistance for hail, wind, tornado, QPE, convection, icing, and turbulence diagnosis. NWS river forecast centers have the capability to ingest MRMS gridded data into MPE within AWIPS to further enhance precipitation estimates. The benefit of utilizing the combined MRMS and AWIPS-2 based MPE is that little cost would be needed to acquire the software. If the MPE/MRMS system were used in Vietnam, there would likely be some software modification required by the contractor who is assigned this task to customize existing core precipitation processing software to the VHMS system. Adoption of the combined US NOAA/NWS MPE/MRMS precipitation processing system in Pakistan for operational hydrometeorological forecasting, would carry with it significant expertise with the system from the United States that would be available as a user community to provide guidance in problem solving and support during periods of system enhancements. Essential to robust precipitation processing is the real-time data from the hydrometeorological network of observations stations, which is used by the bias correction algorithms.

The overall precipitation processing system would take the form shown in Figure 18.5, including raw data collection, transmission, data quality control, gridded QPE field generation, and bias removal. The final step is to use the data as

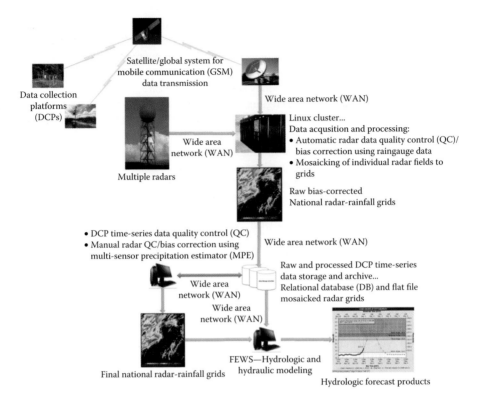

FIGURE 18.5 Precipitation processing data flow from multiple radars and hydrometeorological data collection platforms (DCPs) through automated and manual quality control steps and bias correction for use as the primary hydrologic model forcing. Data flow for observed hydrometeorological data.

the primary hydrometeorological model forcing with the production of hydrologic forecast products. An essential feature of the precipitation processing system is the archival of both observation station time series data in a relational database and storage of both the raw and final, processed mosaicked radar-rainfall grids. These data are needed for research, hydrologic model calibration, and future enhancements of the precipitation processing algorithms.

18.4.2 FLOOD EARLY WARNING SYSTEM

The development of a robust flood forecasting framework is critical; it is the focus of the entire system and the glue that holds it together. The flood early warning system (FEWS) has been implemented in a rough form previously in Pakistan, but it has fallen into disuse by the FFD due to insufficient training and institutional support. Consequently, an entirely re-engineered FEWS configuration is needed for the FFD to work in a nodal, remote server configuration. The remote server arrangement depicts Delft-FEWS central server locations in Islamabad and Lahore, which are synchronized as backups, with a representative remote node in Karachi. Obviously, many other remote Delft-FEWS forecast nodes are possible and would be recommended. This configuration has many advantages, including the following.

1. Facilitates maintenance efficiency with centralization
2. Cost savings with only two redundant (backup) systems in Islamabad and Lahore
3. Promotes cooperation and collaboration
4. Makes FEWS configuration sharing much easier than having locally developed and controlled configurations
5. More easily scalable with centrally located servers

18.4.3 HYDRAULIC AND HYDROLOGIC MODELS

Due to its current use by the PMD, it is recommended that the FEWS-based forecast system, developed by Deltares (The Netherlands) serve as the basis as the hydrologic and hydraulic modeling framework for by PMD/FFD and PID in a physically distributed nodal configuration. This is the same FEWS implementation strategy used the UK Environment Agency and the Australian Bureau of Meteorology. The FEWS is becoming the international standard as a hydrologic forecasting framework, with use in major countries, such as the United States, U.K., Australia, Germany, The Netherlands, and others.

PMD has the use of the Sacramento Soil Moisture Accounting (SAC-SMA) model defined in their current FEWS configuration; continued use of the SAC-SMA is possible and may be desirable. However, by moving to a distributed hydrologic model, such as WRF-Hydro/SAC-HTET, much greater modeling flexibility is possible, including significant support for flash flood modeling and forecasting, and quicker model implementation. So, the use of WRF-Hydro/SAC-HTET is desirable and is recommended.

The SOBEK hydraulic model is implemented for use in PMD's current FEWS configuration; a possible alternative is the US Army Corps of Engineers (USACE) 1D/2D (with version 5.0) unsteady flow HEC-RAS, which is freely available for use in FEWS from the NOAA/NWS. Snow modeling is necessary, especially for water resources planning uses to provide reservoir inflow forecasting for downstream irrigation applications; SNOW-17, developed by NOAA/NWS is also available for use in FEWS. When combined with reservoir modeling using NOAA/NWS and USACE RES-J, RES-SNGL, or RES-SIM models, complete water resources needs for Pakistan focused on irrigation applications can be met. By utilizing numerical weather prediction (NWP) model-based ensemble modeling, it will be possible to utilize the FEWS-based modeling system to provide both short and long lead-time probabilistic hydrologic forecasts to meet risk-based flood forecasting needs and water resources forecasting needs for irrigation. Moreover, the utilization of regional climate models will make seasonal and annual outlooks feasible.

The major components either in use or proposed are as follows:

1. NOAA/NWS Research Distributed Hydrologic Model (RDHM)—SAC-HTET.
2. SOBEK (or HEC-RAS) hydraulic routing model.
3. Reservoir model—RES-J, RES-SNGL, or RES-SIM.
4. Snow model—SNOW-17.

In summary, due to their highly successful use by NOAA/NWS river forecast centers in the United States within the FEWS framework and availability, an appropriate reservoir simulation model (RES-J, RES-SNGL, or RES-SIM) and SNOW-17 snow simulation model are recommended. The SOBEK model has been used by FFD and is currently available for use within their FEWS configuration. It is recommended that the recently developed RDHM/SAC-HTET model be adapted for use within Delft-FEWS. Use of the USACE HEC-RAS model within FEWS is also possible as an alternative to use of SOBEK.

18.4.4 HYDROMETEOROLOGICAL DATABASE SYSTEMS AND SOFTWARE SYSTEM

Development and implementation of an open source relational database system, such as PostgreSQL with independent databases for incoming raw, nonquality controlled data, a live database for quality controlled data, and a long-term archive database. PostgreSQL is used operationally by all NOAA/NWS weather forecast offices and river forecast centers within the AWIPS and has proven to be very robust operationally. A major advantage with the use of PostgreSQL is that it is free, open source, software. Online help is readily available and a very large technical user community exists, while commercial support and training can be secured as well.

Development of the capability to allow users of the FEWS-based hydrologic modeling system to visualize PMD weather and climate forecasts, including rain gauge, radar-precipitation, temperature, NWP model output, and other hydrometeorological datasets in real-time using the telecommunications infrastructure is essential. Such a system is critical for monitoring data quality and for use as a tool for situational awareness, especially during periods of rapidly changing weather, such as for flash flooding events.

As discussed above, a critical element of real-time hydrometeorological activities to support flood forecasting and water resources is real-time precipitation estimation based on rain gauge and radar sensors and satellite-based remote sensing. Extensive research has shown that radar and satellite precipitation estimates must be bias corrected by rain gauges. Moreover, to support real-time operations, the rain gauge, radar, and satellite estimates must be mosaicked seamlessly into a single multisensor precipitation field on an hourly basis, covering the entire hydrologic domain of interest to Pakistan, including areas outside the country because of the upstream river systems that drain into Pakistan. Operationally, both automatic and manual processes are necessary to produce optimal multisensor precipitation estimates.

18.5 INTEGRATED WATER RESOURCES PLANNING AND MANAGEMENT

Fundamental to the economy of Pakistan and its livelihood is the agricultural sector, which relies heavily on an extensive irrigation system that depends both on spring-time snowmelt from the mountainous north and monsoonal rains. Reservoirs to the north capture runoff to the Indus River and major tributaries and are closely managed to provide water for irrigation uses in fertile agricultural regions in the Punjab and Sindh provinces in the south during the summer dry season. The reservoirs also serve a flood control function. Barrages control flow throughout the complex system of major and more minor irrigation canals, which feed irrigation ditches and fields. An extensive system of levees protects villages and cities in the central and southern agricultural areas. This means that Pakistan must balance, sometimes conflicting, water resources demands. Managing the complexity of flood control, irrigation, and water supply needs strongly suggests the necessity of a comprehensive, integrated water resources planning, and management strategy that utilizes a flood and streamflow forecasting system that is closely tied to an integrated DSS. The DSS should be designed to accommodate water resources applications to meet water supply and irrigation requirements as well as having flood forecasting capabilities. The hydrometeorological inputs to the modeling systems should be seamlessly integrated to span the spatial and temporal scales needed for a comprehensive water resources planning and management system. To successfully meet this goal the PMD, FFD, irrigation departments, and central government bodies that regulate reservoir releases must work together utilizing a single integrated system that combines all modeling components, including snowmelt simulation, rainfall–runoff modeling, reservoir simulation, hydraulic channel routing, and irrigation water distribution.

REFERENCES

Abbott, M. B., Bathurst, J. C., Cunge, J. A., O'Connell, P. E., and Rasmussen, J. (1986a). An introduction to the European Hydrological System—Systeme Hydrologique Europeen, "SHE", 1: History and philosophy of a physically-based, distributed modelling system. *Journal of Hydrology,* 87(1–2), 45–59.

Abbott, M. B., Bathurst, J. C., Cunge, J. A., O'Connell, P. E., and Rasmussen, J. (1986b). An introduction to the European Hydrological System—Systeme Hydrologique Europeen, "SHE", 2: Structure of a physically-based, distributed modelling system. *Journal of Hydrology,* 87(1–2), 61–77.

Albrecht, R., Goodman, S., Buechler, D., and Chronis, T. (2009). Tropical frequency and distribution of lightning based on 10 years of observations from space by the Lightning Imaging Sensor (LIS). In *Fourth Conference on Meteorological Applications of Lightning Data* (preprint), pp. 2–12.

Andersen, T. K. and Shepherd, J. M. (2013). Floods in a changing climate. *Geography Compass,* 7(2), 95–115.

Barredo, J. (2009). Normalised flood losses in Europe: 1970–2006. *Natural Hazards and Earth System Sciences,* 9(1), 97–104.

Beven, K. J. and Kirkby, M. J. (1979). A physically based, variable contributing area model of basin hydrology/Un modèle à base physique de zone d'appel variable de l'hydrologie du bassin versant. *Hydrological Sciences Bulletin,* 24(1), 43–69. doi: 10.1080/02626667909491834.

Bogaard, T. and Greco, R. (2014). Preface Hillslope hydrological modelling for landslides prediction. *Hydrology and Earth System Sciences,* 18(10), 4185–4188. doi: 10.5194/hess-18-4185-2014.

Brissette, F. P., Leconte, R., Marche, C., and Rousselle, J. (2003). Historical evolution of flooding damage on a USA/Quebec river basin1. *Journal of the American Water Resources Association,* 39(6), 1385–1396.

Christian, H. J. et al. (2003). Global frequency and distribution of lightning as observed from space by the Optical Transient Detector. *Journal of Geophysical Research,* 108(D1), 4005.

Crawford, N. H. and Linsley, R. S. (1966). Digital simulation in hydrology: The Stanford Watershed Model IV. Technical Report No. 39. Palo Alto, CA: Department of Civil Engineering, Stanford University.

Dirmeyer, P. A. (2011). The terrestrial segment of soil moisture–climate coupling. *Geophysical Research Letters,* 38(16), L16702.

Gourley, J. J., Hong, Y., Flamig, Z. L., Wang, J., Vergara, H., and Anagnostou, E. N. (2011). Hydrologic evaluation of rainfall estimates from radar, satellite, gauge, and combinations on Ft. Cobb basin, Oklahoma. *Journal of Hydrometeorology,* 12(5), 973–988. doi: 10.1175/2011jhm1287.1.

Guha-Sapir, D. and Hoyois, P. (2012). *Measuring the Human and Economic Impact of Disasters.* Brussels, Belgium: Centre for Research on the Epidemiology of Disasters (CRED).

Khan, S. I., Hong, Y., Gourley, J. J., Khattak, M. U. K., Yong, B., and Vergara, H. J. (2014). Evaluation of three high-resolution satellite precipitation estimates: Potential for monsoon monitoring over Pakistan. *Advances in Space Research,* 54(4), 670–684. doi: http://dx.doi.org/10.1016/j.asr.2014.04.017.

Maggioni, V., Vergara, H. J., Anagnostou, E. N., Gourley, J. J., Hong, Y., and Stampoulis, D. (2013). Investigating the applicability of error correction ensembles of satellite rainfall products in river flow simulations. *Journal of Hydrometeorology,* 14(4), 1194–1211. doi: 10.1175/jhm-d-12-074.1.

Nikolopoulos, E. I., Anagnostou, E. N., and Borga, M. (2013). Using high-resolution satellite rainfall products to simulate a major flash flood event in northern Italy. *Journal of Hydrometeorology,* 14(1), 171–185.

Shepherd, M., Mote, T., Dowd, J., Roden, M., Knox, P., McCutcheon, S. C., and Nelson, S. E. (2011). An overview of synoptic and mesoscale factors contributing to the disastrous Atlanta flood of 2009. *Bulletin of the American Meteorological Society,* 92(7), 861–870.

Smith, K. (1996). Natural disasters: Definitions, databases and dilemmas. *Geography Review,* 10, 9–12.

Trenberth, K. E. and Dai, A. (2007). Effects of Mount Pinatubo volcanic eruption on the hydrological cycle as an analog of geoengineering. *Geophysical Research Letters,* 34(15), L15702.

Vergara, H., Hong, Y., Gourley, J. J., Anagnostou, E. N., Maggioni, V., Stampoulis, D., and Kirstetter, P. E. (2013). Effects of resolution of satellite-based rainfall estimates on hydrologic modeling skill at different scales. *Journal of Hydrometeorology,* 15(2), 593–613. doi: 10.1175/jhm-d-12-0113.1.

Wang, J. et al. (2011). The coupled routing and excess storage (CREST) distributed hydrological model. *Hydrological Sciences Journal,* 56(1), 84–98. doi: 10.1080/02626667.2010.543087.

Wigmosta, M. S., Vail, L. W., and Lettenmaier, D. P. (1994). A distributed hydrology-vegetation model for complex terrain. *Water Resources Research,* 30(6), 1665–1679. doi: 10.1029/94WR00436.

Williams, J. R. and Hann., R. W. (1978). Optimal operation of large agricultural watersheds with water quality constraints (pp. Tech. Rept. No. 96.).

World Meteorological Organization (WMO). (2009). Joint meeting of CBS expert team on surface-based remotely-sensed observations, Geneva, Switzerland, November 23–27.

Xue, X. et al. (2013). Statistical and hydrological evaluation of TRMM-based Multi-satellite Precipitation Analysis over the Wangchu Basin of Bhutan: Are the latest satellite precipitation products 3B42V7 ready for use in ungauged basins? *Journal of Hydrology,* 499, 91–99. doi: 10.1016/j.jhydrol.2013.06.042.

Zhao, R. J., Zhuang, Y. G., Fang, L. R., Liu, X. R., and Zhang, Q. S. (1980). *The Xinanjiang Model.* Paper presented at the Hydrological Forecasting Proceedings Oxford Symposium, Oxford University, Oxford.

19 Investigating Satellite-Based Observations to Improve Societal Resilience to Hydrometeorological Hazard in Colombia

Humberto Vergara, Maria Jurado, Paola Gonzalez, Yang Hong, Kenneth Ochoa, and Maritza Paez

CONTENTS

19.1 INTRODUCTION

The geographical location of Colombia makes its territory particularly prone to hydrometeorological hazard. The country is located on the northwest of South America within the Intertropical Convergence Zone (ITCZ). Moreover, Colombia has a complex topography characterized by three mountain ranges and two valleys that are part of the Andes, and lowlands to the north over the Caribbean coast and to the southeast comprised mainly of the plains on the eastern region and the Amazon rainforest. These characteristics of the terrain and the hydroclimatic regime of the region make the population of Colombia subject to impacts from drought, landslides, floods, and flash floods.

Although the country has been recognized for its leadership being a pioneer in risk management with a holistic vision, the early warning systems in Colombia have been noted to have weaknesses: (1) the lack of consistency in the information; (2) a relatively limited historical database (30–40 years); and (3) more importantly, there is insufficient information over more than 40% of the territory (particularly over Orinoquia and Amazon regions) (Cardona, 2007; Mundial, 2012). Therefore, the necessity to explore remote sensing platforms as an alternative to the spatial coverage gap of observational systems becomes clear for this region. Current satellite-based precipitation estimation techniques represent an important advantage for hydro-meteorological applications, especially in sparsely or ungauged regions. The National Aeronautics and Space Administration (NASA) has taken a leading role in research for the development of natural hazards warning systems based on information from satellites such as the NASA Marshall Goddard Space Flight Center internet-based Regional Monitoring and Visualization System (SERVIR) in Mesoamerica (www.servir.net) and Global Flood Monitoring system (http://trmm.gsfc.nasa.gov/) at relatively coarse resolution ($0.25°$ and 3-h).

Following the successful experiences of the aforementioned systems, El Bosque University in Colombia (http://www.uelbosque.edu.co) and the Hydrometeorology and Remote Sensing Laboratory of the University of Oklahoma in the United States of America (http://hydro.ou.edu) have started a collaborative effort to promote the investigation of the utility of satellite-based observations for hydrometeorological applications over Colombia, particularly for disaster mitigation. The development of the research is centered on the design and implementation of a real-time integrated system for the monitoring and prediction of environmental phenomena. The project started with a pilot study on flood prediction, which is presented in the remaining of this chapter.

19.2   A PILOT STUDY ON FLOOD WARNING

Flooding is considered the geophysical hazard with the most devastating effects on society causing more than 20,000 deaths and adversely affecting about 140 million people per year over the globe (Adhikari et al., 2010). In Colombia, flooding causes significant damages to people in different regions of the country every year. Between 1986 and 2008, 10/11 natural disasters with a significant number of affected people (i.e., more than 180,000) in Colombia were flooding events (World Bank, 2010). Therefore, the development of methodologies to mitigate the impacts from this hazard is of critical importance.

Early warning systems have become a standard nonstructural method for flood mitigation due to the advances in information and communication technology, the availability of new hydrometeorological observations through remote sensing platforms and hydrometeorological sciences (Bedient et al., 2008). The present pilot study describes the investigation of a satellite-based hydrologic modeling system for the detection of flooding in the Bogota River basin, where the capital of Colombia is located.

19.2.1 THE BOGOTA RIVER BASIN

The Bogota River basin is located on the eastern Andean mountain range region in central Colombia with a drainage area of approximately 5671 km². Forty six municipalities are located along the river, which use the water resource for economic activities (CAR, 2006). Elevation in the basin ranges between 280 and 3300 m.a.s.l (Figure 19.1). The complex topography of the basin defines a high variability in geomorphology (e.g., 70% of the basin has slopes in the 12%–50% range) and hydroclimatic regimes (400–2200 mm of annual precipitation and average temperatures in the 9°C–15°C range). Other factors such as the movement of the ITCZ and the Niño and Niña phenomena influence the precipitation dynamics of the area. Adequate representation of the spatial and temporal distribution of rainfall in this region is, therefore, an important consideration for hydrometeorological applications.

Panel (b) of Figure 19.2 presents mean monthly series of rainfall. It can be seen that the rainfall regime over the basin is highly variable due to the complex topography. The majority of gauges show the characteristic bimodal rainfall regime of the region. However, the gauges over the far northwestern portion of the basin display series indicating a weak bimodal regime almost indicating a single rainy season.

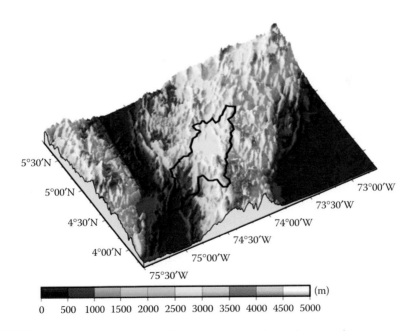

FIGURE 19.1 A map of the Bogota River basin boundaries and topography.

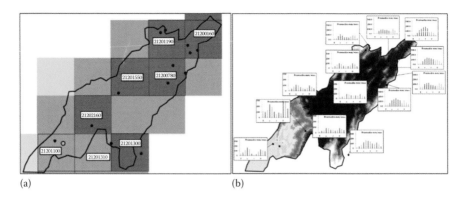

(a)   (b)

FIGURE 19.2 Rain gauge network over the Bogota River basin: (a) 3B42RT pixels covering the basin. The eight-digit numbers indicate the gauge ID associated to each pixel. The large green dot represents the location of the stream gauge; (b) monthly rainfall regime over the basin.

19.2.2 HYDROLOGIC MODELING WITH SATELLITE-BASED RAINFALL ESTIMATES

19.2.2.1 NASA's 3B42RT Product

Rainfall estimates are the main forcing of hydrologic models. NASA's Precipitation Measurement Missions (http://pmm.nasa.gov) offer a way to estimate rainfall using remote sensing platforms from space. In this exercise, the near-real-time infrared and microwave merged precipitation product 3B42RT (version 7) from NASA's Tropical Rainfall Measurement Mission (TRMM) was employed. 3B42RT estimates result from the combination of passive microwave low earth orbit precipitation estimates and infrared precipitation products from geostationary satellites. Estimates are available at the spatial resolution of 25-km and 3-h time intervals (Huffman et al., 2007). NASA's Global Precipitation Measurement (GPM) mission promises new possibilities for flood forecasting with the advent of precipitation products with higher quality and better spatiotemporal resolution (down to 30-min and 10-km). TRMM precipitation products have been heavily used to prototype future GPM-era, high-resolution flood prediction models.

3B42RT data from a 12-year period (2001–2012) were collected for the hydrologic modeling of the Bogota River basin. As a first step, an evaluation of the satellite-based estimates was performed using a rain gage network managed by the hydrology and meteorology authority of Colombia *Instituto de Hidrología, Meteorología y Estudios Ambientales* (http://www.ideam.gov.co). The rain gauges are well distributed across the basin (Figure 19.2), although only eight pixels out of the 3B42RT's grid covering the basin (approximately 15 pixels) have a corresponding rain gauge available for comparison purposes. The locations of the rain gauges span a wide range of elevations (Table 19.1). Satellite-based estimates were evaluated at the annual, monthly and daily scales in this work.

19.2.2.2 Hydrologic Model

The Coupled Routing and Excess Storage (CREST; Wang et al., 2011) distributed rainfall-runoff model was configured with an implementation of the kinematic wave

TABLE 19.1
Geographical Information of the Hydrometeorological Gauge Network Over the Bogota River Basin

Gauge Name	Longitude	Latitude	Elevation (m a.s.l)	Municipality
Australia	74°07 W	04°23 N	3050	Bogotá
Guanquica	73°56 W	05°11 N	2950	Tausa
Hidroparaiso	74°24 W	04°34 N	1600	El Colegio
Lagunitas	73°54 W	05°12 N	3100	Tausa
Lourdes	73°51 W	04°58 N	2750	Gachancipa
Panonia	73°44 W	05°03 N	2800	Choconta
Piscis	73°41 W	05°04 N	2820	Choconta
Potrero Largo	73°46 W	04°55 N	2780	Guatavita
Preventorio Infante	74°16 W	04°27 N	2650	Sibaté
Pte Portillo (stream)	74°36 W	04°27 N	361	Tocaima
Roble El	74°13 W	04°47 N	2560	Madrid
San Isidro	73°53 W	04°51 N	2698	Guasca
Sta. Rosita	73°45 W	05°06 N	2750	Suesca

approximation of the Saint-Venant equations of open channel flow (Chow et al., 1988) for surface routing. CREST models the transformation of rainfall into runoff through the application of a variable infiltration curve (Zhao et al., 1995) and a conceptual partition of water flow into its surface and subsurface components. The hydrologic model was implemented on a 1-km rectangular grid based on digital elevation data from NASA's Shuttle Radar and Topography Mission (http://www2.jpl.nasa.gov/srtm/). Model parameters were prepared using *a priori* information related to geophysical characteristics of the basin. Table 19.2 lists all model parameters and includes the source of the information for each of them.

19.2.3 SIMULATION OF FLOODING EVENTS

19.2.3.1 Selection of Events

Official reports of flooding events are not readily available and are generally dispersed among different agencies. This lack of information consolidation represents a challenge for research on flood modeling, particularly in developing countries. News archives, and other secondary information sources, represent a potentially useful alternative for model evaluation purposes in data-poor regions. Consequently, several events were identified in an examination of local news archives (e.g., El Tiempo; http://www.eltiempo.com) to establish the modeling system's ability to simulate flooding. Table 19.3 presents a list of the selected events found in local news archives. Although these data are limited to the occurrence rather than an observation of a measurable aspect of a flooding event (e.g., streamflow or flood extents), useful information related to spatial extent of the event is available.

TABLE 19.2
CREST-KW Model Parameters and *A Priori* Data Source

Parameter	Description	Source
Wm	Soil water capacity	World soil information—http://www.isric.org
b	Infiltration curve exponent	
ksat	Hydraulic conductivity	
kI	Speed of subsurface flow	
coem	Manning's coefficient for overland routing	UMD vegetation category from 2007 MODIS. At: http://webmap.ornl.gov/wcsdown/wcsdown.jsp?dg_id=10004_32
IR	Imperviousness area ratio	URB_2000—built-up land (residential and infrastructure) From Harmonic World Soil Database (HWSD; Fischer et al., 2008) http://www.iiasa.ac.at/Research/LUC/External-World-soil-database/HTML/LandUseShares.html?sb=9
CET	Linear adjustment factor on potential evapotranspiration	Set subjectively to 1.0
Th	A threshold drainage area value above which a pixel is defined as a stream	Set subjectively to 5.0 km^2
α	Kinematic wave coefficient of the momentum equation	Vergara et al. (2015)
β	Kinematic wave exponent of the momentum equation	Vergara et al. (2015)
α_0	Overland kinematic wave conveyance parameter	Manning's equation using slope derived from DEM and parameter coem

TABLE 19.3
Selection of Events Based on Local News Archives

Date	Affected Municipality
June 8–9, 2002	Bogota, Soacha, Cota
May 3, 2005	Bogota
April 2, 2009	Bogota
November 6, 2010	Cota, Soacha
November 17, 2010	Bogota
November 30, 2010	Zipaquira
December 4, 2010	Bogota, Soacha, Cota
May 14, 2011	Villapinzon, Suesca
December 7, 2011	Bogota,
April 10–11, 2012	Suesca, Zipaquira

19.2.3.2 Flood Detection Approach

Because rainfall-runoff models do not resolve processes depicting the dynamics of flow exceeding bank-full conditions, indirect methods must be used to detect the occurrence of floods. In operational hydrologic modeling systems, a common approach to the detection and prediction of flooding events is based on the use of flood thresholds. That is, a minimum value of streamflow (or water depth) above which flooding is believed to occur (Williams, 1978). Reed et al. (2007) developed a threshold method based on recurrence intervals for distributed hydrologic simulations. The approach is centered on the notion introduced by some studies that bank-full conditions are generally associated to recurrence intervals between 1.5 and 2 years (Carpenter et al., 1999). Therefore, using historical simulations of streamflow it is possible to compute recurrence intervals over the entire basin regardless of the availability of flow observations. Moreover, the method is immune to overall systematic bias in the simulations (although not to conditional biases) because it uses peak flow ranking of the historical series rather than peak flow magnitudes. This threshold-based approach was adopted for the simulation of floods in this work.

The computation of recurrence intervals employed a Log-Pearson Type III (LP3) distribution fit to the simulated annual peak flow series from the 13-year period of study. The LP3 distribution's first three moments (i.e., the mean, standard deviation, and the skewness) were used to estimate recurrence interval for any given value of simulated streamflow. Recurrence intervals fields were computed for the 1-km grid representing the basin's surface. A 2-year recurrence interval was chosen as flooding threshold for this exercise.

19.2.3.3 Evaluation of Detection Skill

The interest in this exercise was to evaluate the ability of the satellite-based flood modeling system to detect the occurrence of events with particular focus on their spatial extent. Hence, the evaluation consisted of an analysis on the ability of the forecasts to detect locations affected by flood. A selection of nine municipalities (Villapinzon, Suesca, Zipaquira, Sopo, Cota, Facatativa, Bogota, Soacha, and Tocaima) within the basin was made based on their location and size (i.e., population) to have a good coverage of the basin's surface. A qualitative assessment was performed based on the available reports.

19.3 DISCUSSION OF RESULTS

19.3.1 A General Evaluation of Satellite-Based Rainfall Estimates

The evaluation of 3B42RT estimates at the annual scale is presented in Figure 19.3. The skill of the satellite-based product to estimate these cumulative values is significant over the majority of the basin. However, significant bias can be observed for pixels 21201310 and 21202160, which are located over the highest gradients of elevation near the outlet. The same observation can be made in the comparison at the monthly scale presented in Figure 19.4. This highlights the challenges in rainfall estimation over complex terrain regions. Despite this limitation, the linear correlation between gauge and satellite-based estimates at the monthly scale is very good in

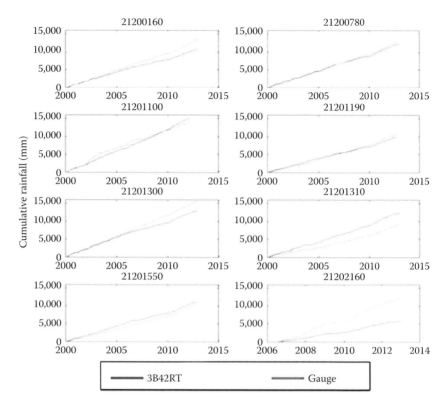

FIGURE 19.3 Comparison of yearly cumulative rainfall estimates between 3B42RT satellite product and rain gauge.

all cases (correlations > 0.6). The latter gives an indication of the satellite product's ability to represent rainfall dynamics relevant to water resources and climate studies.

Figure 19.5 presents scatter plots of gauge versus 3B42RT rainfall estimate at the daily scale. The high dispersion observed in the plots indicates considerable uncertainty in the satellite-based estimates at this scale. This is an indication of limitations of the satellite-based product in resolving short-term variability of rainfall, which represents a challenge for flood forecasting applications. However, because watersheds naturally integrate rainfall as water flows downstream, the impact of the error variability in satellite-based estimates on hydrologic forecasts might be attenuated.

19.3.2 FLOOD DETECTION SKILL

The results from the simulation of selected flood events are presented in Figures 19.6 and 19.7. Three instances for each event are included: the beginning of the flooding (left column); the time at which the maximum recurrence intervals and extent were predicted (center column); and the end of the event (right column). It can be observed that the satellite-based estimates of rainfall result in flooding detection for the majority of the selected events. Moreover, some of the cases show the transition from flash flooding (i.e., over small drainages) to river flooding (i.e., on the main streams).

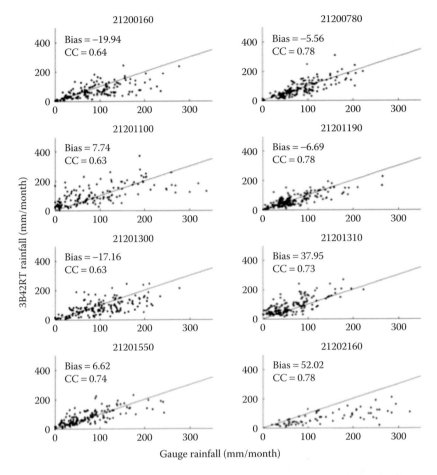

FIGURE 19.4   Scatter plots of monthly totals of 3B42RT satellite product and rain gauge estimates.

Likewise, widespread flooding can be observed in some of the events. These spatial features of the detection of flooding give a preliminary indication of the potential usefulness of hydrologic estimates forced by satellite rainfall for an early warning system over this region. A more detailed description of the skill of the system is discussed in the following.

Some of the identified events were of significant magnitude and of high impact on the society. These events were particularly notorious in the national news reports. The June 8, 2002, event was reported in the local newspapers as one of great proportions. The Tunjuelito river, which is part of Bogota River's stream network, rose to unprecedent levels causing significant economical impact and health problems (López, 2006). The hydrologic model detected this event with flow values exceeding the 5-year recurrence interval over the city of Bogota and the southeastern area of the basin. The April 25, 2005, event was also reported on the news where flooding was noted to cause damages for more than 30 people

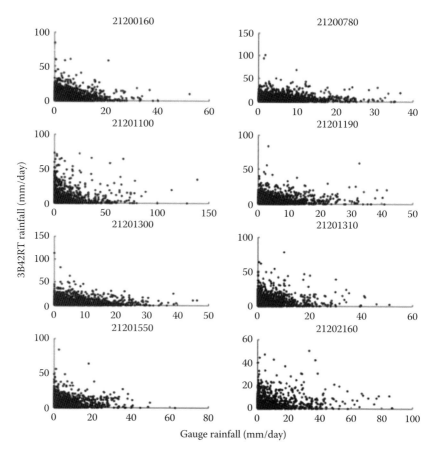

FIGURE 19.5   Same as Figure 19.4 but for daily estimates.

in different areas east of the Bogota city (Tiempo, 2005). The model detected this event with flows exceeding the 3-year recurrence interval over the upper portion of the basin. Likewise, the March 26, 2009, event affected thousands of people over the western area of the basin (Espectador, 2009). The modeling system detected flooding with flows exceeding the 2-year recurrence interval near the municipality of Facatativa west of the city of Bogota.

The rainy season of 2010–2011 in Colombia was particularly wet leading to various flooding events with disastrous consequences. In fact, 5 of the 12 selected events from the 12-year dataset employed herein occurred during this period. For the case of the November 6, 2010, event, the Civil Defense agency of Colombia (Defensa Civil Colombiana, 2010) reported flooding over areas south of the Bogota city and over the municipality of Cota. The model detected flow values with approximately 2-year recurrence interval over those regions. In the November 17, 2010, event, the local news reported intense rainfall over areas west to the city of Bogota, leaving some access roads unavailable (Citytv, 2010). The model detected flows exceeding the 2-year recurrence interval over the middle portion of the basin. For the

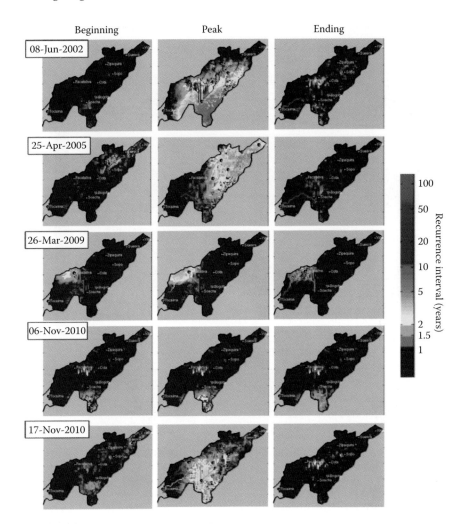

FIGURE 19.6 Streamflow recurrence interval fields over the basin for June 8, 2002; April 25, 2005; March 26, 2009; November 6, 2010; and November 17, 2010, events. Three snapshots of each event are included (from left to right): beginning (first column), peak (second column), and ending (third column).

December 4, 2010, event, news reported flooding caused by Bogota River overflow. This resulted in inundation over entire neighborhoods on the southwestern side of the city. The modeling system indicated flows exceeding the 2-year recurrence interval and propagation of the flood wave downstream in the main stem. In the May 14, 2011, event, half of the population of the municipality of Villapinzon over the northern headwaters of the basin was affected by flooding (Lider, 2011). Likewise, downstream municipalities of Chia, Cajica, and Suesca were reported to be impacted by the same rainfall event (Periodico, 2011). The hydrologic model detected flows exceeding the 2-year recurrence interval that propagated downstream. Finally, the December 1, 2011, event was reported to leave 50,000 people affected

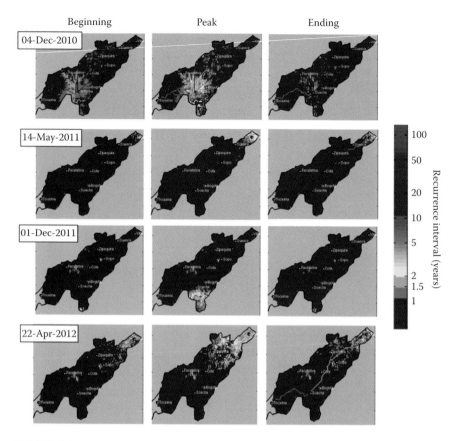

FIGURE 19.7   Same as Figure 19.6 but for December 4, 2010; May 14, 2011; December 1, 2011; and April 22, 2012, events.

over areas south and southwest of the Bogota city (Semana, 2011). The modeling system detected flooding over the southwest area to the city of Bogota with flows nearing the 100-year recurrence interval mark.

19.4   SUMMARY AND CLOSING REMARKS

The purpose of the study presented in this study was to explore the utility of a satellite-based rainfall product for the modeling and prediction of flooding in the Bogota River basin in Colombia. The exercise was a pilot for a collaborative research project between the El Bosque University in Colombia and the Hydrometeorology and Remote Sensing Laboratory of the University of Oklahoma in the United States of America. The basin is located on the Andean mountain range, which makes it of complex topography and particularly challenging for satellite-based estimation of rainfall. The satellite product employed for the hydrologic modeling was NASA's 3B42RT, a near-real-time dataset available every 3 h on a 0.25° grid. A statistical comparison between the satellite-based rainfall estimates and data from the national rain gauge network was performed over a period of 12 years. Likewise, a hydrologic model was configured to simulate streamflow for various flooding events in the period of study. A threshold-based detection methodology that uses recurrence intervals was employed to estimate flooding with the satellite-forced hydrologic simulations. Some of the most relevant remarks are as follows:

- Significant uncertainty was observed in the satellite-based estimates at the daily scale. However, the skill of the product improved at the monthly and annual scales, which indicates the potential utility on water resources and climate change applications.
- The assessment of the flood detection skill suggests that satellite-based rainfall estimates are useful in defining spatial features of flooding. The latter is particularly encouraging for the implementation of early warning systems based on satellite-based observations over this region.

In general, the results of the pilot study motivates the continuation of research and development efforts toward an integrated system for monitoring and prediction of environmental phenomena. A challenge of critical importance is that in the context of data availability for verification purposes. This particular issue will be addressed in future work. Likewise, future endeavors will concentrate on extending this work to different regions of the country and to applications on drought and landslides, which are the other two main hydrometeorological hazard in Colombia.

REFERENCES

Adhikari, P., Y. Hong, K. Douglas, D. Kirschbaum, J. Gourley, R. Adler, and G. Robert Brakenridge, 2010: A digitized global flood inventory (1998–2008): Compilation and preliminary results. *Natural Hazards*, **55**, 405–422 10.1007/s11069-010-9537-2.

Bedient, P. B., W. C. Huber, and B. E. Vieux, 2008: *Hydrology and Floodplain Analysis,* Prentice Hall, Upper Saddle River, NJ.

CAR, 2006: Plan De Ordenacion Y Manejo De La Cuenca Hidrografica Del Rio Bogota. In *Resumen Ejecutivo*, Corporacion Autonoma Regional de Cundinamarca.

Cardona, O. D., 2007: *Información Para La Gestión De Riesgo De Desastres: Estudio De Caso De Cinco Países, Colombia*, Mexico: CEPAL.

Carpenter, T. M., J. A. Sperfslage, K. P. Georgakakos, T. Sweeney, and D. Fread, 1999: National threshold runoff estimation utilizing GIS in support of operational flash flood warning systems. *Journal of Hydrology*, **224**, 21–44.

Chow, V. T., D. R. Maidment, and L. W. Mays, 1988: *Applied Hydrology*. McGraw-Hill, New York.

Citytv, 2014: Pérdidas en el occidente de la ciudad por desbordamiento del río Bogotá, <http://www.citytv.com.co. Recuperado el 25 de Abril de 2014, de http://www.citytv.com.co: http://www.citytv.com.co/videos/287640/perdidas-en-el-occidente-de-la-ciudad-por-desbordamiento-del-rio-bogota>.

Defensa Civil Colombiana, 2010: Reporte Diario de Emergencias—Noviembre 2010. Division Prevencion—Central de Comunicaciones.

Espectador, E., 2009: Créditos de hasta $23 millones para afectados por inundaciones en Bogotá, <Recuperado el 25 de Abril de 2014, de http://www.elespectador.com/noticias/bogota/articulo135597-creditos-de-hasta-23-millones-afectados-inundaciones-bogota.>.

Fischer, G., F. Nachtergaele, S. Prieler, H. Van Velthuizen, L. Verelst, and D. Wiberg, 2008: *Global Agro-Ecological Zones Assessment For Agriculture*. FAO, Rome, Italy.

Huffman, G. J., D. T. Bolvin, E. J. Nelkin, D. B. Wolff, R. F. Adler, G. Gu, Y. Hong, K. P. Bowman, and E. F. Stocker, 2007: The TRMM multisatellite precipitation analysis (TMPA): Quasi-global, multiyear, combined-sensor precipitation estimates at fine scales. *Journal of Hydrometeorology*, **8**, 38–55.

Lider, E., 2011: Villapinzón está inundada por creciente del río Bogotá, <Recuperado el 25 de Abril de 2014, de http://www.ellider.com.co/2011/05/15/villapinzon-esta-inundada-por-creciente-del-rio-bogota/>.

López, V. L., 2006: Comportamiento histórico de la precipitación en la cuenca media y alta del río Tunjuelito y análisis del evento ocurrido entre los meses de mayo y junio de 2002. *Revista Épsilon*.

Mundial, B., 2012: Análisis de la gestión del riesgo de desastres en Colombia: Un aporte para la construcción de políticas públicas. In *Análisis de la gestión del riesgo de desastres en Colombia: un aporte para la construcción de políticas públicas*: Banco Mundial, Colombia.

Periodico, E., 2011: *Varios Municipios en Alerta por Desbordamiento del Rio Bogota, in El Periodico*. Bogota, Colombia: Grupo Editorial El Periódico S.A.S.

Reed, S., J. Schaake, and Z. Zhang, 2007: A distributed hydrologic model and threshold frequency-based method for flash flood forecasting at ungauged locations. *Journal of Hydrology*, **337**, 402–420 10.1016/j.jhydrol.2007.02.015.

Semana, R., 2011: Cerca de 50 mil afectados por inundaciones en el suroccidente de Bogotá, <Recuperado el 25 de Abril de 2014, de http://www.semana.com/nacion/articulo/cerca-50-mil-afectados-inundaciones-suroccidente-bogota/250472-3>.

Tiempo, E., 2005: Inundaciones en chapinero por quebrada las delicias puede ser peor la próxima vez, <Recuperado el 25 de Abril de 2014, de http://www.eltiempo.com: http://www.eltiempo.com/archivo/documento/MAM-1643955>.

Vergara et al., 2016: Estimating a-priori kinematic wave model parameters based on regionalization for flash flood forecasting in the conterminous United States. *Journal of Hydrology* (in press).

Wang, J., Y. Hong, L. Li, J. J. Gourley, S. I. Khan, K. K. Yilmaz, R. F. Adler, F. S. Policelli, S. Habib, and D. Irwn, 2011: The coupled routing and excess storage (CREST) distributed hydrological model. *Hydrological Sciences Journal*, **56**, 84–98.

Williams, G. P., 1978: Bank-full discharge of rivers. *Water Resources Research*, **14**, 1141–1154.

World Bank, T., 2010: *Disaster Risk Management in Latin America and the Caribbean Region: GFDRR Country Notes, Colombia*, Washington, DC: World Bank.

Zhao, R., X. Liu, and V. Singh, 1995: The Xinanjiang model. In: *Computer Models of Watershed Hydrology*, V. Singh (ed.), Water Resources Publications, CO, pp. 215–232.

20 Cloud-Based Cyber-Infrastructure for Disaster Monitoring and Mitigation

Zhanming Wan, Yang Hong, Sadiq Ibrahim Khan,
Jonathan Gourley, Zachary L. Flamig,
Dalia Kirschbaum, and Guoqiang Tang

CONTENTS

20.1 INTRODUCTION

Natural disasters are always dangerous and they have been causing tremendous loss of life and economic damages globally and frequently. According to the International Federation of Red Cross and Red Crescent Societies, there were 3867 reported natural disasters that happened between 2004 and 2013. During this period, natural disasters caused approximately 1 million fatalities worldwide, affected approximately 2 billion people, and led to economic losses totaling approximately $1.6 trillion (Cannon et al., 2014). The significant global impact of recurring natural disaster events leads to an increased demand of having comprehensive disaster databases for hazards studies. There are several existing databases, such as the International Disaster Database, ReliefWeb (launched by the United Nations Office for the Coordination of Humanitarian Affairs), and the International Flood Network. However, there is often a lack of specific geospatial characteristics of the disaster events or a failure to enlist all disaster events due to variable entry criteria. Moreover, these data warehouses lack interactive information sharing with the communities affected by the disaster events. Therefore, it is essential to establish a cloud-based cyber-infrastructure for synthesizing disaster information, visualizing disaster information in different formats, and sharing disaster information with the communities to monitoring disasters as well as mitigating the impact. An example towards this direction was given by Adhikari et al. (2010), who developed a methodology to utilize valuable flood events information from the aforementioned sources. Because the authors' focus was flood, the Global Active Archive of Large Flood Events (created by the Dartmouth Flood Observatory) was taken into account as well. In the study, the data from different disaster databases were synthesized with media reports and remote sensing imagery to provide a record of flood events from 1998 to 2008. The digitized Global Flood Inventory (GFI) gathers and organizes detailed information of flood events from reliable data sources, defines and standardizes categorical terms as entry criteria for flood events (e.g., severity and cause), and cross-checks and quality controls flood event information (e.g., location) to eliminate redundant listings. These characteristics make GFI an appropriate starting point to develop a unified global flood cyber-infrastructure. However, one limitation of this database is that GFI only contains flood events through 2008. Although it is possible that flood events after 2008 can be collected manually, as was done in Adhikari et al. (2010), it can be incomplete and inefficient because this process only involves a limited number of resources and people. Recently, technological advances in social media have tremendously improved data gathering and dissemination, especially with the development of world-wide web technologies. Built on the platform of social media, crowdsourcing has become a versatile act of collecting information from the public.

Crowdsourcing is a term that generally refers to methods of data creation, where large groups of potential individuals generate content as a solution of a certain problem for the crowdsourcing initiator (Hudson-Smith et al., 2009; Estellés-Arolas et al., 2012). In theory, crowdsourcing is based on two assumptions described by Goodchild et al. (2010). First, "a group can solve a problem more effectively than an expert, despite the group's lack of relevant expertise," and second, "information obtained from a crowd of many observers is likely to be closer to the truth than information obtained from one observer." Based on the definition and assumption of crowdsourcing, it has the ability to collect a considerable amount of information from its randomly distributed participants. The nature of crowdsourcing accommodates data collection in numerous forms, including questionnaires, phone calls, text messages, e-mails, web surveys and other paper-based, mobile phone-based, and web-based methods. Moreover, crowdsourcing can be embedded with location-based information by using GPS-enabled devices, IP (Internet protocol) addresses, or participants' awareness of their current locations. Crowdsourcing offers new opportunities to expand the information available to impacted communities and provide a "two-way" street for the same affected populations to communicate with the global community. The crowdsourcing idea of gathering information can be abstracted as utilizing human sensors or citizen sensors (Goodchild and Michael, 2007). Besides human sensors, physical sensors can also be easily applied to monitoring disasters, providing early warning, and supporting decision-making process. Some research has been conducted to investigate using sensor web for disaster management. An example is the Namibia flood SensorWeb infrastructure, which was created for rapid acquisition and distribution of data products for decision-support systems to monitor floods (Kussul et al., 2012). The decision-support system utilizes the Matsu Cloud to store and preprocess data through hydrologic models, eliminating the latency when clients select specific data. Wang et al. (2010) proposed an emergency airship monitoring system by extending the ground-based sensor web to the sky with high mobility to overcome the shortcoming that fixed ground sensors might be out of service due to damages caused by the happening disasters.

The data collected from the sensor web can be used in a cloud-computing framework for information sharing that includes data processing and visualization. Gong et al. (2010) adopted cloud-computing technology in geoprocessing functions to provide elastic geoprocessing capabilities and data services in a distributed environment. Behzad et al. (2011) used cloud-computing in addition to a cyber-infrastructure-based geographic information system to facilitate a large number of concurrent groundwater ensemble runs by improving computational efficiency. Huang et al. (2012) integrated cloud-computing in dust storm forecasting to support scalable computing resources management, high-resolution forecasting, and massive concurrent computing. The study of Sun (2013) presented a collaborative decision-making water management system using a cloud service provided by Google Fusion Table. The author describes the migration of the management system from a traditional client–server-based architecture to a cloud-based web system, revealing the potential to fundamentally change a water management system from its design to the operation. As defined by the National Institute of Standards and Technology, cloud-computing is a model for supporting elastic network access to a shared pool of configurable computing resources (Mell et al., 2011). The nature of cloud-computing assures that it can (a) reduce the time and cost during implementation, operation, and maintenance of the global flood cyber-infrastructure, (b) provide an interface for collaboration at both global and local scales, and (c) conveniently share data in a secure environment. These advantages make cloud-computing an attractive technique in the global flood cyber-infrastructure that can maximize the efficiency and data safety during collaboration, while minimizing time and expense spent on the system.

This study presents the possibilities of adopting cloud-based cyber-infrastructure for disaster monitoring and mitigation. The next section focuses on the methodology of building the cloud-computing system designed for global disaster monitoring, analysis, and reporting. Section 20.3 demonstrates a case study of a cloud-computing service provided by Google to establish the global flood cyber-infrastructure, to share the GFI, to provide statistical and graphical visualizations of the data, and to expand the breadth and content of the GFI by collecting new flood data using crowd-sourcing technology (i.e., CyberFlood). Discussion and conclusion are provided in Sections 20.4 and 20.5 respectively.

20.2 CYBER-INFRASTRUCTURE DESIGN

The global disaster cyber-infrastructure consists of four components: data source, cloud service, web server, and client interface (Figure 20.1). The data source should be preprocessed before being imported into the cyber-infrastructure, as explained later in this section. The cloud service, which significantly improves the performance and decreases the burden on the web server, handles all data queries, data visualization, and data analysis. The web server simply deals with sending requests and responses between clients and the cloud. The client interface is mainly built with hypertext markup language (HTML) and JavaScript. Because all the data are processed before being imported into this cyber-infrastructure, the client side only sends operational requests from users and renders responses from the cloud service.

Data preparation is the very first and important step in designing the cyber-infrastructure. This process directly affects the results users can get from the system. With modern technology support, it is possible to use any different types of data. However, if data conversion is included, it is crucial to keep consistency between the original and converted data. Disasters are usually associated with location information. Judging by the type and scale of disasters, it is possible to illustrate disasters in the form of points (e.g., small-scale events), polylines (e.g., tornado paths), and polygons (e.g., inundation, landslide, and drought affected areas) on maps using well-known formats such as GeoJSON (Butler et al., 2008) and Keyhole Markup Language (KML) (Wilson, 2008). Processed data are then imported into

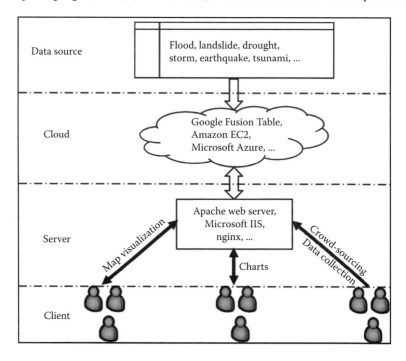

FIGURE 20.1 The global disaster community cyber-infrastructure framework.

cloud service. Elasticity should be considered when selecting an appropriate cloud service. Typically, Google Fusion Table is suitable for data with simple data types, less than 100,000 rows, and light computation. If the cyber-infrastructure requires complex data types, more data rows, and heavy computation, other cloud services are available from major providers such as Google, Amazon, and Microsoft. Because all the computation load is carried in the cloud service, web server only provides the user interface for people to access cyber-infrastructure. Three dominant web servers Apache, Microsoft IIS, and nginx are widely used. The operating system and coding language should also be considered when selecting a web server.

20.3   CYBERFLOOD CASE STUDY

20.3.1   DATA PREPARATION

A cloud-computing service enabled global flood cyber-infrastructure was established to share the aforementioned GFI, to provide location-based visualizations as well as statistical summary of the data, and to expand the breadth and content of current GFI by collecting new flood data using crowdsourcing technology (Wan et al., 2014). GFI standardizes categorical terms as entry criteria for flood events. In other words, every data column in GFI is carefully designed so that each entry strictly follows the criteria of the corresponding data column (Figure 20.2a). GFI was preprocessed before being successfully imported into a Google Fusion Table. Python, which is a cross-platform, extensible, and scalable programming language (Sanner, 1999), was used to write code for data conversion. The purpose is to maintain data consistency, making the converted data readily readable and reducing the data conversion load on the client side. In this process, cells containing −9999, which represent no value in GFI, are removed because they are not consistent with empty cells that also represent no value. Data columns of flood severity, cause, country, and continent are

ID	Year	Month	Day	Duration (days)	Fatality	Severity	Cause	Lat	Long	Country code	Continent Code
2707	2008	12	28	23	25	1	2, 1	−22.92	34.03	140	1
2706	2008	12	26	18	24	1	1	−3.33	103.14	93	3
2705	2008	12	26	3	−9999	1	1	44.66	−123.53	213	6
2704	2008	12	26	3	−9999	1	1	41.04	−89.46	213	6
2703	2008	12	25	12	9	1	1	16.89	107.06	219	3
2702	2008	12	13	31	76	1.5	1	9	−74.23	42	8
2701	2008	12	13	2	2	1	1	51.49	−1.73	212	5

(a)

ID	Year	Month	Date	Duration	Fatality	Severity	Cause	Geometry	CountryCode	ContinentCode
2707	2008	12	12/28/2008	23	25	Class 1	Tropical cyclone, Heavy rain	22.92, 34.03	Mozambique	Africa
2706	2008	12	12/26/2008	18	24	Class 1	Heavy rain	3.33, 103.14	Indonesia	South East Asia
2705	2008	12	12/26/2008	3		Class 1	Heavy rain	44.66, 123.53	United States	North America
2704	2008	12	12/26/2008	3		Class 1	Heavy rain	41.04, 89.46	United States	North America
2703	2008	12	12/25/2008	12	9	Class 1	Heavy rain	16.89, 107.06	Vietnam	South East Asia
2702	2008	12	12/13/2008	31	76	Class 2	Heavy rain	9, 74.23	Colombia	South America
2701	2008	12	12/13/2008	2	2	Class 1	Heavy rain	51.49, 1.73	United Kingdom	Europe

(b)

FIGURE 20.2   Comparison of data tables. (a) Global flood inventory and (b) Google Fusion Table.

filled with numbers to indicate certain meanings in GFI. A look-up table was used to convert the numerical codes into text. For example, "1" means "heavy rain" in the column pertaining to flood causes, whereas it means "Africa" in the column pertaining to continents (Figure 20.2a and b). In other words, if the GFI with numbers are imported into the Fusion Table and used directly by the cyber-infrastructure, the numbers have to be converted to the corresponding texts each time during the refresh on the client side. As a result, text is assigned to severity, cause, country, and continent during this process. Location, the most important information for map visualization in this cyber-infrastructure, is described in two columns representing latitude and longitude in GFI. However, if one flood event involves more than one location, then there will be multiple data records, and only the first data record has shared information such as event ID and date. To improve this data structure and for better visualization, multiple data records representing the same flood event are combined into a single record, while the location is presented as MultiGeometry using KML (Wilson, 2008).

An example of a flooding event in New Hampshire in October 2005 is illustrated in Figure 20.3. Figure 20.3a shows the event as stored in the original GFI covering

ID	Year	Month	Day	Duration (days)	Fatality	Severity	Cause	Lat	Long	Country Code	Continent Code
1859	2005	10	8	10	11	1.5	1	42.9475	−72.2944	213	6
								43.0767	−72.0989		
								43.0839	−72.4317		
								42.8653	−72.555		
								42.8125	−72.5444		

(a)

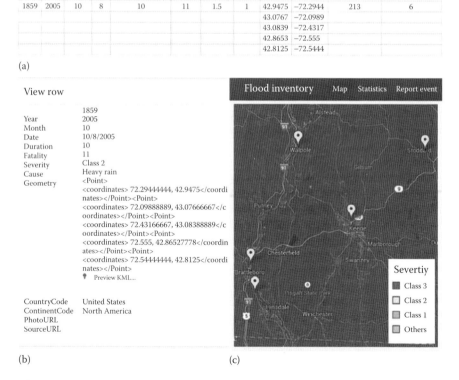

(b)　　　　　　　　　　　　　　　　　　　　(c)

FIGURE 20.3 Flood event over Northeast US in New Hampshire of October 2005. (a) Global flood inventory, (b) Google Fusion Table attributes, and (c) Google Map view.

events from 1998 to 2008. Five locations were associated with this event. Cells are left blank if they share the same record as in the first row. Figure 20.3b shows the same flooding event as in Figure 20.3a but converted into a Google Fusion Table. This table also includes all five locations that are now represented in the geometry column with KML. Figure 20.3c illustrates the visualization of this event, showing the severity as well as the specific locations impacted. Additional layers such as rivers, roads, and topography can also be included during this step to ascertain the spatial extent of inundation.

The processed GFI, now converted to a Google Fusion Table (Figure 20.2b), belongs to a "Software as a Service" (Yang et al., 2011) type of cloud-based service for data management and integration (Gonzalez et al., 2010). Google Fusion Table was created to manage and collaborate with tabular datasets in which geospatial fields can be included to provide location information. These geospatial fields can be in the form of latitude and longitude in two separate columns, latitude and longitude pairs in one column, or KML strings in one column. Fusion Table accepts many different tabular formats of files as its data source. Any text-delimited files such as comma-separated values (CSV) files, KML files, and spreadsheets can be imported directly into a Fusion Table. Because Google Fusion Table is a part of Google Drive, users can simply select an existing spreadsheet from their Google Drive and import it into a Fusion Table. Cloud computing is embedded to provide rapid responses to requests from users for data querying, summary, and visualization. Moreover, data security and sharing is already implemented in Google Fusion Table.

The steps required to import data into a Google Fusion Table are straightforward. First, the data must be in one of the supported formats (tabular or text-delimited data such as CSV files, excel spreadsheets, and other similar types). A wizard then provides easy-to-follow instructions describing how to upload the data. Fusion Table looks like a common table in a spreadsheet, whereas it supports structured query language (SQL) to operate the table. Keywords, such as *SELECT, INSERT, DELETE*, and *UPDATE*, can be used to manipulate Fusion Table, which is similar to how a table is handled in a database. Fusion Table provides application programming interface (API) to programmatically perform SQL-based, table-related tasks through using hypertext transfer protocol requests (Google, 2013). By combining with other Google-provided APIs, the capability of Fusion Table can be extended to not only manipulate the data in the table, but also to visualize the data through thematic mapping and analytic charts.

Fusion Table, which plays an important role in this global flood cyber-infrastructure, provides data storage, data sharing, and fast data access. However, because the infrastructure is functioning from the backend, users cannot benefit from this service unless a traditional server and client components are included for interaction. Because all the computing loads are on the cloud, the web server only serves as a "middleware" dealing with requests and responses between the cloud and clients. The web server also protects the Fusion Table on the cloud from being accidentally modified by clients. Google provides two kinds of API keys for programmers to develop applications. One of the keys is a string, which grants permission to applications to select items from the Fusion

Table. The other key is a special file that should be stored securely with the application on the web server. This type of key grants permission to the application from the specific web server to insert, update, or delete items from the Fusion Table. The client side is programmed with HTML and JavaScript, along with several APIs from Google, to send requests through the server to the cloud, receiving responses for location-based and analytic visualization.

20.3.2 DEMONSTRATION

The global flood cyber-infrastructure is currently running at http://eos.ou.edu/flood/ (Figure 20.4). An Apache web server is deployed to host the frontend web interface. Google Map has been integrated to map the locations of flood events after querying the Fusion Table using the Google Map API. All the points representing locations of flood events are color coded by severity or fatalities associated to the flood event. Severity is classified into classes 1, 2, and 3, with "Class 1" being least severe and "Class 3" the most severe. Fatalities are categorized into four classes based on the value. Users are allowed to select a range of years and causes of flood events from the provided controls. Each selection will lead to a new query from the Fusion Table, which means that the desired data will be plotted on the map and can include event details that have just been uploaded in real time. In addition to visualization of the data using the information stored in the Fusion Table, a Google Chart API is utilized

FIGURE 20.4 The map visualization of global flood cyber-infrastructure. The top and bottom maps are color coded by severity and fatalities, respectively.

FIGURE 20.5   The statistical chart and table of global flood cyber-infrastructure.

to create analytic charts for the statistical analysis of the flood events (Figure 20.5). Variables such as the year, month, severity, cause, continent, and country, can be analyzed in a chart and a table. Variables can be summarized by the count of the variables, sum of fatalities, or average of fatalities. For instance, Figure 20.5 demonstrates the summary of flood events by year and severity. Flood events with Class 1 severity are in a blue color on the chart, with about 270 of the flood events in 2003 occurring with such a severity class.

To expand and update the existing GFI, now stored as a Fusion Table, crowdsourcing from public entries is implemented in this cyber-infrastructure by providing a flood events observation report form (Figure 20.6). Most of the fields are the same as the existing GFI. However, photo URL and source URL fields are appended to the Fusion Table to store additional details about the submitted flood event. This means that users are able to upload one photo per submission and provide a URL of the web source as a proof or supplemental information of that flood event. The current date will be retrieved from the users' operating system by default to submit present flood events. Users can also select any date between 1998 to present if past events are reported. Because reported events will be displayed on the map in real time immediately following submission, location is a required field in the report form. The location will be automatically retrieved if a location service is allowed by the client's browser or the uploaded photo is geo-tagged. This report form is submitted directly

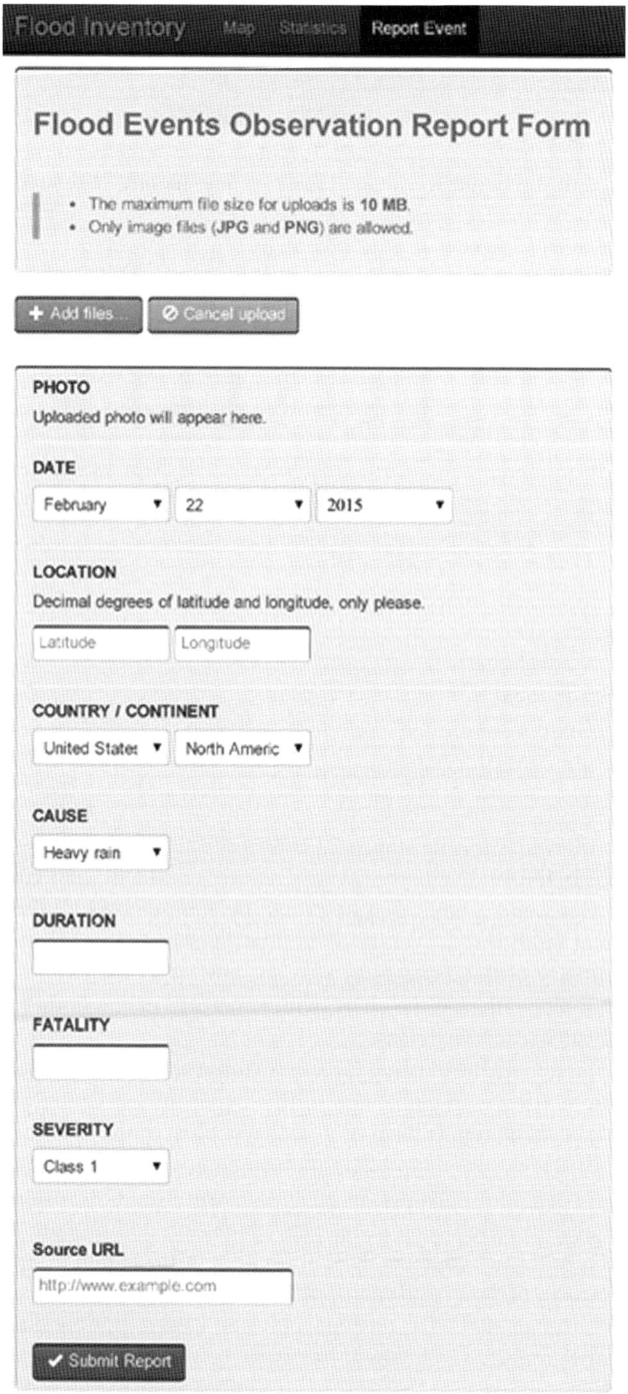

FIGURE 20.6 The flood events observation report form.

into the Fusion Table through the server, and this process is protected by Google Account Authentication and Authorization Mechanism to secure data on the Fusion Table. A two-way quality control approach of data from crowdsourcing is implemented. First, when a user submits a report of flood events, the system will automatically check if each field is correctly formatted. For example, fields of latitude and longitude can only be numeric values. Fields of day, month, and year are restricted to certain numbers which can only be selected by users. Instructions have also been created for first-time users and they can learn what each field means and how to retrieve current location to help them submit correct information. Second, after submission, the data will be manually checked with different sources, including news reports, flood reports from other major disaster data sources, and satellite imagery. Checking data sequentially is not an efficient way of quality control. However, it is effective in this case because the number of data received so far is limited. Newly submitted events following postprocessing will be assigned IDs according to the number of milliseconds from January 1, 1970, to the time of the submission. For example, a flood event reported on December 18, 2013, 23:35:15.199 will be assigned an ID of 1387431315199. Sequential IDs will be assigned to newly submitted data after quality control is complete. If crowdsourced data submissions increase in frequency in the future, automated data quality control procedures will be developed to check the spatial and temporal consistency with other flood reports. Other automated procedures can cross-check the reports with global flood forecasts available from http://eos.ou.edu/Global_Flood.html. A crowdsourcing way to control the quality of crowdsourced flood events reports are under consideration. A mechanism could be established to grant permission to qualified users and students who have expertise in flood monitoring and validation to check the data quality in the Fusion Table.

20.4   DISCUSSION

20.4.1   ADVANTAGE

Although CyberFlood does not directly solve flooding problems, this work is expected to be able to help advance flood-related research areas such as hydrologic model evaluation, flood risk management, and flood awareness. Both the public and research community can use the resources provided by this cyber-infrastructure to analyze retrospective flood events and submit their witness accounts of previously unreported flood events. Therefore, this approach is useful for flood monitoring and validation research. The long-term database could also help generate flood climatology of occurrences and damage and therefore could potentially lead to better flood risk management for zoning and other flood-related decision-making purposes. Public engagement using crowdsourcing and cloud-based techniques could potentially raise flood awareness around the globe and provoke citizen-scientists to consider careers in the natural sciences, engineering, and mathematics.

CyberFlood has been created to be used by anyone with internet access. To access the flood resources, a web-based interface is provided and is becoming accessible through iOS apps for mobile users. As CyberFlood becomes more accessible through these apps, more people will use it to view retrospective flood events, monitor current flood events,

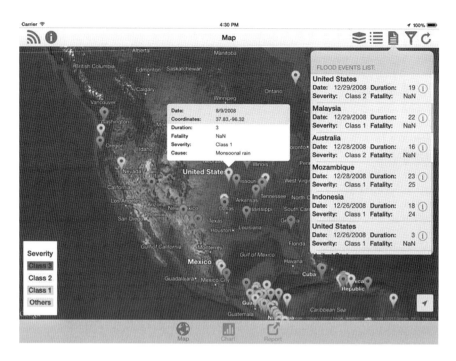

FIGURE 20.7   iPad app demonstration of CyberFlood.

and contribute to the flood community by submitting their reports of flood events. CyberFlood has been created to adapt the idea of Volunteered Geographic Information, which is described as tools to create, assemble, and disseminate geographic information provided voluntarily by individuals (Goodchild et al., 2007). The CyberFlood app is able to facilitate the process of compiling flood events by involving map-based visualization and utilizing human sensors to collect useful data globally (Figure 20.7).

Compared with the traditional server–client structure, the cloud-computing service provided by Google Fusion Table enhances the performance of the global flood cyber-infrastructure in terms of the speed during data query and data visualization. By providing a Fusion Table API, the complexity of the global flood cyber-infrastructure is significantly reduced. This benefits both programmers and clients because they are able to focus more on the actual functions they need to implement and use, not on the logistics with the cloud itself. Rather than using the traditional server–client-based structure, this simplified cloud-based framework makes it easier to develop scalable applications. Furthermore, taking into consideration data sharing and collaboration, Fusion Table provides a comprehensive solution to keep data secure while making seamless communications between collaborators and Google servers for data updates, queries, and visualization.

20.4.2   Performance Experiment

An experiment was developed to compare the speed of reading data and geographically displaying data using a Google Map API with a Google Fusion Table and a

TABLE 20.1
Performance Comparison Results

Test Order (Records)	Google Fusion Table (google.maps. FusionTablesLayer)					
	1	2	3	4	5	Average
1000	17	8	8	7	8	9.6
5000	9	7	6	7	9	7.6
10,000	12	7	8	6	7	8.0
50,000	8	9	8	7	7	7.8
100,000	14	10	9	8	8	9.8
Test Order (Records)	MySQL (google.maps. Marker)					
	1	2	3	4	5	Average
1000	1052	1039	1041	1049	1048	1045.8
5000	1128	1138	1143	1111	1116	1127.2
10,000	1202	1194	1230	1233	1240	1219.8
50,000	1842	2145	1915	1867	2211	1996.0
100,000	3050	3332	2938	2895	3123	3067.6

Note: Units are in ms.

MySQL database, respectively, both of which contain the same dataset. A Google Map API provides two ways to display markers on Google Map. The traditional way is by using the google.maps. Marker class. The more efficient way is to utilize the google. maps. FusionTablesLayer class that can only be employed by data from the Google Fusion Table. As a result, the data in the Google Fusion Table is visualized by google. maps. FusionTablesLayer class while the data in the MySQL database is visualized by google.maps. Marker class in this experiment. The query speed of both Google Fusion Table and MySQL database are rapid, taking a few milliseconds. However, the speed advantage becomes predominant when using data from the Google Fusion Table with the google.maps. FusionTablesLayer class. Table 20.1 demonstrates the results of this performance experiment. The first 1,000, 5,000, 10,000, 50,000, and 100,000 records are retrieved from the dataset. The average time of reading and displaying different sizes of data is calculated from five consecutive measurements. When data records increase from 1,000 to 100,000, the average elapsed time for using the Google Fusion Table with the google.maps. FusionTablesLayer class is always low (<10 ms) while the average elapsed time for using the MySQL database with the google.maps. Marker class is much higher (>1000 ms) and increases significantly to more than 3,000 ms when displaying 100,000 records.

20.4.3 LIMITATION AND SCALABILITY

The Fusion Table has some limitations on storage and usage. Each user can import data files no more than 100 MB into each Fusion Table, and each Google cloud account can contain data no more than 250 MB. The Google Fusion Table is an experimental product, which does not have a payment option for increasing the storage space. However, the data inside the Google Fusion Table is text-based that takes up very little space. When data are inserted into the Google Fusion Table, efforts have been made with additional code/scripts to save space by normalizing each field and trimming unnecessary spaces. Currently, there are 2730 records in the Fusion Table, which takes up 657 KB out of 250 MB. This means approximately 1 million similar data records can be stored with just this one Google cloud account. Furthermore, photo submissions are uploaded to a separate server with a terabyte-level shared storage space and only the URLs linked to the photos are stored in the Fusion Table.

The situation when the dataset grows beyond the limit of approximately 1 million records has also been taken into consideration. One solution is to have the data stored in multiple Fusion Tables of multiple Google accounts and perform a cross-table query. Another way is to use other cloud-based services, such as Google Cloud SQL and BigQuery, Amazon EC2, and Windows Azure. Google services will be our first choice because it is usually straightforward to develop applications with other Google products, such as Google Maps/Earth and Google Chart.

When inserting a data record into the Fusion Table, the record should be less than 1 MB, and a maximum of 25,000 requests per day can be sent to one Google account with free Fusion Table API access. However, the number of maximum requests per day can be increased by request through Google.

As a result, there is a trade-off between using Fusion Table resources directly and consuming a small portion of the resources from clients. To reduce the times in querying the Fusion Table, data from the prior queries are stored on the client side in the global flood cyber-infrastructure. If the next operation from the client side returns the same result as the previous operation, no request will be sent to the Fusion Table. It will use the stored data instead.

20.4.4 Data Sharing

Although the Google Fusion Table API does not provide a way to download raw data programmatically, as a shared cyber-infrastructure, the data in the Fusion Table of CyberFlood is free to download. A link can be provided to the actual Fusion Table from where users can view raw data and download them as a CSV or KML file. After the raw data have been made accessible, it is possible for users to adapt the raw data to visualize flood events in their own way and gain more discovery.

20.4.5 Sustainability

To involve people, some poster presentations about CyberFlood have been given at several conferences. Meanwhile, iOS apps for iPad and iPhone are under development, providing functions for people to view map and chart visualization of flood events and submit their witness accounts of flood events. Plans are made to advertise the CyberFlood through nontraditional media, such as social media Facebook and Twitter. We have also developed the mPING (Meteorological Phenomena Identification Near the Ground: http://www.nssl.noaa.gov/projects/ping/) app that includes flood entries (four levels of severity) and uses the crowdsourcing technique to obtain data. Given that the mPING has more than 200,000 active users today, this app will also be utilized to advertise our CyberFlood system. Because only limited entries from crowdsourcing during the 2009–2013 period are obtained, locally recruited students are compiling flood events from multiple sources for that period with manual quality control. Data for these years will be available in CyberFlood.

20.5 CONCLUSION

The global flood disaster community cyber-infrastructure (CyberFlood), with cloud-computing service integration and crowdsourcing data collection, provides on-demand, location-based visualization, as well as statistical analysis and graphing functions. It involves citizen-scientist participation, allowing the public to submit their personal accounts of flood events to help the flood disaster community to archive comprehensive information of flood events, analyze past flood events, and get prepared for future flood events. This cyber-infrastructure presents an opportunity to eventually modernize the existing methods the flood disaster community utilizes to collect, manage, visualize, and analyze data with flood events. In the future, data describing the flood reports in this cyber-infrastructure will be linked to real time and archived satellite-based flood inundation areas, observed streamflow, simulated surface runoff from a global distributed hydrologic modeling system, and precipitation products. These datasets will be beneficial both as method to validate the crowdsourced flood events and to help educate, motivate, and engage citizen-scientists about the latest advances in satellite remote sensing and hydrologic modeling technologies. Given the elasticity of a cloud-based infrastructure, this cyber-infrastructure for global floods can be applied to other natural hazards, such as droughts and landslides (Li et al., 2013), at both global and regional scales.

All kinds of disasters have significant impacts on the development of communities globally, often causing loss of life and property. It is increasingly important to create globally shared cyber-infrastructure to collect, organize, and manage disaster databases that visually provide useful information back to both authorities and the public in real time. By utilizing cloud-computing services and crowdsourcing data collection methods, it is practicable to provide on-demand, location-based visualization as well as statistical analysis and graphing capabilities. Cloud computing can be integrated into cyber-infrastructure by using cloud services that effectively accelerates the speed during data processing and visualization over the Internet. As a cloud-based cyber-infrastructure, people can access this infrastructure from all over the world through the Internet or mobile phones. The involvement of public participation allows the public to submit their entries of disaster events to help the community to archive comprehensive information of happening disasters, past and present. The crowdsourcing approach, which enables web-based data entry for the public to report or record their personal accounts of local flood events, can be employed in cyber-infrastructure to expand and update the existing disaster inventories. This step is also intended to engage citizen-scientists so that they may become motivated and educated about the latest developments in satellite remote sensing and hydrologic modeling technologies. The shared vision is to better serve the global disaster community by providing essential information aided by the state-of-the-art cloud-computing and crowdsourcing technology, and present an opportunity to eventually modernize the existing paradigm used to collect, manage, analyze, and visualize natural disasters (e.g., floods, landslide, and droughts).

REFERENCES

Adhikari, Pradeep, Yang Hong, Kimberly Douglas, Dalia Bach Kirschbaum, Jonathan Gourley, Robert Adler, and Robert Brakenridge. 2010. A digitized global flood inventory (1998–2008): Compilation and preliminary results. *Natural Hazards* 55(2):405–422.

Behzad, Babak, Anand Padmanabhan, Yong Liu, Yan Liu, and Shaowen Wang. 2011. Integrating CyberGIS gateway with Windows Azure: A case study on MODFLOW groundwater simulation. In *Proceedings of the ACM SIGSPATIAL Second International Workshop on High Performance and Distributed Geographic Information Systems*, pp. 26–29. ACM, New York. http://dx.doi.org/10.1145/2070770.2070774.

Butler, Howard, Martin Daly, Allan Doyle, Sean Gillies, Tim Schaub, and Christopher Schmidt. 2008. *The GeoJSON format specification*. http://geojson.org/geojson-spec.html.

Cannon, Terry, and Lisa Schipper. 2014. *World Disasters Report 2014 Focus on Culture and Risk*. International Federation of Red Cross and Red Crescent Societies, Switzerland, Europe.

Estellés-Arolas, Enrique and Fernando González-Ladrón-de-Guevara. 2012. Towards an integrated crowdsourcing definition. *Journal of Information Science* 38(2):189–200.

Gong, Jianya, Peng Yue, and Hongxiu Zhou. 2010. Geoprocessing in the Microsoft Cloud Computing Platform-Azure. In *Proceedings the Joint Symposium of ISPRS Technical Commission IV & AutoCarto*, Orlando, FL, p. 6. http://www.isprs.org/proceedings/XXXVIII/part4/.

Gonzalez, Hector, Alon Halevy, Christian S Jensen, Anno Langen, Jayant Madhavan, Rebecca Shapley, and Warren Shen. 2010. Google fusion tables: Data management, integration and collaboration in the cloud. In *Proceedings of the 1st ACM symposium on Cloud computing*.

Goodchild, Michael F. 2007. Citizens as sensors: The world of volunteered geography. *GeoJournal* 69(4):211–221.

Goodchild, Michael F and J Alan Glennon. 2010. Crowdsourcing geographic information for disaster response: A research frontier. *International Journal of Digital Earth* 3(3):231–241. doi: 10.1080/17538941003759255.

Goodchild, Michael F, May Yuan, and Thomas J Cova. 2007. Towards a general theory of geographic representation in GIS. *International Journal of Geographical Information Science* 21(3):239–260.

Google. 2013. [Accessed June 28, 2013]. Available from https://developers.google.com/fusiontables/.

Huang, Qunying, Chaowei Yang, Karl Benedict, Songqing Chen, Abdelmounaam Rezgui, and Jibo Xie. 2012. Utilize cloud computing to support dust storm forecasting. *International Journal of Digital Earth* 6(4):338–355.

Hudson-Smith, Andrew, Michael Batty, Andrew Crooks, and Richard Milton. 2009. Mapping for the masses accessing Web 2.0 through crowdsourcing. *Social Science Computer Review* 27(4):524–538. doi: 10.1177/0894439309332299.

Kussul, Nataliia, Dan Mandl, Karen Moe, Jan-Peter Mund, Joachim Post, Andrii Shelestov, Sergii Skakun, Joerg Szarzynski, Guido Van Langenhove, and Matthew Handy. 2012. Interoperable infrastructure for flood monitoring: Sensorweb, grid and cloud. *IEEE Journal of Selected Topics in Applied Earth Observations and Remote Sensing* 5(6):1740–1745. doi: 10.1109/JSTARS.2012.2192417.

Li, Weiyue, Chun Liu, Zhanming Wan, Yang Hong, Weiwei Sun, Zhiwei Jian, and Sheng Chen. 2013. A Cloud-based China's Landslide Disaster Database (CCLDD) Development and Analysis. In *1st International Workshop on Cloud Computing and Information Security*. Atlantis Press, Amsterdam, the Netherlands. http://dx.doi.org/10.2991/ccis-13.2013.49.

Mell, Peter and Timothy Grance. 2011. The NIST definition of cloud computing (draft). *NIST Special Publication* 800:145.

Sanner, Michel F. 1999. Python: A programming language for software integration and development. *Journal of Molecular Graphics & Modelling* 17(1):57–61.

Sun, Alexander. 2013. Enabling collaborative decision-making in watershed management using cloud-computing services. *Environmental Modelling & Software* 41:93–97.

Wan, Zhanming, Yang Hong, Sadiq Khan, Jonathan Gourley, Zachary Flamig, Dalia Kirschbaum, and Guoqiang Tang. 2014. A cloud-based global flood disaster community cyber-infrastructure: Development and demonstration. *Environmental Modelling & Software* 58:86–94. doi: 10.1016/j.envsoft.2014.04.007.

Wang, Fei and Hongyong Yuan. 2010. Challenges of the sensor web for disaster management. *International Journal of Digital Earth* 3(3):260–279.

Wilson, Tim. 2008. *OGC KML, Version 2.2.0: Open Geospatial Consortium*. http://www.opengeospatial.org/standards/kml.

Yang, Chaowei, Michael Goodchild, Qunying Huang, Doug Nebert, Robert Raskin, Yan Xu, Myra Bambacus, and Daniel Fay. 2011. Spatial cloud computing: How can the geospatial sciences use and help shape cloud computing? *International Journal of Digital Earth* 4(4):305–329.

Index

Note: Page numbers followed by f and t refer to figures and tables, respectively.